Restoring the Balance

Restoring the Balance

WOMEN PHYSICIANS AND THE
PROFESSION OF MEDICINE, 1850–1995

ELLEN S. MORE

HARVARD UNIVERSITY PRESS
Cambridge, Massachusetts
London, England

Copyright © 1999 by the President and Fellows of Harvard College
All rights reserved
Printed in the United States of America

Second printing, 2000

Library of Congress Cataloging-in-Publication Data

More, Ellen Singer, 1946–
    Restoring the balance : women physicians and the profession of
medicine, 1850–1995 / Ellen S. More.
        p.   cm.
    Includes bibliographical references and index.
    ISBN 0-674-76661-X (alk. paper)
    ISBN 0-674-00567-8 (pbk.)
    1. Women physicians—United States—History 19th century.
2. Women physicians—United States—History 20th century.
3. Women in medicine—United States—History 19th century.
4. Women in medicine—United States—History 20th century.
I. Title.   [DNLM: 1. Physicians, Women—United States.
2. Feminism—history—United States.
3. History of Medicine, 20th Cent.—United States.
WZ 80.5.W5 M835r   1999]
R692.M645   1999
610.69'52'0820973—dc21   99-38185

This book is dedicated with love to my husband, Micha Hofri, and to my daughter, Betsy More. Without their love and support—and humor—I surely would have lost my own sense of balance. I dedicate it, too, to the memory of my parents, Ben and Dorothy Cooperman Singer.

Contents

Acknowledgments

This project was intended to be a small, highly focused study of the professionalization of women physicians in one region of the country, upstate New York, during the years between 1880 and 1920. It was inspired by my students' persistent questions about women in medicine during an experimental seminar on the history of the American medical profession, given at the University of Rochester in 1978. Dr. Edward Atwater's generous tip that the Edward G. Miner Library at the University of Rochester medical school held the extraordinary, forty-year correspondence of Dr. Sarah Dolley and her son sent me into the archives in Rochester and at the Medical College of Pennsylvania. (This was just the first of many acts of generosity by Ed and Ruth Atwater over the years, for which I am immensely grateful.) An unexpected telephone call from Dr. Leah Dickstein (whose New York accent made me nostalgic) to my office at the Institute for the Medical Humanities at University of Texas Medical Branch in Galveston resulted in an invitation from the American Medical Women's Association to research the history of AMWA. At that point, my "focused" study began to expand. Regina Morantz-Sanchez, whose works have illuminated so much of our understanding of nineteenth-century American women physicians, read my original book prospectus and presciently advised me to expand my perspective and, particularly, to incorporate the history of African American women into my narrative—excellent advice.

I am grateful to the National Endowment for the Humanities for two grants for Travel to Collections (Nos. RV-21087-85, FE-22611-88); to the Susan B. Anthony Center for the Study of Women in Society; to the National Institutes of Health National Library of Medicine for a two-year Publication Grant (No. 5 RO1 LMO4980-02); to AMWA for research support; to the Office of the Dean of Medicine, UTMB, for research support; and to the Institute for the Medical Humanities and its director, Ronald A. Carson, for that most important resource, dedicated research time.

Among the many librarians and archivists who provided crucial assistance, I wish to thank specifically Philip Maples of the Baker-Cederberg Archives of Rochester General Hospital; Christopher Hoolihan, Lucretia McClure, and Janet Brady Joy, currently or formerly of the Edward G. Miner Library at the University of Rochester School of Medicine and Dentistry; Karl Kabelac and Mary Huth, Rush Rhees Library Department of Rare Books and Special Collections, University of Rochester; Sandra Chaff, Margaret Jerrido, and Jill Gates Smith, formerly of the Archives and Special Collections on Women in Medicine of the Medical College of Pennsylvania (now MCP Hahnemann University); Adele Lerner, Dan Cherubin, and Stephen Novak of the New York Hospital–Cornell Medical Center Medical Archives; and the archival staff at the Schlesinger Library of Radcliffe College. The archivists at the University of Pennsylvania library, the Rochester Public Library, and the National Archives Research Center all provided excellent help. Closer to home, the reference librarians at the Moody Medical Library of UTMB regularly resolved bibliographical mysteries and tracked down wayward sources. Linda Zuber of the University of Rochester Alumni Office provided data on medical graduates' employment patterns. I enjoyed the research assistance of University of Rochester and UTMB students Laura Graham, Corinne Sutter (Brown), and Victoria Neidell, as well as research and graphics support from Dr. Kayhan Parsi and Dr. Sara Clausen, respectively. To all of them, I am most grateful.

I want to thank, especially, the many individuals, cited throughout the book, who agreed to be interviewed or to provide uncataloged research materials for this project. They include Mrs. Georgia Gosnell and Dr. Marion Craig Potter (granddaughters of Dr. Marion Craig Potter), Janet Bickel, Phyllis Kopriva, Francesca Calderone-Steichen (daughter of Dr. Mary S. Calderone), Eileen McGrath, and Drs. Sally

Abston, Mary Ellen Avery, Judith M. Cadore, the late Mary Steichen Calderone, Leah Dickstein, Suzanne Hall, Ruth Lawrence, Eugenia Marcus, Carol Nadelson, Bertha Offenbach, and the late Mary Saxe.

The following journals allowed me to make use of materials previously published: *The Bulletin of the History of Medicine, American Quarterly, Journal of the American Medical Women's Association,* and *Rochester History*. I am grateful to them for this courtesy. Archival materials, including photographs, have been used with permission from the following sources: Edward G. Miner Library of the University of Rochester School of Medicine and Dentistry; Baker-Cederberg Archives of Rochester General Hospital; the Rochester Public Library; the Department of Rare Books and Special Collections, Rush Rhees Library, University of Rochester; the Schlesinger Library, Radcliffe College; New York Hospital–Cornell Medical Archives; the Archives and Special Collections on Women in Medicine of MCP Hahnemann University; and the National Archives Research Center.

The following family, friends, and colleagues read and provided useful criticism of all or part of the manuscript: Micha Hofri, Tom Cole, Chester Burns, Ron Carson, Steve Peitzman, Kenneth Ludmerer, Regina Morantz-Sanchez, Gert Brieger, and Warren Carrier. (Unfortunately they refuse to accept any responsibility for mistakes.) Many of my graduate students, including Laura Kickleiter, Toni Schossler, Heather Campbell, David Stevenson, and Craig Klugman, also read the manuscript and gave insightful critiques.

The editorial staff at Harvard University Press has been a pleasure to work with. I have particularly benefited from the professionalism of Editor-in-Chief Aida Donald, my acquiring editor, and Ann Downer-Hazell, editor/diplomat extraordinaire. Copyediting by Julie Ericksen Hagen has greatly strengthened the manuscript although (again) whatever mistakes remain must be credited to the author.

At the Institute for the Medical Humanities, crucial secretarial assistance was provided efficiently and sympathetically by Donna Vickers, as was expert editorial guidance, by Sandy Sheehy. I extend my deep appreciation to my colleagues at the IMH, who have always provided the rich, multidisciplinary environment needed to sustain intellectual risk. But my deepest, most profound gratitude is reserved for my husband and daughter, who have contributed to this book in ways that cannot be cataloged.

Introduction:
Restoring the Balance?

In all departures of health of body, mind, or spirit, I believe there is a loss of balance. [Though] we may have other terms, harmony, equilibrium, etc., the point and principle of getting righted . . . must be to restore that balance.

Sarah Adamson Dolley, M.D., 1896

If we women can be more honest in dealing with conflicts between family and career, we can lead our male colleagues to also be more open and flexible in balancing personal and professional lives.

Anonymous quotation in Association of American
Medical Colleges Project Committee Report,
Increasing Women's Leadership in Academic Medicine, 1996

FROM DRS. BERNADINE HEALY, Frances Conley, Antonia Novello, Joycelyn Elders, Ruth Kirschstein, Vivian Pinn, Susan Love, Joyce Wallace, and Nancy Dickey to *Dr. Quinn, Medicine Woman,* American women physicians were a notable public presence during the 1990s.[1] As government officials, researchers, clinicians, and reformers, women physicians appeared prominently in media coverage of the medical profession. This is a far cry from the mixture of public condescension and admiration accorded the nineteenth-century pioneer Dr. Elizabeth Blackwell, or the anger and resentment directed at so many of her successors until well into the 1970s.

Not all cultural indicators of women's progress in medicine are so encouraging, however. For women struggling to succeed in the medical professions, indeed in all fields, books and articles about sexual harassment, the "mommy track," and the "second shift," as well as the persistence of sex stereotyping, glass ceilings, unequal pay, and unreliable day care, indicate that much work remains to be done. Women in medicine

1

today, like their predecessors, still must work hard at balancing the many demands of their composite role: woman/physician.

It took me a long time to recognize that the theme of "balance" winds through the entire history of women in medicine. Most cultures find it difficult to reconcile equality and difference, and ours is no exception—especially in the domain of gender. For more than a century and a half, American women physicians have grappled with the dilemma of how to be a woman and a physician, how to be different from yet *equal to* their male colleagues. Legal scholar Martha Minow framed the problem as the "dilemma of difference."[2] The many achievements of women doctors have been accomplished in the face of this long-standing problem. The ideal of balance has been a virtue and a conscious goal throughout their history in the American medical profession, the hoped-for solution to the dilemma of difference. Indeed the metaphor of balance that binds together my account of their history not only has maintained currency during the past century and a half but also has taken on additional layers of meaning, resonating powerfully with the realities of life for women in medicine today—as for women in the workforce in general.

Like my fellow historians, I am acutely aware of both the ethical and the epistemological importance of the ideal of objectivity in the construction of narrative. Yet scholars from Wilhelm Dilthey to Paul Ricoeur, Evelyn Fox Keller, and Joan Scott remind us of the limited reach of our truth-seeking ambitions. Historians frame questions according to their own lights, guided (self-consciously, one hopes) by the social, psychological, and historiographical moment in which they find themselves. As I began the process of shaping this narrative, I could not help but acknowledge that, as a woman, a professional, a wife, and a mother, I was writing about women whose values and concerns were in many cases much like my own. When Amelia, the pediatrician-protagonist of Dr. Perri Klass's novel *Other Women's Children*, finds herself unbearably stretched between responsibilities to her family and her profession, she agonizes: "This is a moment when things are badly out of balance for me. How can I go back and forth, how can I hold these two realities in my mind at the same time?"[3] I recognize this dilemma. When I first encountered similar accounts in the historical record, I felt as if I'd uncovered an ancient, partially encrusted mirror. Rather than obscuring the underlying significance of such accounts, or distorting my historical perspective, my own experience has deepened my awareness of the meaning of my sources.

Women physicians, of course, like women in general, are no less diverse than their male colleagues. Many, for example, have forgone childbearing; some live alone, while others live with a same-sex partner. Still, in the face of what philosopher Iris Marion Young termed humanity's "inexhaustible heterogeneity," generalization can be a tool of understanding as well as prejudice.[4] Thus the metaphor of balance—balancing work and home, family and career—although it appears in many guises, nevertheless recurs throughout the history of women physicians. With the title of this book, *Restoring the Balance*, I highlight three facets of the balance metaphor: first, the process by which American women physicians fought for professional equality in medicine; second, their steadfast resistance to a one-dimensional conception of professionalism by pursuing a judicious balance of personal, community, and professional interests; and third, the attentiveness of many women physicians to interactions among the psychological, social, and physiological dimensions of their patients' lives.

Although it is well known that women have practiced as healers and midwives since the days of ancient Greece, it is less commonly recognized that they have also practiced as full-fledged physicians since ancient times. Women may have practiced medicine and surgery—not midwifery alone—at least from the fifth century B.C.E., a tradition extending through the fifteenth century, when Costanza Calenda of Naples became the first woman known to have received a doctorate in medicine. In England, some women practiced as surgeons, licensed and unlicensed, well into the seventeenth century.[5] As historians are discovering, however, female participation in the profession was increasingly suppressed between the Renaissance and the nineteenth century when, first in the United States and soon afterward in Europe, women were finally readmitted to medicine's formal educational and professional institutions.[6]

This book takes up their history with a close reading of the life of one pioneering nineteenth-century medical graduate, Dr. Sarah Adamson Dolley, and then examines the various strategies employed by succeeding generations to balance professional equality and feminine difference. With the exception of a few glittering tokens, female physicians were consigned to the margins of the profession until quite recently.

Only since 1970, in the wake of complex social, political, and legal changes, have significant numbers of women doctors been able—and willing—to move toward the center of professional authority in the United States. If access to the profession was the major obstacle confronting the pioneer generations—something that previous historians have written about extensively and well—the challenge for their successors has been to solidify their standing, and to flourish on their own terms.[7]

Even during the centuries when women were absent from the roster of medical doctorates, however, they made signal contributions to the practice of medicine. From the earliest years of the American colonies, women such as New Englanders Martha Ballard and Joanna Cotton served their communities as skilled midwives and healers. During the early nineteenth century, when the role of physician began to supplant that of midwife, a few women, such as Harriot Hunt of Boston, completed apprenticeships and enjoyed successful careers as physicians. Yet it was not until 1849 that Elizabeth Blackwell became the first woman in America to graduate from a college of medicine. In the second half of the nineteenth century women won gradual acceptance from male colleagues and the general public, primarily as physicians to women and children. By 1920 women represented approximately 5 percent of American doctors.[8]

Prior to the American Revolution, African Americans—medicine's *other* minority—also participated in the healing arts, primarily as informally trained slave practitioners, herbalists, and midwives but also, in rare cases, as apprenticeship-trained physicians (like most white physicians). By the nineteenth century, according to historian Herbert Morais, "Negro women engaged in the general practice of medicine were frequently listed in plantation inventories as 'Doctor.'"[9] Plantation practitioners used the same techniques of bleeding, blistering, purging, and emesis as their freeborn colleagues. Morais also found that a "handful of free Negroes" practiced in the northern states before the Civil War. Some were largely self-taught; some had trained as apprentices. A few, like Dr. David J. Peck, who graduated from Rush Medical College of Chicago in 1847, were graduates of a college of medicine.[10] By 1860 at least nine northern medical colleges had begun to admit African American men. The first African American woman received a degree from a college of medicine in 1864. That year, Dr. Rebecca Lee graduated from the New England Female Medical College; three years later Rebecca J. Cole graduated from Woman's Medical College of Pennsylvania. (By the end of the nineteenth century, a total of 12 black women had graduated from Woman's Medical College.) In 1890, according to historian Darlene Clark Hine, 909 black physicians were in practice, including 115 women.[11]

White and black, most female physicians mainly treated women and children. But by World War I, women's medical schools, medical societies, hospitals, dispensaries, houses of refuge, and settlement houses— the institutional settings of the nineteenth-century "woman's sphere" in medicine—began to decline. The growth in numbers of women enrolling in medical school and the increasing proportion of female physicians nationwide also lost momentum in the early years of this century. African American women may have been affected even more than white women; compared with the 3,885 black male physicians listed by the U.S. Census for 1920, only 65 black women were listed as practicing medicine (just 1.6 percent of all women physicians). More than two-thirds of them are estimated to have graduated from Howard University and Meharry Medical College; 9 others were graduates of Woman's Medical College of Pennsylvania.[12]

The proportion of all physicians who were women did not exceed the modest level of 7 percent until 1970. Since then the number has continued to rise. Women comprised 11.6 percent of the profession in 1980, 16.9 percent by 1990. This represents an increase of 92 percent, although the overall rate of increase in numbers of physicians during the same decade was 32 percent. By 1995 women physicians numbered more than 149,000: 20.7 percent of doctors were women. By the year 2010 they are expected to represent nearly 30 percent.[13]

Statistics reveal less progress, however, in achieving racial and ethnic diversity. African Americans accounted for 5 percent of medical graduates in 1975; by 1995, they accounted for just 6.6 percent. As for social class, women physicians of all races and ethnic categories, like men, are still predominantly middle to upper-middle class in origin. From the beginnings of women's formal medical training in the late 1840s, most white women medical graduates were from the middle class; often they had spent some years as a teacher to earn the money for their medical education. The first generation of African American women physicians, too, were the daughters of prominent black families or had acquired

middle-class standing on their own as teachers or civil servants in post-
bellum Washington, D.C., prior to matriculating in medical school.
Not all, however: for example, Dr. Virginia Alexander (1899–1949), a
leading African American physician in Philadelphia, was able to attend
the University of Pennsylvania and Woman's Medical College through
the help of scholarships and by working twenty hours a week.[14] As one
recent study commented, "Considering the economic ramifications of
medical school attendance, it is not surprising that the major changes
have not been in social class but in gender; it appears that the daughters
of the middle and upper middle classes are joining their brothers in
medical school in greater numbers."[15]

The careers women doctors are constructing, however, do differ in
important ways from those of their male colleagues. The differences
result both from personal agency (choice) and sociocultural imperatives
(necessity). For example, recent research on the career patterns of contemporary women physicians suggests that the majority willingly combine the multiple responsibilities of work, family, and community involvement.[16] Women's specialty choices continue to cluster in fields
that largely are defined as primary care medicine, a trend that dates
from the nineteenth century. As of 1991, more than one-third of
women residents had trained in internal medicine or pediatrics. About
one-fourth of the remainder specialized in obstetrics-gynecology, family practice, or psychiatry. These choices result from their own interests and because, in settings such as health maintenance organizations
(HMOs), physicians in these fields have the opportunity to structure
reasonably regular work hours during their years of child rearing—
something many women doctors consider important. In addition, certain specialties, such as orthopedics and neurosurgery, have only recently begun admitting more than token numbers of women into
residencies. The question of specialty choice is a complex one, however.
Within fields like internal medicine and pediatrics, many more women
are entering and practicing subspecialties such as oncology and hema-
tology. In their professional style and setting, these practice areas afford
doctors the opportunity to combine aspects of primary care and subspecialty practice.[17]

To understand the evolving situation of women in the medical profession, I have found an explanatory model developed by sociologist
Rosabeth Moss Kanter particularly useful. Kanter describes the effect

of what she terms "skewed" sex ratios on the internal dynamics of occupational groups. She argues that the status and prestige of a minority subgroup, in this case women, will vary according to its level of representation in the group. In groups with a *skewed* sex ratio, defined as female representation of at most 15 percent, the majority overwhelms the minority. Under such circumstances members of the minority, often seen as "tokens," will have virtually no opportunity to shape the group's culture. In groups with a *tilted* ratio, Kanter hypothesizes, where women constitute no more than about one-third of the membership, they have more opportunity to form alliances and affect decision making. (A group that is *balanced* would be composed of approximately equal numbers of the subgroups.)[18]

Although Kanter was primarily interested in the movement of women into corporate leadership, her hypothesis also helps describe the evolving status and prestige of women in the American medical profession. At the outset of the nineteenth century the profession's sex ratio was entirely *uniform*—in other words, all male. From the 1850s to 1985, fewer than 15 percent of American doctors were female (a skewed sex ratio), and those women generally occupied the margins of professional power. Since 1985, as women have become a more visible presence in the profession, their number has risen to more than 20 percent, a sex ratio that Kanter's model would characterize as tilted. The profession will reach balance, according to this model, when the proportion of women approaches 40 percent—something not projected to occur for decades.

What Kanter's model describes, historians must attempt to explain: Why is it taking so long for women physicians to reach the higher levels in their profession, both in numbers and in power? Having finally moved beyond tokenism since the 1970s, will women capitalize on their increasing numbers by moving into leadership roles that reshape the profession to reflect their perspective? Here the work of historians and sociologists of science, such as Gerald Holton and Gerhard Sonnert, offers insight by pinpointing two possible explanatory models. One, which they label the "deficit model," presumes that women have been the targets of external, structural, formal and informal mechanisms of discrimination that kept their numbers down. I refer to this dynamic as "necessity." A second explanatory model claims that differences in women's preferences and values persuade many to choose scientific or medical career paths with fewer opportunities for major success. This they label the "difference" model, what I refer to as "choice."

Detailed historical investigations of the careers of women in American science and medicine by Margaret Rossiter, Regina Morantz-Sanchez, and others do not fully support either model by itself. As this book will argue, both necessity and choice, social deficits and individual preferences, have shaped the majority of women physicians' career opportunities and selections. Probably the answer lies closer to what Sonnert and Holton, following sociologists Jonathan R. Cole and Burton Singer, call the "kick-reaction" model, or what I, following Harriet Zuckerman, term the model of "cumulative advantage or disadvantage," according to which individual careers are shaped both by external social forces (especially early on) and by the individual's response to those factors.[19]

So, what cultural forces, professional traditions, and individual values have informed the experiences of women in American medicine? I interpret their history as a careful attempt to fulfill expectations that originally were characteristic of mainstream professional culture in nineteenth-century medicine. As historian Judith Leavitt has observed, the majority of nineteenth-century American practitioners combined general medical practices with the routine details of domestic, agricultural, and small-town life. Unavoidably, "they wove together their domestic experiences with their perceptions of their own medical practices . . . [T]hey saw the parts as an integrated whole."[20] Not the least of women physicians' achievements was their ability to sustain into the twentieth century this traditional professional culture, firmly situated in the context of community life, by exploiting its many points of congruence with feminine gender norms.

Although all physicians, male and female, are charged to be both empathic and expert, the historically dichotomous identity imputed to women physicians in Western culture—to be womanly as well as scientific—has complicated this challenge, while situating it at the center of their professional life. Furthermore, the duality of their own experience, particularly the intersection of professional and feminine cultures, has fostered a continued appreciation for the complicated interplay between a patient's mode of life and state of health. Throughout their history, women physicians have attempted to integrate professionalism with civic and personal life, to sustain an older model of "civic profes-

sionalism."[21] The Association of American Medical Colleges' anonymous respondent (quoted in the epigraph) is part of a long line of
women doctors who have tried, however imperfectly, to put things back
into "balance" for their patients and for themselves. These efforts, re-
affirmed by each generation, lie at the heart of my central question,
What does it mean to be a woman physician?

Previous works on women physicians in America, notably Regina
Morantz-Sanchez's *Sympathy and Science: Women Physicians and Ameri-*
can Medicine, explored the tension between two typologies of female
professionalism: separatist perfectionism and collegial assimilation.
Morantz-Sanchez concretely represents these types through two emblematic nineteenth-century figures, Elizabeth Blackwell and Mary
Putnam Jacobi. Blackwell disavowed marriage, family, and even the
daily care of patients to devote herself instead to the sociomedical evangelism of popular health reform and to preventive medicine. She has
been taken, rightly enough, to represent a distinctly feminine type,
inhabiting the homosocial world of female separatism, reigning supreme over women physicians' own separate sphere. Mary Putnam Jacobi managed to combine medical practice, medical research, marriage,
and motherhood into a rich but volatile compound. Her life, in contrast
to Blackwell's, was a deliberate attempt to demonstrate that women
physicians' careers need be little different from men's. Her own writ-
ings attest to a vision of professionalism that, while acknowledging the
"peculiar destiny" of many women physicians (i.e., marriage and childbearing), attempted nevertheless to minimize women's distinctiveness
and to claim full equality for them.[22]

The careers of Sarah Adamson Dolley (1829–1909) and many other
women who figure prominently in my analysis represent a third and
intertwining strand in the complex braid of traditions of women in
American medicine. Indeed, their lives in medicine represent a model
that continues to the present as the life plan and implicit value structure
of the majority of American women physicians: the quest for a balanced, well-integrated life for their patients and for themselves. As
Perri Klass's Amelia puts it, "The strength you draw from home and
family is the strength that lets you turn outward."[23] For example, Dr.
Dolley's efforts to reconcile the values and responsibilities of professional and private life, described in Chapter 1, epitomized her professional generation's etiological and therapeutic worldview. As her letter
quoted at the start of the introduction makes clear, balance and har-

mony were, for her, essential ingredients of both personal health and professional competence.

The values expressed by Sarah Dolley's lifework were widely shared by many of her professional contemporaries, male and female.[24] Principally derived from the Hippocratic-Galenic tradition, the metaphor of balance was firmly embedded in the etiology and therapeutics of medicine from ancient times through much of the 1800s. The nineteenth-century physician's conception of professionalism also was shaped by the metaphor of balance in terms of the balanced character undergirding good professional judgment. Like the Galenic philosophy of health, Victorian professionalism strongly echoed the Aristotelian doctrine of the golden mean. It linked the achievement of the good life to the union of virtue and practical wisdom, the ability to make sound, or balanced, judgments.[25] Such judgments depended also on the deep familiarity with one's patients and community that came from the integration of professional, civic, and personal responsibilities. During the late nineteenth century the ideal of professionalism thus was intimately bound up with the personal character and judgment of the practitioner. And underlying both professional and therapeutic values was the principle of balance.[26]

The continuity of this ideal among many women physicians from the nineteenth century to the present day is one of the themes of this book. Thus, although I focus primarily on the period from the 1890s to the 1990s, I have chosen to begin, in Chapter 1, with Dr. Sarah Dolley's emblematic nineteenth-century career. Dolley's activities carried her from her own household (and those of her patients) into the center of professional and civic life in her community. Her career stood squarely in the mainstream of mid-nineteenth-century medicine. Moreover, her special vantage as a wife, then a widow, and as a mother, gives us a particularly clear view of the professional choices and constraints increasingly faced by her women colleagues and, even more, by their successors. I have chosen not to focus on the experiences of other pioneer women physicians, such as Elizabeth Blackwell, Marie Zakrzewska, or Mary Putnam Jacobi, about whom others have written at length or are currently completing extensive studies. Nor have I devoted many pages to the effort to establish women's right to formal medical education. This, too, has been closely examined by others.[27] Chapters 2 and 3 explore the ways in which women physicians, at a time when their proportion in the profession was radically skewed, carved out a place for themselves. Chapter 2 describes the all-female professional institutions created to enhance women physicians' professional development. Chapter 3 moves into the twentieth century, tracing several Progressive-era "maternalist" health care initiatives, such as homes for unwed mothers, settlement house clinics, and public health bureaus dedicated to the needs of women and young children, through which many women doctors hoped to publicly enshrine the ideals of "scientific" motherhood and to establish themselves as professional authorities on its medical aspects. As historians Seth Koven and Sonya Michel observe, "maternalist" ideologies "exalted women's capacity to mother and extended to society as a whole the values of care, nurturance, and morality."[28]

From Chapter 4 onward, the narrative takes on a more familiar character—familiar, that is, to twentieth-century sensibilities steeped in the politics of modern professionalism. Chapter 4 examines the effect of the rise of professionally run hospitals and the growth of specialties and postgraduate residency education on the career opportunities open to women physicians. Women's overall exclusion from these new professional institutions effectively slowed their diffusion into positions of professional leadership for decades. And women physicians were fully aware of their marginality. Chapter 5 relates the stratagems they adopted to combat the isolation entailed in their minority status, such as founding the Medical Women's National Association in 1915 and, in 1917, its affiliate, the American Women's Hospitals service which organized wartime overseas hospitals staffed by women physicians and surgeons, who were formally excluded from the military reserves until World War II.

Chapter 6 analyzes the 1921 Sheppard-Towner Act—a culmination of women reformers' "maternalist" effort to improve health care for mothers and young children—and the demise of maternalist medicine in a tide of private enterprise rhetoric from politicians and physicians. Yet despite the decline of their maternalist niche during the 1920s and 1930s, some women physicians, principally those involved in children's health, began to demonstrate the potential of medical careers for women who acquired a first-class education and a good mentoring network. By the beginning of World War II, as Chapter 7 details, women physicians were able to marshal support even among members of Congress for their participation in the Army Medical Reserves. Following the war, however, along with most other professional women, they were relegated to the domestic domain described in Betty Friedan's *The Feminine Mystique*. Even a threatened shortage of physicians did not evoke unequivocal support for women in the profession. Women unquestionably would have remained on the margins of medicine had not medical education been swept up in the legal battles inspired by the civil rights and women's movements of the late 1960s. Chapter 8 assesses the effect of "equal opportunity" legislation on the gender and racial profile of the medical profession in recent decades, and the status of African American women physicians since the start of the modern civil rights movements. Finally, I conclude with a consideration of women physicians' current attitudes toward balancing work, family, and civic responsibilities, and the obstacles they still encounter in a profession in which the models of productivity frequently conflict with the ideals of balance and health.

Women practitioners express their ideal of professionalism through the medium of their own gendered experience of the world. Rather than force their history to oscillate between the poles of separatism and assimilation, femininity and professionalism, we should heed the words of Simone de Beauvoir or more recent theorists of gender such as Joan Scott and Martha Minow. It is true, as they have noted, that for centuries gender functioned as a signifier of "difference." Today, however, they urge us to resist the discouraging inflexibilities of a world of "either/or." We are trying to rid future generations of the need to choose *either* to play by the old rules *or* not to play at all. Many women physicians today, like Sarah Dolley before them, seek professional equality expressed through feminine difference.[29]

By dint of dual citizenship in the public world of medicine and the private domain of the household, women physicians have tried to become expert at balancing the values and customs of both. Setting a pattern for generations of professional heirs, physicians like Sarah Dolley not only endured but positively thrived on the creative tension between their professional and their feminine roles. The insistent effort of women in medicine today to integrate public and private, civic and professional interests aims for the inner balance they have long sought for themselves and their patients. Dolley was a fighter; she knew a struggle when she saw one. We are still learning. We cannot afford to become complacent.

1

The Professionalism of Sarah Dolley, M.D.

> I have often, while trying to develop my intellectual faculties, forgotten, that to cultivate the moral faculties, was equally necessary, and by this neglect, have lost much. I do think that perfection, will never be personified, without physical, intellectual and moral cultivation; and when the intellectual is cultivated at the expense of the physical or moral, there is a loss of that symmetry which was designed by nature.
>
> *Sarah Adamson (Dolley), February 3, 1850*

THREE MONTHS after beginning medical school in Syracuse, New York, twenty-year-old Sarah Read Adamson wrote these lines to her cousin Elijah back home in Philadelphia.[1] In 1851 she became the third woman in the United States to graduate from a chartered college of medicine. Adamson's ideal of professionalism—the art of good judgment expressed as a balance of professional and moral values—animated the practices of most American physicians of the time. But in the long run, this professional (rather than scientific) ideal proved especially characteristic of women physicians, linking Sarah Dolley's pioneer generation and its successors. As she wrote her son after nearly half a century in practice, "In all departures of health of body, mind, or spirit, I believe there is a loss of balance."[2] Dolley's call for balance was both a literal reference to physiological systems and a metaphor for personal integration. Seeking balance for herself as a physician, a wife, a mother, and an active citizen of her community, Dolley forged a career that, from a late twentieth-century perspective, characterized the careers and life choices adopted by many of her successors. Increasingly, this model has become an ideal for men as well as women.[3]

Long before her death in 1909 at the age of eighty, Dolley was acknowledged as one of the "eminent ones," laden with honors and the respect of peers and patients alike. The third woman medical graduate in America she also, in 1851, was the first to complete a year's internship in an American hospital.[4] After marrying one of her medical school professors, Dr. Lester Dolley, she settled into a productive and pioneering career in Rochester, New York. At first Dolley and her husband shared a practice, but after his death she established her own practice, sometimes in partnership with another woman physician. She also spearheaded the establishment of a woman's medical society, a dispensary for women and children, the Women's Medical Society of New York State, and numerous political initiatives to improve conditions for women in her city and state.

For Dolley, personal and professional values acted as mutual supports rather than mutual antagonists. Middle-class Victorian culture was built on the foundation of domestic life. For the typical practitioner, male or female, personal and professional life intersected in the household.[5] Judith Leavitt convincingly argues that rural nineteenth-century male physicians, whose typically part-time practices complemented the daily work of running a farm, were as steeped in the domestic details of general practice as were women. But for male doctors—especially those in cities and towns—the cross-fertilization of private and public identities slowly declined as medicine became a full-time, office-based occupation toward the end of the century.[6] Female physicians, however, having retained dual citizenship in the public world of medicine and the private domain of the household, sustained their commitment to balancing the values and customs of both worlds.[7]

Dolley was one such practitioner. Yet she was no stranger to the excitement and promise of medical science. Like the physicians of the generation following hers, she valued laboratory science as one of the underpinnings of practice, avidly reading the latest medical journals, keenly anticipating the effects of scientific advance on medical care. Unlike her contemporary Elizabeth Blackwell, she had few qualms about the impact of bacteriology or vivisection on clinical practice, though she shared Blackwell's commitment to keeping the "whole" patient uppermost in mind. But she rarely undertook scientific investigations herself. Instead of pursuing research as an explicit complement to clinical practice, as Drs. Mary Putnam Jacobi and Mary Dixon-Jones were doing, Dolley strove where possible to integrate the results of others' research into her own daily activities as a clinician or, at the least, into her understanding of the nature of health and disease.[8]

Dolley's professional generation expressed its commitment to integrating the values of professional and private life through the governing metaphor of balance. Drawn from a larger belief system principally derived from the Hippocratic-Galenic tradition, the concept of balance was deeply rooted in ancient medicine and philosophy. For more than a millennium ancient doctrine had held that health depended on a relationship of internal and external forces acting on the individual. Good health was understood as not only the homeostatic equilibrium of forces internal to the body but also the harmonious balance between one's internal and external environments. Internally, the individual's unique balance of humors, infused by the "vital force" necessary to life, provided a foundation for health. But health also required control of the external factors affecting the constitution, including air, food and drink, motion and rest, sleep, evacuation and retention, and the passions of the soul. Health depended, therefore, on a balanced life: a moderate regimen that avoided extremes, a middle way.[9]

Through three-quarters of the nineteenth century, the ideal of balance was ubiquitous in the literature of medicine, physiology, and religio-moral prescription. Despite the decline of the Hippocratic-Galenic model of the humors and its replacement by the mechanical models of seventeenth- and eighteenth-century physiology, the body continued to be understood as a holistic, interactive system.[10] Lesions in one region of the body could act sympathetically, it was believed, on another part of the body.[11] By midcentury, when Dolley began to practice medicine, "equilibrium" and a balance of "energy" had become dominant metaphors in natural science.[12]

The metaphor of balance also shaped the Victorian physician's concept of professionalism, defined as the sound character required for good medical judgment. Reflecting the Galenic conception of health, nineteenth-century medical professionalism fully affirmed the Aristotelian doctrine of the "golden mean." In the *Nichomachean Ethics*, Aristotle ascribed to *phronesis*, or "practical wisdom," the secret to living well. Practical wisdom was defined as "wisdom in action," or the skill of deliberating well.[13] Practical wisdom produced good judgment, an essential quality of good medical practice.[14] The Victorian conception of medical professionalism closely resembled the ideal of practical wisdom.[15] In the face of nineteenth-century social volatility, integrity rather than gentility became the sine qua non of professionalism. Balanced judgment, the outward sign of inner character, was the hallmark of the true professional. Professionals, moreover, demonstrated their character not only at the bedside but also in the arena of civic affairs, particularly in matters relating to public health and morals.[16]

Nineteenth-century medical professionalism, the basis for a physician's claim to be trustworthy, thus depended on an alliance of competence, virtue, and practical wisdom—the ability to make sound, or balanced, judgments. The linchpin of the mid-Victorian physician's professional identity was the concept of therapeutic "specificity," a diagnostic approach that emphasized the physiological, psycho-social, and geographic particulars of a patient's life. Therapeutic skill was wedded to judgment, and judgment revealed inner character.[17] When patients summoned their physicians to the bedside, they sought maturity, judgment, and character as much as good breeding and an elite education or high social standing.[18] Practical wisdom, then, characterized the goals of Victorian medical practice and defined the role of science within everyday practice. Underlying medicine's physiological, professional, and therapeutic values stood Sarah Dolley's cardinal principle of balance.

Medical Education for Women: "Sarah, Thee Must Not Fail"

Sarah Adamson was born in 1829 to one of the old Quaker families of Chester County, near Philadelphia. She grew up at the family farm and country store, located in Schuylkill Meeting near Phoenixville, site of the largest nail factory in the country.[19] Quakers, staunch believers in a democracy of the "inner light," produced many of the earliest female reformers in America, including the first generation of women doctors. The Adamsons were Hicksite Quakers, a group that separated from the main body of the Society of Friends in 1827 largely out of a commitment to an active political engagement against slavery. Some of the staunchest women's rights leaders, such as Lucretia Mott of Philadelphia and Susan B. Anthony of Rochester, New York, were Hicksite Quakers. They also played an important part in the movement for women's higher education, especially medical education. Philadelphians such as Joseph Longshore and Ann Preston, who helped found the Female Medical College of Philadelphia in 1850 (renamed Woman's Medical College in 1867), were affiliated with the Hicksite Quaker community.[20]

The young Sarah Adamson benefited from these liberal influences. Her parents, Charles and Mary Corson Adamson, took seriously their obligation to educate Sarah as well as her two brothers. (Later, they supported her right to marry "out of meeting.") In 1844, when Sarah was fifteen, she began attending the "select school" in Phoenixville run by her cousin, Elijah Pennypacker. Instruction in astronomy and the natural sciences, especially botany, was carried out by Graceanna Lewis, a pioneering educator and, later, a fellow of the Academy of Natural Sciences of Philadelphia. It was Lewis who gave Sarah a physiology text to study at home, one source of her early interest in becoming a doctor. When Adamson transferred to the Friends Central School of Philadelphia for her last two years of secondary education, she began to pursue her goal actively, reading everything from a copy of Wistar's *Anatomy* (borrowed from the public library) to works on midwifery and diseases of women borrowed from family friends.[21]

In 1847, the same year in which Elizabeth Blackwell left Philadelphia to attend Geneva Medical College in upstate New York, the eighteen-year-old Adamson gravely informed her parents that she wished to become a doctor. She recalled her father replying, "If it was not a mere freak of fantasy and I would make a success (he meant by that was I in dead earnest and would I diligently pursue and never go back on my intention), then he would do just as he would for [brothers] Thomas or Charles." But he also insisted, "Sarah, thee must not fail." Soon after, Sarah wrote to Elizabeth Blackwell for advice. Blackwell answered her letter kindly but added, realistically, "'If you can endure vulgar ridicule and reproach I bid you Godspeed'" (October 17, 1904).[22]

Her most valuable ally proved to be her uncle, the eminent Philadelphia physician Hiram Corson. Understanding her need for preliminary medical education before applying to medical school, Adamson wisely asked Corson to be her preceptor. That same year, 1847, Corson had helped found the Montgomery County Medical Society. Six years later he was elected president of the State Medical Society of Pennsylvania. Corson is remembered for defending women's medical education and championing their admission to the State Medical Society in the 1860s. In 1879 he sponsored a bill requiring that women physicians care for the women inmates of the state's mental hospitals. He was also one of several Quaker physicians around Philadelphia who served as preceptors to a small circle of women students. Besides Sarah Adamson, the group included Ann Preston, future dean of Woman's Medical College.[23]

At first, as she wrote to her son many years later, Corson "kindly wrote me a discouraging letter . . . Now I say this was *kind* to bring before my mind all the difficulties I had to face, but I said to my parents, 'I will never coax anybody for anything.'" Concluding prematurely that Corson had refused her request, Adamson then appealed to Dr. Bartholemew Fussel, another preceptor to Ann Preston, who quizzed her in reading, writing, and grammar and agreed to take her on. Dolley went home "elated; but Mother said, 'I have two brothers doctors and I will see Hiram.' So Father and Mother went to Plymouth and when they came home, a skeleton and a lot of books were brought in and Father raised the east wing of our house one story for my study" (October 17, 1904). Corson had no illusions about the road ahead for his niece. What she wanted to do, he said, would be

> a new step. We must, therefore, look to the motives and objects . . . The end or object is professedly to practise. But for what? Money? honor? a name? . . . Or, is it rather to introduce a new custom, one more consonant with common sense, and one which will be to the mutual advantage of the sexes, and of course of general society? This . . . is perhaps the strongest inducement, for one who loves her kind, to engage in a new pursuit in which she will most certainly meet with obstacles of great magnitude, and contempt and scorn in abundance.[24]

Soon Adamson began studying with Dr. Corson and "riding with him betimes [to see his patients], but not much till after I had read some months" (October 17, 1904). After a year or two she moved to Philadelphia to study anatomy with Dr. Edwin Fussel (Bartholemew Fussel's son), who was soon to lecture on anatomy at the new Female Medical College. Meanwhile, with the help of Hiram Corson, Adamson applied to—and was rejected by—every medical college in Philadelphia. She was even turned down by Geneva Medical College, whose faculty and students were just recovering their composure—and barring women matriculants—following Elizabeth Blackwell's graduation.[25] But Corson soon learned of a new *co*educational medical college in upstate New York, the "eclectic" Central Medical College of Syracuse. After satisfying his own doubts (whether over the school's curriculum or its coed student body is not clear) he agreed that Sarah should enroll for the first session of 1849–50, beginning on November 5.

Gender discrimination, indeed universal exclusion, was of course the overriding concern of women seeking a medical education at midcentury. But once a few medical colleges opened their doors to women during the 1850s, two questions demanded attention from prospective students—even if, in reality, they would find little opportunity to act on their preferences: whether or not to choose a coeducational institution, and whether to seek a sectarian or a "regular" medical education. Adamson opted for coeducation (a choice implicit in her decision *not* to remain in Philadelphia for the opening of the Female Medical College). Neither she nor Corson seemed put off by Central Medical College's sectarian curriculum.

Central Medical College, in fact, has the distinction of being the first explicitly coeducational medical college in the United States, having decided, three weeks before the beginning of its first term, to admit women students. But it was only the first of dozens of sectarian schools to adopt coeducation—nearly two decades before most "regular" medical colleges attempted it.[26] Prior to the 1870s, the majority of coeducational medical colleges were founded by sectarian physicians. In 1849 there was no other place for Adamson to get a complete medical education, and certainly not a coeducational one. Not until 1894 did the majority of women medical graduates, regular and sectarian, graduate from coeducational schools.[27] This was a significant issue for her. She chose Central Medical College, thereby rejecting the psychological comfort of study in an all-female setting, at a time when plans for the new Female Medical College in Philadelphia were well known.[28]

Two other women were already in attendance by the time Adamson arrived: Lydia Folger Fowler and Rachel Brooks Gleason. Fowler, wife of well-known phrenologist, hydropathist, and publisher Lorenzo Fowler, graduated on June 5, 1850, becoming the second woman in America to receive a medical degree. Adamson and Mrs. Rachel Brooks Gleason graduated (in alphabetical order) in 1851, respectively the third and fourth women in America to be granted medical degrees.[29] Folger traveled widely in England and America as a popular health lecturer. Gleason authored the widely read home medical text *Talks to My Patients* and for many years was coproprietor, with her physician-husband, of a hydropathic sanatorium in Elmira, New York. (The Gleasons' daughter, Adele, eventually became a physician too.) The three alumnae remained lifelong friends and correspondents.[30]

The faculty of Central Medical College, including Sarah Adamson's future husband, were followers of Dr. Wooster Beach, the founder of "eclectic" medicine, one of several sectarian therapeutic systems competing with "regular," or allopathic, medicine for the public's allegiance.[31] The movement for sectarian medical reform dated from the earliest years of the century. Distaste for regular medicine's harsh therapeutics was the most readily given reason for the growth of alternative systems such as Thomsonianism, homeopathy, hydrotherapy, and eclecticism. Sectarian doctors drastically reduced or eliminated the use of bloodletting, leeching, mineral compounds, and other depletive therapies relied on by so-called heroic medicine. Many other factors contributed to the flowering of sectarian medicine between the 1820s and the 1870s, including its claim to represent a uniquely American, homegrown, plainspoken therapeutics. Sectarian medicine's individually empowering therapeutic style also resonated with the evangelical revivals sweeping northeastern and frontier towns after the 1820s.[32] By the 1840s, moreover, in the wake of heroic medicine's failure to combat yellow fever and cholera, even mainstream practitioners began broadcasting their doubts about depletive interventions. Beach, a graduate of a regular medical school, rejected heroic medicine in favor of milder botanical therapies. Because he resisted adopting one exclusive therapeutic method, as he considered them to be marks of medical dogmatism, his system was soon labeled "eclecticism."[33]

The beginnings of women's medical coeducation in America depended in large part on the institutionalization of sectarian medicine. The National Eclectic Medical Association, for example, officially sanctioned coeducation as early as 1852.[34] The combination of sectarians' democratic ethos (especially the admission of women to medical schools) and their emphasis on preventive health care proved highly attractive to many women.[35] By 1862, approximately two hundred fifty women had managed to obtain a medical degree from a chartered college of medicine, and over half were graduates of sectarian schools. The majority of the more than five hundred women physicians listed in the 1870 census were "irregular" practitioners, although this number undoubtedly included some who were not graduates at all. After 1871, however, the number of women graduating from regular medical schools began to rise as new land-grant universities, such as the Universities of Michigan and Iowa, received charters requiring the admission of women to all departments. (Syracuse University, perhaps reflecting its Methodist evangelical tradition, also began graduating women from its medical department by the mid-1870s.) Still, as late as 1917 a high percentage of the women physicians who had been practicing more than twenty years were graduates of sectarian medical colleges.[36]

This heavy reliance on "irregular" medical training by the first generation of women practitioners played into the hands of their critics. The stigma of an "inferior" education was difficult to dispel.[37] William Rothstein writes, for example, "The scientific and intellectual content of eclecticism was inferior . . . Most eclectic medical colleges were undistinguished and unsuccessful." Yet Rothstein agrees that until the 1870s, eclectic schools' curricula differed from the regulars' only in "materia medica and the administration of remedies."[38] In fact, at mid-century a common approach united all schools of therapeutics.[39] As for the teaching of anatomy (including dissection), physiology, chemistry, obstetrics, and surgery, they were essentially the same at the early eclectic schools—such as the Penn Medical University of Philadelphia, the New York Medical College and Hospital for Women, the Eclectic Medical Institute of Cincinnati, and Central Medical College—as at the regulars. Not until the 1870s and 1880s could a meaningful distinction be made between regular and sectarian schools.[40]

In later life Sarah Dolley always minimized the potential embarrassment of her eclectic education. She maintained—perhaps disingenuously—that she had had no idea such distinctions as "regular" and "irregular" even existed. According to her account, written five years before her death, she discovered that the school was "not 'regular'" only after her arrival in Syracuse; saying, "I had never known of any but regular physicians. . . ." Even her textbooks, she insisted, were the "same standard works" in use at medical colleges everywhere.[41]

Nevertheless, once she arrived at Central Medical College, she certainly learned of the competing therapeutic systems. After all, her professors, including Lester Dolley, were members of state and national eclectic medical societies and editors of the *Eclectic Journal of Medicine*. During her first semester she wrote her cousin that she was not yet "converted to any system of medicine, without it is the system of *Pre-*

vention." Even after she and her husband shucked off the more obvious
traces of their eclectic training, she always remained loyal to its commitment to moderation.[42] Throughout her career, she drew on its legacy.[43]

By 1850 dissension among the faculty led to the division of the college into factions and the removal of the school from Syracuse to Rochester, where Adamson and Gleason received their degrees on February
20, 1851. By 1852 it closed for good, having graduated forty-seven
students.[44] Their diplomas announced them "qualified to discharge all
the highly responsible, important and complicated duties attached to
the office of PHYSICIAN, SURGEON, and ACCOUCHEUR and to sustain
a good moral character."[45]

After graduation, Adamson's education continued in Philadelphia.
Midcentury medical schools offered scant clinical experience. Only an
elite minority acquired postgraduate clinical training in hospitals, but
Adamson was determined to be among them. Her best chance lay in
securing a position on her home territory. Whereas the most promising
midcentury male graduates would have traveled to the clinical "para-
dise" of Paris, women found themselves greatly restricted at the Paris
hospitals.[46] Aware that Elizabeth Blackwell had attended Philadelphia
Hospital for one summer following her graduation in 1849, Adamson,
too, returned to Philadelphia.[47] Sponsored by her two influential un-
cles, Dr. Isaac Pennypacker (Elijah's father) and Dr. Hiram Corson, and
invoking "the most determined perseverence," she applied to the Board
of Guardians of the Poor of Philadelphia for a "situation" at Philadelphia Hospital. The hospital, commonly known as Blockley, had evolved
from the medical wing of the city almshouse into a large municipal
hospital (it was officially renamed Philadelphia Hospital in 1835). Always overburdened with the poor and outcast, Blockley's staff regularly
saw those patients for whom no other refuge was available.[48] Fortunately for Adamson, the hospital was currently without a senior medical
board. Since 1845 it had been run by a chief resident, with consultations
provided by one surgeon and one physician. Her application was judged
not by a group of senior physicians but by laymen, the Guardians of the
Poor. Only two members opposed her application, and on May 25,
1851, Blockley's chief resident was directed to "assign her to such position as will best enable her to obtain the knowledge she desires without
detriment to the institution." She remained in attendance for eleven

months, the first woman to complete a hospital internship. Certificate in hand, Adamson left Philadelphia Hospital several weeks before her marriage on June 9, 1852.[49]

The Question of Marriage

> The question of marriage . . . which complicates everything else in the life of women, cannot fail to complicate their professional life. It does so, whether the marriage exist or does not exist, that is, as much for unmarried as for married women.
>
> *Dr. Mary Putnam Jacobi, 1880*

> Surely this is a little difficult. We were told that the great objection to women entering professional life was that they would not care to marry, hence the desire to keep them out. Now conditions are reversed and the objection is that they do. Really it is not easy to please.
>
> *Woman's Medical Journal, 1893*

Sarah Adamson became engaged to Lester C. Dolley, M.D., professor of anatomy at Central Medical College, during her last year of study. A native of upstate New York, Dolley later served as health officer for Rochester and was also a physician in private practice. At Sarah's insistence (firmly supported by her parents), they delayed marrying until she completed her internship in Philadelphia. As her parents wrote to their future son-in-law, the internship would "give her a better chance of improving [*sic*] herself in the practice of medicine than perhaps years of common practice." For most of their married lives, the couple practiced medicine in Rochester. They had two children, Loilyn, who was born on April 19, 1854, and died in 1858 of "typhoid pneumonia," and a son, Charles Sumner (called Sumner by the family) who was born in Elyria, Ohio, on June 16, 1856. He lived until 1948. After her husband's death on April 6, 1872, Sarah Dolley practiced for another thirty years.[50] Her letters to Sumner during this period provide the chief basis for our knowledge of her increasingly prominent career.

Achieving success was no easy task for men or women in the overpopulated, competitive world of nineteenth-century medicine. For women, the question of marriage complicated matters considerably. Marriage was not then the norm for most women professionals—at least, not while they continued in their chosen professional calling. The Victorian idealization of home and motherhood seemed to leave little room for the pursuit of any outside activities other than the voluntary charitable work that was seen as a legitimate expansion of the "woman's sphere." At the same time, until the end of the century and in some cases even beyond, nurses and teachers either were forbidden or actively discouraged from continuing to work once they married. Undoubtedly, as one woman doctor wrote in 1894, many professional women had become "convinced of their ability to take care of themselves . . . We have no desire to reject our new-found liberty for what would handicap and hamper us in our individual vocations."[51]

Still, marriage was undertaken by a surprisingly large proportion of nineteenth-century women physicians. (Among the first four women medical graduates, only Elizabeth Blackwell chose to remain unmarried.) A significant percentage, ranging from 25 percent to 35 percent in the nineteenth century and 30 percent to 40 percent in the first quarter of the twentieth century, chose not only to marry but also to continue practicing their profession. Of a sample consisting of the fifty-nine members of the Blackwell Medical Society of Rochester between 1887 and 1927, 41 percent were married. At least eight, including Dolley, also had children.[52] Likewise, although only two of the six women graduates of the Cleveland Medical College before 1860 ever married, at least five were biological or adoptive mothers. In this they expressed a significant difference from other professional women of the period, reflecting both the greater ease of maintaining both a medical career and a family and a strong belief in the desirability of doing so. Their determination to balance love and work, family and career, sprang from a tolerance for unconventionality coupled with a belief that medical practice and professionalism must be grounded in community life.[53]

Many exceptionally successful women physicians, such as Mary Putnam Jacobi, her protégée Emily Dunning Barringer, and Bertha Van Hoosen, cofounder of the Medical Women's National Association, gave a central place to familial matters in their busy, productive lives, regardless of their marital state.[54] Such women may have seemed unconventional by Victorian standards of domesticity, but their lives were reasonably congruent with the expected norms for nineteenth-century medical practitioners, whose work was rooted, after all, in both the household and the community. (Sarah Dolley's only surviving journal, for the year 1860, routinely intermingled case descriptions with comments on her family's health.)[55] From the perspective of the Victorian patient, practitioners like Dolley, who allied professional authority to a close attention to domestic detail, represented a desirable blend of professional strengths.

Fortunately for Dolley, her parents and husband all supported her goals. In addition, Rochester at midcentury provided ample scope for her development both as a practitioner and as an active participant in the city's civic culture. Rochester was settled during the "New England diaspora" of the early nineteenth century (although its first settlers had migrated from Maryland).[56] The city's civic culture was complicated but enlivened by the religious, racial, social, and feminist reform movements that shaped Victorian society. Indeed Rochester lay in the heart of the "burned-over district," that swath of northern America engulfed in the evangelical revivals ignited by the Reverend Charles Grandison Finney in the 1820s. It later became the home of abolitionist Frederick Douglass and suffragist Susan B. Anthony; neighboring villages were home base for Anthony's fellow activist Elizabeth Cady Stanton, Frances Willard of the Women's Christian Temperance Union, and Clara Barton, founder of the American Red Cross. Dolley's own Quaker-forged commitments to abolition, suffrage, and women's rights flourished among these overlapping circles of like-minded friends.

By midcentury Rochester was also beginning to feel the effects of immigration and industrialization. By the time of the Dolleys' arrival, it was a city of nearly 40,000, of whom about three-fourths were native born. It stood on the threshold of a period of rapid growth and diversification. Between 1855 and 1890, the primary years of Sarah Dolley's medical career, Rochester's population increased 200 percent, and it ranked twenty-second in size among all American cities in the 1880s. By then its population base had shifted slightly toward the typical makeup of late nineteenth-century cities—poor immigrant industrial workers—yet the city never experienced the crush of impoverished tenement dwellers characteristic of larger urban centers. Of its 27,000 families in 1890, 24,000 lived in separate units of housing. With the notable exceptions of 1873 and 1893, years of disastrous economic contraction and widespread unemployment throughout the country, the city prospered. Most of Rochester's public health problems, such as impure water, inadequate sewage disposal, impure milk and other foods, were the result of rapid growth, political corruption, and public indecision.[57]

These were good years in which to launch a general medical practice. St. Mary's, the city's first hospital and one where Dolley's ties were strongest, was established in 1857; the Rochester City Hospital was founded in 1863. Surgeon-in-chief Dr. Edward Mott Moore, a Quaker, headed St. Mary's medical staff.[58] Moore, who became the second president of the American Surgical Society, president of the American Medical Association, and president of both the Rochester and the New York State Boards of Health, was a lifelong friend of Lester and Sarah Dolley. Like Sarah, he was born into a progressive Quaker family in Philadelphia, and he had capped his medical training with an internship at Blockley from 1838 to 1840. He settled permanently in Rochester shortly after the Dolleys, soon becoming the city's leading medical citizen. After Lester Dolley's death, Moore became Sarah Dolley's chief mentor until his own death in 1902. St. Mary's was the hospital where Sarah Dolley first attended patients, occasionally assisting Moore in surgery. The two founded the Rochester chapter of the American Red Cross; he served as president, Sarah Dolley as secretary. Dolley succeeded him as president after his death.[59]

Although Lester Dolley's tenure at the Central Medical College was cut short when the school closed, he did well in general practice and as a city health officer. The couple's lives seemed full of promise. But Lester Dolley died suddenly of cerebrospinal meningitis in 1872 at the age of 47, leaving his desolate wife and teenage son with several rental properties in the city but little liquid capital. No longer could Sarah rely on her husband's counsel in her medical practice. Her journal suggests that they had maintained separate panels of patients (Sarah's were mainly women and children). She wrote, "I always see my patients alone, it is my habit, though I am willing to see and hear from their friends[,] first though I must make out their story from their own lips." But Lester's support was crucial on occasion. When a patient disputed her billing, for example, Lester had been the one to settle the matter.[60] Plainly she was devastated by her husband's death. At no other time was her confidence so shaken. "I seemed 'cribbed, cabined and confined,'" she remembered, "and I knew not how it was possible that I could bring to pass that which I had pledged myself perhaps to do" (February 12,

1888). The next two decades tested her determination to succeed as a physician and mother.

Dr. Dolley's letters from this period reveal her many responsibilities: to educate Sumner, nurture an independent medical practice, and manage the real estate holdings that were her late husband's main tangible asset.[61] During a restorative visit to her family in Philadelphia two months after Lester's death, she paused to write a letter to Sumner on the occasion of his sixteenth birthday, on June 16, 1872. Weighing their future—his responsibilities as a young man beginning his preprofessional years, hers as a single mother and solo practitioner—she began:

> My dear son, I remember this is thy birthday and wish very much I could be at home with thee. It seems so long since I left home and I am so anxious to return that I can scarcely wait. I can hardly realize here how my little boy will be a man if his life is spared. I want so much thee should have more rest and so grow and consolidate into a stronger physique than thy nervous, active temperament will allow without the greatest care to counterbalance constitutional tendencies. Now is the time to lay the foundation for good health and everything else good.

Blending the gently authoritarian tones of maternal "influence" with the language of medical expertise, Dolley laid claim to the role of both parents. Yet her authoritative tone in this brief letter belied her anxiety. Three months later she made the painful decision to send Sumner to boarding school in Geneseo, about thirty miles from Rochester. Her loneliness permeated her letters. In October she wrote, "I can't wait 'til Thanksgiving to have thee come, so thee may come this week or next as thee can best arrange with thy studies . . . If the teachers know that our horse is sick and that I want thee to come home, they won't care, I guess" (October 29, 1872). Only a month after Sumner's departure, she fretted over her decision to send him away. Many years later she still recalled "how many times I was told 'You should never have allowed your son to leave Rochester . . . You should have had him be a help to you'" (October 18, 1904). In a typical note, she alerted Sumner that she would soon send him a new blanket, chided him for forgetting his umbrella, and added, "When thee writes me, put the M.D. on the superscription of the envelope" (September 4, 1872).

Scholars have discovered many examples of letters to young adult children from nineteenth-century parents attempting to prepare them for a world of competition, unpredictability, and psychological unease. Most often, fathers wrote to sons and mothers to daughters. Typically, mothers took responsibility for modeling the moral virtues of conscience and self-restraint. Fathers, figures of worldly wisdom and familial authority, instructed sons in the virtues of perseverence, thrift, ambition, and industry. Boys learned boldness, entrepreneurialism, and other combative virtues mainly from their companions outside the household, often in defiance of maternal authority. Such lessons, of course, were never intended for Victorian girls.[62]

Sarah Dolley's correspondence never conformed to this Victorian maternal script. Her professional status, coupled with her widowhood, structured an atypical and frequently tense relationship with her son. Just below the surface of her letters lay a complex and unfamiliar dilemma. Dolley was one of the first women physicians to face a challenge that has helped define the personal meaning of female professionalism: practicing medicine while fulfilling her responsibilities (as she and her social peers defined them) to her son. Her letters reveal the clash of roles as she struggled to make sense of her compound status as physician and single parent. Her lifelong call to "restore the balance," certainly a fitting motto for a general practitioner, also shaped her understanding of how the disparate elements in her life might make up a coherent whole. Still, as her correspondence with Sumner attests, a genuine sense of balance often eluded her.

For example, preparing Sumner for a professional career was no simple matter, and she did not take this duty lightly. Unfortunately, Sumner felt no such obligation. Within a few months of leaving for boarding school, he sneaked home to attend a "masked ball," to the bewildered exasperation of his mother. By the end of January, less than halfway through his second term at school, she was reconsidering his preparation for college. "I want to know some things definitely," she wrote him. "One is . . . whether, if I had thee a tutor here to teach thee and get thee ready for college, thee would like that, as well as for me to go with thee wherever it seems best for thee to go . . . I want to do what is best for thee. I can read or study and be content anywhere that thee can be under good influences and have good advantages. Now where will that place be found? We must learn and must go and I am ready to cross oceans or continents or do anything that may be for thy good" (January 27, 1873).

In truth, Dolley neither would nor could afford to jeopardize her own career, even for Sumner. She briefly considered their living in Edinburgh, where she could brush up on her surgical skills while Sumner prepared for college. However, a letter from Dr. Lydia Fowler, now living in London, put an end to this idea. As Dolley explained it to Sumner, "no medical advantages [for me] are now at Edinburgh. [It has] closed to women. If thee was prepared for the university, I might sacrifice opportunities to employ my own time to advantage, but as thee will have yet to take either a year or eighteen months to be ready for almost any college, I think we must seek preparatory schools where climate and medical advantages may be found" (August 8, 1873).[63] Four days later she returned to this theme: "I think I could get a good practice [in Rochester] after a little, but I need rest very much from care and responsibility." In addition, that year the economic difficulties of the 1873 panic were never far from her thoughts: "The amount of labor and effort required to supplement our income to enable us to live here as we have been accustomed to live is a great matter, I don't want to give up professional work, only to take a rest and a change that will enable me to do better work in the higher and better paying departments of practice . . . I want to do credit by my scholarship as well as by my practical tact to my professional position" (August 10, 1873).

By the fall of 1873, Dolley had found her solution in a temporary appointment to the faculty of Woman's Medical College of Pennsylvania. Her friend the respected gynecological surgeon Dr. Emeline Horton Cleveland was temporarily unwell. Dolley agreed to take over Cleveland's teaching load for two terms, but only on condition that she not be offered a permanent post at the college. Dolley was no stranger to the leading women of her profession. Over the years she was the beneficiary of referrals from many of them. Accustomed to acting as preceptor to female and male students, she nevertheless dreaded all occasions requiring a public presentation. In giving her first lecture at the college, she confessed, "I was some embarrassed as I knew I would be . . . had control entirely of my voice, but shook some. I was awkward at first in handling my notes and had to confine myself to them . . . for fear of being disconcerted"(October 5, 1873). Within a week, though, she regained her confidence: "I see, when I am quite elementary with my class, and repeat and talk very slowly, they think it is 'perfectly splen-did.' So, as this will suit them, it will not be near so great exertion for me and I will profit by a little experience" (October 11, 1873). As it turned out, Dolley intensely disliked her brief academic career. She later wrote that if she had stayed on past the year, she "would have gone crazy" (October 11, 1887).[64]

The following September found Dolley back in Rochester, refreshed, retrained, and ready to take up her practice with vigor. Her son, who had spent part of the spring with her in Philadelphia, left for his first year at Syracuse University. Instead of devoting himself to scholarship, however, Sumner was busy courting his future wife, Elizabeth Gilman. In December Elizabeth received some unsolicited advice from Sum-ner's mother. Concerned about the effects of love's "first flush," Dolley feared her son might leave college and start out in business for the sake of an early marriage. "*Now I know*," she wrote, underlining her words, "that in these times of sharp rivalry if a man is going to have any *desirable position in society* and particularly if he were to be a professional man he must have the most thorough and careful preparation" (Decem-ber 14, 1874, emphasis in original). She urged Elizabeth to encourage Sumner to graduate. She knew her headstrong son too well to approach him directly on the subject.

Sarah Dolley pursued advancement for Sumner as assiduously as she pursued it for herself. As he wrote her in 1889, "I feel more and more as I progress in my life work that I owe a larger debt to thy teaching than to all the teachers I have had" (March 4, 1889). Steeped in the culture of mid-nineteenth-century clinical medicine, Dolley was intent on social-izing her son into its professional values. When Sumner rashly married Elizabeth right after his freshman year, Dolley wrote her new daughter-in-law, "Men must learn to endure hardness and to work conscien-tiously and thoroughly and while I think Sumner is conscientious and has worthy purposes, he is young and being an only son is [in] danger through motherly tenderness and fondness . . . of being too much made of and so being rendered less equal to grappling with that which is hard or distasteful" (November 12, 1875). Ultimately Sumner did leave Syracuse University without his bachelor's degree. Even so, he was admitted to the medical department of the University of Pennsylvania and graduated in 1882. After an unhappy year practicing medicine with his mother in Rochester, Sumner began graduate work in biology at

Johns Hopkins, intent on pursuing an academic career. He joined the faculty of Swarthmore College in 1885, and in 1886 he became a member of the University of Pennsylvania Biology Department, where he was a protégé of his mother's old friend, the paleontologist Joseph Leidy.[65]

Sarah Dolley was well aware of the increasing prestige of the emerging field of experimental biology; she hoped her son would establish himself as a specialist in some branch of research by devoting less time to his museum and laboratory curatorial duties and more time to his own investigations. She also appreciated a scientist's need to keep up with new work through current publications and overseas study. During the controversy over Robert Koch's claimed discovery of a tuberculosis vaccine, for example, Dolley herself and her medical partner, Dr. Anna Searing, spent "four full hours" reading Koch's account in the German periodicals rather than trusting the press reports.[66]

Yet Dolley's professional counsel was a double-edged sword to her son. Despite her informed enthusiasm for his academic career, her concept of professionalism differed markedly from the intensely specialized approach of the new generation of academic researchers.[67] From her perspective as a clinician and mother, Dolley still kept faith with the ideal of individual well-being that carefully balanced civic, personal, and professional life. She also feared the effects of overwork on her son's constitution. Thus, an ambivalent message ran through her letters of advice: "Broken down scholars are of not much more use than other invalids," she once declared (July 28, 1873).[68] To Sumner's wife she confided, "I am distressed as I have been for months at the intensity with which Sumner works. *No man of his age and build can work to the amount he does after his intense fashion and not be physically wrecked [,] if he lives*" (March 15, 1887, emphasis in original).

Sumner concurred. In his view, the University of Pennsylvania was overburdening its faculty with teaching, leaving little time for research. Even while proudly announcing his election to the American Philosophical Society, he admitted, "I feel quite set up, but also discouraged at the lack of time I have [for] original work." According to his calculations, he had had "more or less" a nervous breakdown every year since college.[69]

Although he published a respectable seventeen articles in six years, established the school's undergraduate biology laboratory, cofounded a marine biological laboratory at Sea Isle, New Jersey, and was a popular professor, Sumner was dismissed in 1892 as part of a general "retrenchment." He spent the rest of his long life doing independent research and consulting on industrial and commercial biological problems.[70]

In a revealing letter to his mother written during a combined onslaught of "nerves" and "the grippe" (influenza) a few months before his dismissal, Sumner described his state of mind and laid some of the blame, ironically, on his mother's too-strenuous professionalism, which he apparently experienced as insufficiently nurturing or maternal:

> The fact is, I am wearing out nervously, too rapidly, and must condition myself somewhat differently or the machine will stop. Speaking from a mechanical standpoint—my bearings will heat. The best I can do, they seem to wear and gut more than is right. Thee knows in making a bearing for a machine, both hard and soft metals are used . . . It doesn't do to put too much of the same kinds of metal in making up a man—a father and mother, both of hard metal, both adapted for the active moving parts of life's machinery, furnish too little proper material in their offspring for actually holding the machine together and preventing friction. I've just had a hot box . . . and now I am afraid to start out again for fear I can't pull through. (December 29, 1891)[71]

"The Medical Brothers Are So Polite and Appreciative"

Establishing a successful practice was never easy in the nineteenth century. It must have been especially daunting after the 1873 panic, when Sarah Dolley returned to Rochester from the Woman's Medical College of Pennsylvania. If not for the extra income from the rental properties acquired by her late husband, she and her son would have had little to live on. Besides her office and residence on East Avenue and a summer cottage, Dolley's properties included a house in Nebraska and three others in Rochester, all of which required oversight and continued full occupancy to offset tax bills that came due quarterly (letter of March 10, 1876). Writing to her son at college, she asked that he "save the pennies and I will try to earn the dollars. Practice is looking up a little and I am well and courageous although everything seems demanding means and little [is] coming in" (September 21, 1875). In November she purchased the horse and carriage she required for her practice, but she denied herself the dinner following the Medical Society of Central New York State's annual meeting because she could not afford it.[72] Her "medical brothers" did not make it easy for her during Dolley's first years in solo practice. During one "trying time . . . Dr. Whitbeck's son tried to get a case from me. Refused to consult with me." Fortunately the family called in another consultant who "sustained" her course of treatment and persuaded them to continue in her care (December 7, 1875).

At about this time Dolley began a long professional association with Dr. Anna Searing. She and Dr. Searing shared an office in Dolley's house on East Avenue, shared their meals, and even shared a bedroom in the house in order to create room for live-in patients. Dolley agreed to pay Searing a monthly salary, but sometimes her life insurance premiums, tax bills, and the unceasing requests for money from Sumner put her in arrears. Still, the two women thoroughly appreciated each other. Dolley often included news of Dr. Searing in her letters, and Searing herself occasionally wrote to Sumner of Dolley's physical or financial health. Nevertheless, her relationship with Searing could not replace the pleasures of her own family. The rare domestic notes in Dolley's letters spoke volumes. In one, headed "Thanksgiving Day, early candlelight," she admitted, "In all the past fifteen years, it has usually been that I felt more bereft on Thanksgiving Day than on any other day in the whole year" (November 24, 1887).

In reality, the house rarely was empty. Soon after Dr. Searing moved in, they began renting the extra bedroom to patients who required constant attention. "We have a patient in the house, a lady from Chicago," she wrote, "on whom I performed a somewhat painful operation, and a week from today one of my patients is to have the operation for ovariotomy which is one of the most serious. So you see how crowded I am" (July 19, 1876). Bill collecting continued to trouble Dolley. A letter to Sumner confessed, "This morning I was almost hopeless when I wrote thee and I sat by the fire and *meditated and calculated* what ways were most likely to bring [the needed money]. Wrote five letters [i.e., bills] and had an answer to two in the afternoon bringing the money" (January 24, 1876). Although most cases mentioned in her letters were located within the Rochester city limits, a fair number took her to outlying towns such as Brockport or the district out at Long Pond where she had her summer cottage. And of course she and Dr. Searing also saw patients at their office.

Dolley was proud of the collegiality she eventually came to enjoy with the male physicians in the community. A little more than a decade after her return from Philadelphia she wrote, "It is certainly gratifying that the medical brothers are so polite and appreciative." One even called her in as a consultant for his father.[73] She also conducted dissections with male colleagues, hardly a traditional activity for a woman physician in those years. One letter to her son at college began, "Today [Dr. Edward Angel and I] had a very interesting dissection . . . It was the case of 'Professor,' that is, Mrs. [Jane Marsh] Parker's favorite dog." After describing the painless procedures by which the dog was "dispatched," she marveled, "His olfactory lobes were wonderfully developed" (December 7, 1877). When in the 1880s vivisection in medical research became notorious among the members of Dolley's all-woman's discussion group, the Fortnightly Ignorance Club, she firmly supported the practice. As she wrote Sumner, "When I put the life of one child by that of a million cats . . . I expect I would give the baby life" (January 31, 1881).[74] Nor did her distress over the loss of a cherished patient, a young woman who died "the death of the righteous" in Dolley's own bed, deter her from conducting an autopsy assisted by three male colleagues: "She had requested it be done. Her heart and lungs were much diseased" (January, 1873, n.d.).

Dolley's therapeutics reflected the growing tension between an older generation's diagnostic holism and the "physiological" therapeutics of a younger generation newly introduced to cellular pathology and specific etiology. Mirroring the discordance within medical doctrine of the period, she viewed the patient as a holistic organism within a complex external environment, but also understood that diseases could arise from specific microorganisms or a breakdown of specific bodily systems at the level of cellular physiology. Although she continued to conceptualize diagnostics as knowing "the condition wherein the balance is disturbed," by the 1880s she routinely combined systemic stimulants and tonics for the nerves and blood—a supportive approach common by the 1860s—with antimicrobial antiseptics. Earlier in her practice she had favored nonspecific remedies such as "Fowler's solution," an arsenical compound believed to have tonic properties for the blood, nerves, and general metabolism. By 1889, prescribing for her grandchildren's colds and sore throats, she relied on both stimulants and antiseptic lozenges (December 25, 1889). In 1891 she dosed herself for an attack of "la grippe" with Quinia (an antipyretic), an iron tonic, and Thymol, an antiseptic (December 17, 1891). In 1896, while Dolley was recovering from an attack of diphtheria, her physician, Dr. Sarah Perry, prescribed alternate doses of whiskey and milk; Dolley herself added an "antiseptic wash" (November 19, 1896).[75]

Her practice precluded no aspect of general medicine. Indeed she never revised her lifelong suspicion of specialization—not merely, as Mary Putnam Jacobi did, for the sake of advancing women in all branches of medicine, but also on account of the specialists' high fees. Given her obstetrics training at Blockley and the urban middle class's increasing desire for physicians in the birthing room, it is likely she handled some obstetric cases during the years she practiced with her husband; in later years she rarely mentioned childbirths in her letters. Like her old friend Edward Mott Moore, she had a particular interest in surgery, especially orthopedics. As a surgeon she was relatively fearless (though not invariably successful). Besides obstetrics and gynecology, Dolley took on ovariotomies, strangulated hernias, and ovarian tumors. In her old age she decried what she saw as the younger generation's overcautious calculus of decision making. She told the story of a traveling salesman with a double hernia, a Mr. Powell, who went to see Dr. Richard Moore, son of surgeon Edward Mott Moore: "When the doctor found that he was a travelling man [he] said that he must change his occupation. Well, he went to keeping a real estate office, got so reduced that now they cannot have needed food or fire . . . I think if I could have known of Mr. P.'s case at the time he went to see Richard Moore and had him sent to the old doctor [Moore] and gotten a permit for him to St. Mary's [Hospital], he would never have been where he is today. He would not have been the first man that I saved after the doctor . . . had not found anything could be done but to give up his work. And I might add *starve*" (December 28, 1902, emphasis in original).[76]

Dolley's fearlessness did not extend, however, to the practice of abortion. Perhaps because of its negative impact on the reputation of women doctors, Dolley always viewed abortion with intense distaste. Women physicians were sometimes accused of being abortionists by hostile competitors, whether male physicians, midwives, or healers without formal training. Dolley recalled that "when I was a young physician a low woman said that I would commit abortions. I lost no time to go around, thy father going with me, and also engaging counsel and my determination could hardly be stayed not to prosecute the party with the utmost rigor of the law. But finally was prevailed upon to let it drop" (April 29, 1892). Late in life, she was outraged by charges of "a vituperative empiric" known as Madame Davenport that "'women doctors are abortionists.'" Dolley's colleagues in the Practitioners' Society declined to lodge a public protest, since the woman "was unworthy of notice" (undated fragment, ca. December 1904). Dolley's description of one of her own tenants, a married mother of several children whose attempted abortion Dolley managed to prevent, displayed none of the sympathy that usually characterized Dolley's accounts of impoverished patients. She also apparently declined to attend the daughter of a poor workman who previously had been attended by a male physician widely rumored to be an abortionist.[77]

"Moral Power and Moral Purpose"

After a decade on her own, Dr. Dolley could write with satisfaction of her well-rounded life practicing medicine, engaging in civic reform, and running a household that, for a time, included her medical partner, her daughter-in-law, two grandchildren under the age of six, a nurse, a cook, and the occasional live-in patient. "With a family of eight," she wrote her son, "I feel pretty tranquil to keep all 'agoing and pay a little bill each week . . . [T]he good example and instruction and mental and physical sustenance of the babies are the matters that take most of my thought" (December 7, 1884). Two years later she proudly reflected, "It is in practical medicine alone that I have done anything, and as the shoemaker must stick to his last, I know without any wavering that this is my work" (January 29, 1887).[78] At a time when women physicians constituted fewer than 3 percent of practitioners, one can appreciate the value she placed on both prudence and perseverance.[79]

Dolley's substantial successes rested uneasily on two distinct traditions: a commitment to Quaker modesty and to evangelical Protestantism. Dolley's letters contain many references to the enduring influence of her Quaker heritage despite her subsequent separation from the Society of Friends. As she wrote to a friend in 1890, "Friends have principles which are fundamental" (October 27, 1890). Nevertheless, from Dolley's days as a medical student, and especially after her husband's death, evangelical Presbyterianism provided the more satisfactory outlet for her strenuous spirituality.[80] The day after her much-dreaded first lecture at the Woman's Medical College of Philadelphia, for example, she attended "a good Presbyterian sermon." Its message, "to look to God for grace *to do, bear, and to suffer* as we all . . . must," gave her comfort. This, not the equable Quaker doctrine of humankind's inborn sanctity, strengthened her resolve during times when she felt "alone without anchorage or direction" (October 5, 1873). The concept of Darwinian evolution, she also believed, was "painfully inadequate to give any solace in affliction." She agreed with those who saw the doctrine as proof of the "theory of design."[81]

It was Dolley's Hicksite Quaker heritage, however, that drew her to Rochester's staunchest women activists. She and Frances Willard, leader of the Women's Christian Temperance Union, were good friends. And despite occasional disagreements with them, she had close ties to Susan B. Anthony and her circle.[82] In 1872, she was one of fifty or so women who attempted to register to vote. Although only Anthony and several others succeeded, Dolley recorded in a letter to Sumner that "the election triumph was a glorious one . . . sixteen women voted in the city" (November 1872, n.d.).[83] Even though her professional concerns took precedence, Dolley was truly committed to women's rights—especially the right to equal protection under the law. Thus she favored legal changes that would establish women's right to manage their own property, to vote, to sit on school boards and on the boards governing such other public welfare institutions as houses of refuge, and to gain some redress against drunken or abusive husbands. In this context she supported a moderate version of her friend Frances Willard's temperance program. Dolley was no teetotaler. She did favor, however, a drastic increase in the tax on alcoholic beverages as a deterrent against excess. She did not expect human nature to change.[84]

Ingrained Quaker modesty precluded Dolley's seeking public office. She considered a potential offer to sit on the board of the Western House of Refuge because of the additional salary it might provide, but she refused to run for the Rochester School Board in 1883 when her friends first began campaigning for a woman board member. And unlike younger colleagues such as Mary Putnam Jacobi, she ceased to participate openly in the suffrage movement. She lent her prestigious name to the invitations to plan the Women's Educational and Industrial Union of Rochester during the depression and widespread hardships of 1893, but only as president of the Fortnightly Ignorance Club, not in her name alone. Finally, in 1894, Susan B. Anthony insisted that Dolley and Searing sign the call for a public meeting on the suffrage question so that the public would know her position on the matter. Anthony also insisted, despite Dolley's reluctance, that they sit on the platform during the rally.[85]

Over the years, Dolley directed most of her attention to professional and family concerns. After advising Susan B. Anthony to "lay aside all weights for a time and rest a bit," Anthony replied, "'Well, you are a wise woman and you have as persistently stuck to your profession as I have to suffrage'" (January 8, 1897). Dolley was a conscientious and successful mentor to younger women colleagues. At least from the 1860s, Dolley was a preceptor to both male and female students, but it was the women students who took precedence in her mind.[86] When young women graduates, such as Dr. Marion Craig, opened a practice in Rochester, they found Dolley at the heart of their women's professional community, their link to that wider society of male physicians with which she was uniquely well connected. She always chose women physicians for herself, and she often expressed pride in their prosperity and growing success. If necessary, she would put her prestige on the line to support their professional interests. When her friend and colleague Dr. Harriet Turner, a graduate of Woman's Medical College, cofounder of the Practitioners' Society, and one of the most experienced women in the city, lost her nomination for city physician in a close ballot during a caucus of the ruling Republicans, Dolley spent days lobbying Rochester's Republican boss, George Aldridge, to overturn the vote. She told her son that if those tactics failed, she would go straight to the Democratic leader. Dolley seems to have won the skirmish.[87] Three years later she also successfully defended the position of Dr. Kathleen Buck, a recent graduate of Trinity College Medical Department in Toronto, whom she recommended over an older male colleague as physician to the nuns at Nazareth Convent. The convent's elderly sister superior at first insisted on the older physician, but she came around after private negotiations with the sisters, all of whom wanted Dr. Buck.[88]

In medical politics Dolley also played an active hand—locally and farther afield. In December 1886 she was instrumental in organizing

Dr. Sarah Read Adamson Dolley (1829–1909), photographed in 1904 in honor of her seventy-fifth birthday, after presentation of a diamond brooch by fellow members of the Practitioners' Society of Rochester, New York. (Courtesy of the Edward G. Miner Library, University of Rochester School of Medicine and Dentistry, Rochester, New York.)

both the all-women's Practitioners' Society, and the Provident Dispensary for Women and Children, which was run by Practitioners' Society members.[89] In 1887, for the first time, Dolley ventured to Albany for the annual meeting of the Medical Society of New York State, where she heard remarks by pediatrician Abraham Jacobi, someone she deeply respected. In 1889, at the age of sixty, she was a delegate of the Monroe County Medical Society at the annual meeting of the American Medical Association (AMA) in Newport, the same year Edward Mott Moore became AMA president-elect.[90] Dolley's presence was less significant as an indication of the rising status of women in medicine than as recognition of her standing within her county society. Although the appearance in 1876 of the first woman AMA delegate, Sarah Hackett Stevenson of the Illinois State Medical Society, had been met with stiff resentment, by 1889 women delegates were rare but no longer controversial. As Paul Starr makes clear, "By the late 1800s the number

of [AMA] delegates had become unmanageable; virtually anyone showing up at the AMA's annual meeting could take part in its business, as it
was impossible to check credentials."[91]

Dolley's appearance at the meeting also attested to the vitality of her
"Philadelphia connection" and her concern over the status of homeopathy in the profession. Much to her chagrin, the Rochester-based majority of the Monroe County Medical Society had followed the lead of
other large-city delegates to the State Medical Society by voting to
allow consultation with homeopathic physicians, an action that violated
the AMA's Code of Ethics and cost the state society its standing as a
constituent member of the AMA. Dolley, Dr. Moore, and the others
chosen to attend the AMA's meeting were elected, in part, in hopes that
their adherence to the AMA consultation clause might placate AMA
officials. Dolley deeply respected the experimental mentality of scientific medicine. Despite—or because of—her eclectic medical schooling,
she could never abide what she viewed as homeopathy's unscientific
therapeutic dogmatism; moreover, she suspected homeopaths of actively misleading their patients.[92] She hoped to use her position to try to
heal the rift between the New York society and the AMA by calling on
the intercession of her uncle, Hiram Corson (June 4, 1888). This was an
ambitious plan even for an elder stateswoman like Dolley.[93]

Dolley's letters testify to the contradictory effects of old age on such
an accomplished and independent woman. During her late seventies,
her handwriting took on the spidery traces of advancing age and declining vigor. Yet in 1908, she became founding president of the Women's
Medical Society of New York State. Her presidential address drew on
the same philosophy (and some of the same language) as her address to
the graduating class of Woman's Medical College thirty-five years earlier. Invoking her own inner rallying cry, "the strength to suffer and the
will to serve," she urged her colleagues not only to hone medical skills
and scientific knowledge but also "to bring moral power and moral
purpose to permeate your whole professional life. At all cost keep the
vantage for the soul."[94] The following year, when she was just short of
her eightieth birthday and recovering from a broken leg, Dolley prepared to attend the society's first annual meeting, held on March 11 in
honor of her birthday. Despite being *"weary,"* she wrote that she was
delighted by the evening birthday festivities replete with flowers, con-
gratulations, and a cake with eighty candles (March 13, 1909, emphasis

in original). In August of that year, four months before her death, she proudly wrote to Sumner, "I am being heralded all over the country as honorary chairwoman of the [Public Health Education Committee] authorized by the American Medical Association at Atlantic City."[95]

Despite the recognition, Dolley lived her last years in Spartan fashion. During a brief stay as a resident in a boarding house that catered primarily to young single men, she felt as lonely as if she had "retired to a convent." She then moved into a room in one of her own houses. As always, she was "cash poor," and she even borrowed small sums from friends. Her letters portray her as owning little more than "one plate . . . two good teacups and several saucers" (September 5, 1909). She managed to invite friends for tea now and then, but she could not keep loneliness at bay. Dolley's real estate holdings could have provided financial independence, but she resisted selling them off so that Sumner could inherit them intact. Three months before her death she completed the final revision of her last will and testament, leaving the bulk of her estate to Sumner. Since he was out of the country, she named Dr. Kathleen Buck, her personal physician and friend of many years, as executor.[96] Dolley died on December 27, 1909.

Sarah Dolley's career epitomized those of many in her generation and set a pattern for even more of her twentieth-century successors, women physicians who strove to balance the demands of personal and professional, private and public life, to reconcile femininity, feminism, and professionalism. She probably never fully achieved the ideal of balance she sought for herself, her women colleagues, and her patients. Yet Sarah Dolley not only endured but positively thrived on the creative tension between her professional and her feminine identities.

2

Gendered Practices: Late Victorian Medicine in the Woman's Sphere

> The fact of a medical society conducted by women physicians is an innovation . . . We are members of a learned profession of which the opposite sex are as the sands of the sea compared with us in numbers. They hold precedence by right of occupation of the premises and also by reason of general prejudice of the public.
>
> *Mary Stark, M.D., Practitioners' Society annual meeting, 1889*

THE ESTABLISHMENT of women's medical societies and dispensaries (outpatient clinics for the indigent and working poor) was one of the legacies of pioneer women physicians like Sarah Dolley.[1] Such institutions provided professional settings in which the next generation of women doctors could prepare themselves to gain professional acceptance in previously all-male medical institutions. Just as important, they also satisfied the special interest in women's and children's health that had motivated many of them to enter the profession in the first place. Primarily, therefore, their earliest professional institution building was accomplished within the "woman's sphere."

Although the etiologic and therapeutic philosophies of male and female physicians were essentially similar, their practices did differ in important ways. The settings in which women practiced medicine, their deep empathy for and interest in the problems of women and children, and the meaning they and society at large accorded their activities bracketed women's medicine as a special domain until nearly the end of the nineteenth century.[2] At a time when the majority of women physicians still received their medical education in all-female colleges, it is not surprising that they firmly upheld the value of feminine solidarity.[3] The rise and decline of women's medical societies and dispensaries loosely parallels the course of overt gender separatism in American medicine.

Because the early communities of women physicians were primarily regional in character and only gradually were absorbed into national professional networks, this chapter draws on the history of the influential (and well-documented) women's medical community of Rochester, New York. For purposes of a case study of late nineteenth-century women's medical institutions, Rochester's all-women's Practitioners' Society and the Provident Dispensary for Women and Children were typical in many ways. They exemplify the settings in which many women physicians first enacted their conception of professionalism.

Women physicians' organization building therefore was influenced as much by changes within their profession as by the goals of social reform. Institutions such as those in Rochester were effective instruments of professional legitimation and, eventually, integration. They also allowed women physicians to fulfill their sense of obligation to the health care needs of their own sex. Any analysis of women's medical institutions must come to terms with their professional significance. They combined both what Estelle Freedman has called "separate institution building" and an intermediate stage of female integration into a male-dominated profession.[4] Such institutions flourished because they introduced women physicians to the professional community while accommodating their long-held and highly valued ties to the community of female social reform.

In the late nineteenth century, professionalism in general, and professional societies in particular, formed a vital link in the process by which middle-class Americans redefined their relationship to the world around them. As older sources of identity dependent on local traditions and values gave way before the complexities of urban, industrial America, identification with one's profession offered a satisfying substitute for community ties defined by geographic locale. Professional societies also functioned as gatekeepers, controlling entry into practice, and as referees deciding questions of legitimacy and professional ethics. They thus conferred a powerful sense of collegiality and allegiance on their members. Members routinely spoke out on the health-related, civic concerns of their home cities, and, increasingly, they did so in the capacity of specialized professionals guarding the public's health, not merely as interested citizens.[5] Their faith in scientific progress and professional authority supported a set of common values that increasingly diverged from those of their neighbors.[6]

Male or female, regular or homeopath, anyone in medical practice around the turn of the century faced the challenges of medicine's professionalization. The impact of these changes, however, was experienced somewhat differently by women physicians. Most women apparently held on to the older professional norms longer than many of their male colleagues—and for good reason. For them, the shift in professional identity toward the collegial and away from the local community proved relatively thankless and far less rewarding than it did for men. They could not easily cast aside the moral, civic, reformist component of nineteenth-century professionalism. After all, those were the qualities that had brought many women into medicine in the first place. Several excellent studies have demonstrated that a strong infusion of the reforming zeal common to "social housekeepers" of the late nineteenth century colored the professional aspirations of the early generations of women physicians.[7] By the turn of the century, feminine professionalism buttressed a commitment to "maternalist" health care reforms that would extend needed social services to impoverished mothers and children.[8]

In short, women doctors demonstrated a diverse combination of interests and activities, some directly professional, some not. These included the opening of women's medical colleges, hospitals, and dispensaries; concern for the health of women and children, especially the poor; a commitment to women's higher education and the suffrage and temperance movements; and membership in countless local women's clubs. With the exception of suffrage, these ideals helped pave the way for widespread public acceptance of women physicians. Feminist professionalism, maternalist social reform, and social feminism, though distinct, were thoroughly compatible through the early 1900s.[9] When these broad-based concerns began to diverge from twentieth-century collegial professionalism, women's medical societies, dispensaries, and other such medical institutions prolonged their compatibility into the twentieth century.

Women's Medical Societies

The proliferation of medical societies was common to all sectors of the physician population—male, female, allopathic, homeopathic—at the end of the century. Most tried to fulfill three related objectives: the advance of professional interests, as, for example, through lobbying state legislatures; the presentation of clinical cases, both as a forum of clinical research and as "continuing education"; and, finally, the opportunity to socialize with colleagues. In the years before medical school hierarchies emerged as rival professional communities sufficient unto themselves, general and specialized medical societies stood at the apex of the professional community. Women's medical societies frequently were modeled after the exclusively or predominantly male medical societies. All physicians in the late nineteenth century grappled with the uncertainties of scientific standards, professional status, and etiquette; medical societies for women, like those for their male colleagues, acted as arbiters of professional conduct, assisting their members to sort out their roles and responsibilities as professionals.

Women's medical societies, however, fused more than one set of values by combining the ideals of professional medicine, feminism, and maternalist reform. Thus they made a unique, if short-lived, contribution to their members' sense of professional identity. The decision to found such societies was a positive strategy rather than a negative reaction to the exclusionary policies of male-dominated societies. Indeed, by 1881 seventeen state medical societies had opened their doors to women, although only a fraction of either male or female physicians ever actually joined their state society. (In contrast, local medical societies, whether restricted by gender or not, enjoyed large and loyal memberships.)[10] The real obstacle faced by women doctors was not exclusion but marginality within existing professional societies.

For more than thirty years, the Practitioners' Society of Rochester, New York, sustained a balance between two increasingly distinct hierarchies of value, one technical and scientific, the other reformist, moralizing, value laden. The first society of regular women physicians in Rochester, it was organized in 1887 and renamed the Blackwell Medical Society in 1906 in honor of Elizabeth Blackwell. It appears to have been the earliest such society in the United States not restricted to the alumnae of a single institution. When it first met in January 1887, the society had eight members.[11] Over the next forty years its membership rolls included a total of fifty-eight women plus an additional half-dozen honorary members. Its membership peaked in 1910 at more than 7 percent of the total physician population of the Rochester region, including nearly all of the women "regulars" in the city. By 1915 the group comprised 6 percent of the local physician population, and by 1920 it had declined further, to 5 percent, a figure more in line with the percentage of women in American medicine at the time.[12]

The heyday of the society coincided with the beginning of the decline of sectarian medicine. Its admissions policies attested to its members' identification with medicine's mainstream, limiting membership to "regularly qualified physicians" as defined by the Medical Society of the State of New York. Given the recent decision of the state society to admit homeopaths, homeopathic graduates technically would have qualified for membership. Nevertheless, whether because they instead joined the local homeopathic medical society, which was open to women members, or were discouraged by the regulars, only Anna Searing, one of the original eight members, was a graduate of a homeopathic medical college. Because she was one of the most senior medical women in the city and an early University of Michigan graduate (two years before women were admitted to its regular medical department), the society applied a kind of "grandmother clause" to her eligibility. Similarly, Dr. Sarah Adamson Dolley, one of the founders and the society's most illustrious member, had graduated from an eclectic medical college but gradually abandoned her identification with its teachings. Only one of the later members was a homeopath.[13] Professional relations with other female homeopaths of Rochester were cordial but distant. Only occasionally were the latter invited to attend Practitioners' Society meetings. In the summer of 1902, for example, two homeopathic physicians cooperated in the efforts of the female regulars to raise money for a medical missionary to China, Li Bi Cu. Several months later, when the president of the New York State Medical Association sent the society a general letter urging the union of regular and homeopathic medical societies, the request was "tabled" without further discussion. Nor did the various hydropaths, osteopaths, magnetotherapists, and Christian Scientists of the city play any part in the society's history.[14]

Unlike its all-male counterpart in the city, the Pathological Society, which was known to blackball prospective members, the Practitioners' Society may have discouraged inappropriate applications for admission so as to avoid such confrontations. Once a prospective member was proposed, procedures were courteous and even casual. Prospective members generally were proposed informally, though procedures were tightened in 1906, when the society reincorporated and changed its name to the Blackwell Medical Society. In most cases a candidate would be elected to membership at the meeting following her initial consideration. Since meetings were held only once a month, there was plenty of time for second thoughts. Yet often a new member would pay her dues or be nominated to a committee even before she was formally elected. No doubt an informal screening of candidates prior to their application made explicit blackballing unnecessary.[15]

The educational, personal, and career histories of the members reflect the extent of their integration into the professional world of their male colleagues. Of the society's fifty-eight active members, forty-one (71 percent) had received coeducational training, the largest single group (fifteen) having graduated from the University of Michigan. Buffalo and Syracuse Universities accounted for nine members and five members, respectively. Those three schools alone had graduated half the society's membership. Seven members had graduated from Woman's Medical College of Pennsylvania and seven from Woman's Medical College of the New York Infirmary.[16] Half the group were native New Yorkers; twenty-five members hailed from Rochester or western New York. Twenty-four women, or 41 percent, were married, a higher figure than the 32 percent to 35 percent marriage rate estimated for female physicians nationwide in the same period. At least one-third of those who were married had married a doctor.[17]

Superficially, the career patterns of the Practitioners' Society members fit the general pattern for women physicians of the era. Of the fifty-two for whom such information was found, forty-two, or more than 80 percent, were engaged in general practice for part of their career; ten of those in general practice, however, became de facto specialists in obstetrics, gynecology, psychiatry, or anesthesiology by the end of their career. Thus, a total of twenty (38 percent) eventually became specialists: ten in psychiatry, six in obstetrics and gynecology, and the remaining four in public health or anesthesiology. (These figures

The Practitioners' Society of Rochester, New York, 1899. Dr. Sarah Dolley (second row) has been deliberately seated in the center. Seated in the first row, left to right, are Drs. Marion Craig Potter, Kathleen Buck, and Mary E. Dickinson. (Courtesy of the Edward G. Miner Library, University of Rochester School of Medicine and Dentistry, Rochester, New York.)

may be conservative.) As late as 1929, according to the well-known Chicago obstetrician-gynecologist Bertha Van Hoosen, women "were not regarded as eligible for membership" in the national obstetrical and gynecological societies.[18] Since women were excluded from such groups, obituaries and other indirect sources are the only means of determining de facto specialization before 1930.[19]

Despite such obstacles, the society's members made every effort to forge strong personal and professional ties to the male medical community of Rochester. Sarah Dolley, as we have seen, was a friend to many of the city's leading male medical figures. Equally comfortable in male professional circles were Drs. Eveline Ballintine, Marion Craig, and her sister, the future Sara Craig Buckley. Marion Craig's marriage to Dr. Ezra Potter, assistant physician at Rochester State Hospital and the future assistant superintendent there, allied two of the leading physicians in the city. A number of the elite medical institutions of Rochester sanctioned the membership of a small number of well-connected women physicians. The Rochester Academy of Medicine, for example, was intended at its founding in 1899 to be the most prestigious medical society in town. Unlike the much older Pathological Society, formerly the most exclusive medical society in town, the academy was not closed to women. The academy's only female charter member, Dr. Eveline Ballintine, was a colleague of Ezra Potter's at Rochester State Hospital and a staunch supporter of the Practitioners' Society.[20] Even more important for the integration of women into formerly all-male medical institutions was the decision of Rochester City Hospital in 1898 to employ its first female physician, Marion Craig Potter, on the outpatient staff. A year later, Potter recommended that her fellow Practitioners' Society member Evelyn Baldwin be appointed to the staff as well.[21] A third society colleague, M. May Allen, was appointed at Potter's and Baldwin's request to fill in during their summer vacations. A fourth, Cornelia White-Thomas, was later added to the permanent staff, but as White-Thomas's husband was already on the hospital staff, it is difficult to know by whose influence this appointment was made.

Practitioners' Society members cultivated collegial ties to Rochester's medical elite individually and collectively. Leading male physicians in general surgery, orthopedics, laryngology, and X-ray technology all read papers before meetings of the society. The society's members also

cooperated with the president of the Pathological Society in a campaign
to strengthen the city charter's public health provisions; they cooperated with the Pathological Society again in lobbying against state licensure of osteopaths and opticians. While only about one-third of the
Practitioners' Society's members joined the New York State Medical
Society, by 1900 at least one member was sent every year to represent
the group at the state society's annual meeting in Albany. In 1895 the
Practitioners' Society heard the account of Dr. Alice Brownell's trip to
Baltimore for the annual meeting of the AMA.

Thus, by the turn of the century the society's minutes record an
increasing number of references to the wider world of American medical practice. In 1900, for example, a trial of cotton gloves for surgery
was described. Sometimes the society heard reports of on-site visits, as
when Marion Craig Potter reported on her trips to the New York
Polyclinic and the hospitals of Chicago. Returning from Chicago in
1905, Potter reported "great attention to asepsis, constant use of rubber
gloves and enveloping sterile garments." That same year Dr. Ida Porter
described an operation performed by Johns Hopkins gynecologist Dr.
Howard Kelly. In 1906, a report to the society included the news that
the Johns Hopkins Hospital was experimenting with the use of paper
"napkins" for patients with chronic cough. Members occasionally reported on the papers presented at the Rochester Academy of Medicine or read to the meeting from the *American Journal of Clinical Science*,
the *Journal of the American Medical Association (JAMA)*, and the *New
York State Journal of Medicine*. Rochester's pioneer roentgenologist, Dr.
Louis Weigel, reported on the status of X-ray technology and orthopedic surgery to both the Practitioners' Society and the Pathological Society. Roswell Park of Buffalo addressed only the Pathological Society in
person, but three months later the Practitioners' members heard an
extensive secondhand report on his work. Their colleague, Eveline
Ballintine, the senior woman physician at Rochester State Hospital,
often reported to society members on the current laws regarding the
institutionalization of the insane and the classification of the various
types of mental disease, reports that were similar to those given by the
hospital's superintendent to the Pathological Society.[22]

A first glance at the proceedings of the two societies suggests a strong
resemblance between the groups. In recent years historians have won-

dered about the ideological and therapeutic differences between male and female doctors. Did male physicians hold a paternalistic view of the female nature and feminine social role, and, if so, did they employ a deliberately punitive, controlling therapeutic regimen with their women patients? Were women physicians different from men in their view of female patients, in their therapeutics, or in both? Although some investigators concluded initially that the two groups differed radically in both respects, detailed investigation has led to a more complex conclusion. While the prescriptive literature by male physicians has been found to be generally more traditional and conservative in its implications for women than the comparable literature by women physicians, few corresponding differences have been found in the therapeutics of the two groups. A closer comparative look at the Practitioners' Society and the Pathological Society may provide some insight into the character and extent of their similarities and differences.[23]

Like those of the Pathological Society, the Practitioners' constitution and bylaws called for "mutual improvement in all scientific subjects pertaining to . . . medicine" and the promotion of "social intercourse among [the Society's] members".[24] But whereas the men's group met at the Rochester Whist Club and followed its scientific program with a manly repast of stewed oysters at a public eatery, the women met at each other's homes or offices and did their own catering. (Only a woman's society, one imagines, would include a formal injunction in its bylaws against serving more than six dishes to the membership.) Similarly, Practitioners' Society members cheerfully departed from the masculine model by exchanging Valentines whenever a meeting fell on February 14.

In their training, therapeutics, and approach to clinical science, however, Rochester's female regulars were virtually indistinguishable from their contemporaries in the Pathological Society. One-third of the Practitioners' Society's active members between 1900 and 1910, for example, were graduates of the University of Michigan, which had one of the best medical departments in the country, with a strong tradition of preclinical science and a clinical clerkship hampered only by the inadequate size of its teaching hospital.[25] Of the ninety-seven members of the Pathological Society in 1910, more than one-third were graduates of either the College of Physicians and Surgeons of New York City (it became Columbia University Medical School in 1896) or the University of Pennsylvania, schools engaged in serious attempts to upgrade their preclinical training but still hampered by inadequate facilities for clinical clerkships. Between one-fifth and one-fourth of the members of both societies were graduates of the upstate medical colleges at the Universities of Buffalo and Syracuse. Half the members of both societies in 1910 had graduated since 1890. Thus, a considerable number of members in both groups were trained to know the value of basic and clinical research and had been given at least some exposure to both. Yet the realities of everyday practice, the unavailability of research facilities, and the extreme limitations of the clinical training available before 1910 everywhere but at Johns Hopkins took a similar toll on local practitioners everywhere: the papers and reports of both groups performed as much a collegial as a scientific service, affording a briefly shared sense of participation in the world of medical science.[26]

Members of both societies would have agreed with Dr. Mary Stark's presidential address to the Practitioners' Society in 1889, in which she noted that "professional competition increases each year and we must keep pace with medical science or we will go to the wall."[27] Perhaps Stark's warning helps to explain Marion Craig Potter's anger at what she viewed as inexcusable backsliding by society members during 1895 and 1896. The minutes tell a tale of declining attendance, perfunctory committee reports, and a persistent unpreparedness by the scheduled speakers. After months of witnessing such behavior, Potter had had enough. During a meeting at Sarah Dolley's house in 1896, Potter "severely" berated the president and secretary for failing to secure scientific papers for the monthly meetings. According to the minutes, "a somewhat prolonged discussion followed." In the end, President Dolley and Secretary Harriet Turner "promised to mend their ways."[28] Similarly, in 1888, one of the senior members of the Pathological Society was appointed to investigate ways to improve meetings. His committee urged that "the discussions be carried on with more spirit."[29]

Papers presented to the Practitioners' Society and to the Pathological Society were cut from the same cloth. Few differences were discernible in either the format employed or the conclusions reached in their scientific and clinical presentations. Generally a discussant would describe the historical progression of opinions surrounding a subject, summarize the best current ideas, and offer a description of clinical cases drawn from her or his own practice.[30] A comparison of the therapeutic values of the two societies reveals their overall similarity of clinical judgment. In obstetrics, for example, concerning the complex matter of the appropriate use of forceps, one discussant at the Pathological Society in 1888 favored their use directly after the "cessation of progress" in labor; in the case of "anaemic women with weak, flabby muscles," he urged the application of forceps "even before cervical dilatation is complete." The speaker's paper provoked considerable disagreement during a lengthy discussion among ten other member physicians. The majority held that forceps should be used only as a last resort. Even those supporting the speaker justified their position on the grounds of "the right of parturient women to relief from pain," in some cases recommending the administration of chloroform in forceps deliveries. On the (quite) related question of whether to repair lacerations of the perineum, a discussion held several years later revealed that most members favored immediate repair. A similar discussion at the Practitioners' Society in 1900 found the majority also in favor of immediate repair. On the use of forceps, Dr. Kathleen Buck's survey of sixteen consecutive cases in 1901 indicated no disagreement with the men. Buck distinguished between "normal" and forceps deliveries; the latter had occurred in one-fourth of the cases, all of which were described as "slow" or "difficult." However, perhaps because of its inherent risks and its apparent tendency to slow down contractions, Buck had administered chloroform only twice, once for a difficult forceps delivery and once for a case of miscarriage at four months.[31]

In gynecological cases, the evidence suggests that the women initially were more willing to operate than the men, although scattered evidence of this period suggests that wide variation existed among all physicians practicing gynecological surgery.[32] In 1901, Drs. Evelyn Baldwin and Marion Craig Potter, partners at the time in the Rochester City Hospital's Outpatient Department Clinic on Diseases of Women, reported three cases of clitorectomies performed on young women who practiced, as the minutes report, "self-abuse." Dr. Frances Hulburt-White reported the same year on a case of hysteria in which an ovariotomy failed to have any effect.[33] Examples of gynecological surgery such as these are not to be found in the minutes of the male physicians' society.

In fact, within two years the Practitioners' Society also explicitly disavowed any presumed link between pelvic and nervous diseases, as did their male counterparts. On the related subject of nervous diseases, by 1903 both societies had heard reports reassessing the etiology of hysteria as a "mental disease" requiring "mental therapeutics," discounting the older notion that nervous disorders in women originated in pelvic or uterine disorders. Both societies also discussed S. Weir Mitchell's rest cure for neurasthenia.[34]

Differences in the topics these societies chose to discuss may be attributable to the different choices of specialties among their members. These choices, in turn, reflected the various factors—social, intellectual, and professional—influencing male and female physicians' decisions on whether to specialize and what specialty to choose. For example, the women's society had a consistently higher concentration of papers on insanity and nervous diseases, reflecting the particular contributions of Dr. Eveline Ballintine, the local importance of Rochester State Hospital, and the general importance of psychiatry to women physicians of the period. New York, after all, was one of a handful of eastern states between 1880 and 1900 requiring that all public mental hospitals have a woman physician on staff.[35] Among the Practitioners' membership, half of those who eventually became specialists chose to work in psychiatry. The women demonstrated a higher interest in obstetrics and gynecology, too. Many women doctors, like their male colleagues, increasingly accepted specialization as a mark of expertise rather than of quackery or commercialization. Moreover, late nineteenth-century medical practice was marked by wide variances in the understanding of what specialization actually signified—whether it meant concentrating on a particular type of technique or type of case within a general practice, or devoting oneself exclusively to one highly focused area of interest.

Gender issues also played a role in the choices made by women and men physicians. An earlier view, expressed by Mary Putnam Jacobi in 1880, equated female specialization in obstetrics, gynecology, and pediatrics with narrow training and incomplete preparation. By the end of the century, this notion appears to have been abandoned as more prestige accrued to specialized practice. Yet while women physicians could be seen as taking advantage of their reputed expertise in obstetrics and gynecology, they also tended to cluster in fields where social acceptance and individual interest coincided.[36] In contrast to the women's emphasis on psychiatry, obstetrics, and gynecology, the Pathological Society included many more discussions of ophthalmology, otolaryngology, and orthopedic surgery, reflecting the importance of these new specialties to its members. Thus, specialization was increasingly characteristic of the careers of both male and female physicians, but a narrower range of choices characterized the specialization of the women. And specialization for women doctors long remained a matter of unofficial concentration, because most of the national specialty societies refused to admit women as members until well after the creation of the specialty boards during the 1930s. In this regard, women's medical societies did serve a compensatory function as de facto specialty societies.

The question remains whether the women of the Practitioners' Society pursued professional advancement at the expense of feminism and social reform. The answer must be no. The most successful female physicians in Rochester were precisely those who cultivated the strongest and most enduring ties to local, state, and national women's organizations, medical and otherwise.[37] At least until World War I, they never flagged in their commitment to women in medicine nor, indeed, in their commitment to more general concerns—such as temperance, social hygiene, and suffrage—inherited from nineteenth-century female reformers and suffragists. The history of the Practitioners' Society between 1900 and the end of World War I reveals its leaders' deliberate attempts to sustain members' loyalties to the values of professionalism, social reform, and women's rights. In contrast, a study of women physicians in Washington, D.C., by Gloria Moldow found a declining interest in social reform and feminism and a correspondingly increased concentration on strictly professional advancement in graduates of the 1880s and later.[38]

What accounts for the Rochester society's sustained social—and feminist—commitment? The answer may lie in its members' experiences prior to their decision to become doctors. One-third of the members had attended a normal school, a female seminary, a women's college, or a nurse training school prior to turning to medicine for a career. Moreover, among graduates from the 1880s, five of whom were founders of the society, the percentage who spent their premedical years at predominantly female institutions was double the rate for the group as a whole. One can reasonably speculate that their social feminism was rooted in an early exposure to the feminine culture of female-dominated educational institutions in the late nineteenth century. In some cases, notably Sarah Dolley's, an egalitarian, Quaker background prepared the way for a lifetime of social activism. Prior experience, therefore, may help to explain why, in spite of a commitment to professional integration, the group remained true to its original values and expectations for more than twenty years.[39]

More important, perhaps, was the influence of the society's remarkable founders. Sarah Dolley, Mary Stark, and Marion Craig Potter—all founders and past presidents of the society—brought to it a strong tradition of social activism. In the 1880s, Dolley, Potter, and other women colleagues joined Susan B. Anthony and her circle in founding the prosuffrage Political Equality Club (PEC); Potter also allied herself with Anthony in a variety of court cases taken up in the cause of women's suffrage. When in 1899 Anthony invited seventy-three women's groups to join in a Rochester Council of Women, Stark proposed that the Practitioners' Society accept the invitation.[40] Dolley, Stark, Potter, and the rest insisted from the beginning that the society support and encourage a broad conception of feminine professionalism—a later generation would call it feminist—encompassing medical expertise, maternalist social reform, and women's rights. For example, prior to World War I, many of the most active members of the society were staunch, prosuffrage members of the PEC. The Political Equality Club served as the mainstay of suffrage activity in Rochester until the passage of the Nineteenth Amendment in 1920.[41]

The Provident Dispensary

By the turn of the century, women's medical societies provided the same services as their male counterparts: continuing education, professional advancement, and collegial socializing. What set them apart was their additional commitment to the goals of maternalist social reform and feminism. Their professional philosophy, as Dolley put it, "to make the prime motive the cure, relief, or alleviation of human ills and [make] the money consideration secondary," added an element of social responsibility to their professional goals. The Practitioners' Society, like other women's medical institutions of this period, should be understood in this context.[42] Like many female physicians in other cities, the members of the Practitioners' Society established an outpatient dispensary for the worthy poor (i.e., the working, or potentially self-supporting, poor). In fact, the society itself was developed in conjunction with the dispensary. Unlike the all-male Pathological Society's Free Dispensary, which catered to men, women, and children, the Practitioners' Provident Dispensary limited its services to women and children.

The dispensary movement first began in late eighteenth-century England. Between 1786 and 1800, American dispensaries were founded in Philadelphia, New York, Boston, and Baltimore.[43] These health care clinics were affiliated neither with almshouses nor hospitals, but were intended to supply the needs of the urban working poor at minimal cost to all concerned. Usually they were financed through philanthropic donations and, in some cases, governmental allocation. Almost always, as Charles Rosenberg writes, they were "shoestring operations."[44] Often they combined provision of on-site office visits (including, most importantly, the dispensing of prescriptions) with home visits for patients unable to travel. By the last quarter of the nineteenth century they were intended to fulfill both their social welfare function and the need of an increasingly self-conscious medical profession for clinical education and experience. At a time when only the most fortunate of medical students and recent graduates were given access to hospital bedside training, dispensaries served a valuable function by rounding out what would otherwise have been a theoretical medical education largely devoid of hands-on, clinical experience. Older, well-established physicians volunteered their services as members of the consulting staffs, while younger, recent graduates—often former students of the consultants— donated a predetermined number of hours per week to staffing the clinic. For those fortunate enough to establish themselves on a dispensary staff, the experience, the opportunity to develop areas of specialization, and the contact with influential physicians and laypersons vastly increased their chances of building a successful practice in the community. By 1900 approximately one hundred dispensaries existed in the country, of which three-fourths were general dispensaries. The remainder specialized in the treatment of specific classes of disease.[45]

Women physicians, who generally were excluded from all hospital staffs except those founded by and for women physicians, also were excluded from the staffs of most dispensaries. As medical educators became increasingly convinced of the importance of hospital-centered education for their students, women physicians' exclusion from clinical educational settings took on ever greater significance. While women's hospitals were the preferred remedy, often the cost of establishing one was prohibitive. Women's dispensaries offered a reasonable alternative.

By the 1880s, women physicians had founded dispensaries in New York, Boston, Philadelphia, Baltimore, Buffalo, Rochester, and Washington, D.C., as well as other cities. In New York, for example, Elizabeth Blackwell opened a one-room dispensary for women and children in 1854 that was expanded into the New York Infirmary for Women and Children in 1857.[46] In Washington, three dispensaries for women and children were in operation between 1883, when the Women's Dispensary opened, and 1894. Many of these institutions, established toward the end of the 150-year history of the dispensary movement, lasted only a brief time. Washington's Dorothea Dix Dispensary, for example, closed after only three years.[47] In Baltimore the Evening Dispensary for Working Women and Girls lasted from 1891 to 1910. Along with dispensing medical care to women and children, the women on its staff gave lectures on hygiene; distributed clean milk to impoverished children; organized a social service department; and conducted studies of midwifery, birth registration, and tuberculosis mortality.[48] Their multifaceted interest in women's health exemplifies the complex and distinctive goals of many dispensaries run by women physicians. At the same time, the Baltimore dispensary's interest in tuberculosis and clean milk foretold a new development in dispensary organization: a move from concern with the general care of patients to a specialized interest in particular organ systems or diseases.

Dispensaries, like hospitals in the same period, struggled to achieve rational and efficient organization. Dispensary doctors also made every effort to avoid the appearance of competition with physicians in private practice. Private physicians, especially those without elite professional connections, regularly accused dispensaries of luring away patients who were perfectly capable of paying fees for service. Usually they expressed this objection as a fear of "pauperizing" the working class. One physician described dispensaries as "'vast schools of pauperism, demoralizing the poor, educating them in improvident habits, and teaching them, in one of the most vital departments of life, to be thriftless and improvident.'"[49] At a time of physician oversupply and generally low incomes, freestanding dispensaries subsidized by local philanthropy were an easy target for physicians' resentment. Soon they were subjected to increasing state control. Beginning in 1895 in New York State, for example, dispensaries were required to submit a quarterly report to the State Board of Charities listing the number of patients, treatments prescribed, and receipts from all sources, with a possible fine of $100 for failure to comply. By 1899, these requirements were expanded to include the presence of a registrar at each dispensary for the purpose of improved record keeping.[50]

Unlike turn-of-the-century hospitals, dispensaries could not deflect the resentment of private practitioners or the accusations of zealous state regulators by catering to the middle class and raising their fees. This dilemma, as well as an increased understanding of infectious disease etiology, prompted a new emphasis on careful case management, efficient record keeping, economies of scale, and, in general, rationalized organizational management. By the end of World War I, dispensaries began to change their philosophy. Mid-nineteenth-century dispensaries subscribed to "a broad conception of the patient as a man [*sic*] and a citizen." By the early twentieth century, at least in the larger cities, they also addressed themselves to the health needs of entire populations. In general the changing climate of health care delivery encouraged the survival of large institutions and hindered the survival of smaller ones. Further, patient intake procedures were reorganized for greater efficiency and rationality, with dispensaries increasingly organized along lines of specialization by disease, and with many specialties accommodated within a single clinical setting.[51] As the history of the Provident Dispensary shows, none of this boded well for smaller clinics devoted to the general care of women and children.

Planning for the Provident Dispensary for Women and Children began on November 23, 1886, with a meeting at Dr. Mary Stark's house. Only the regular women physicians of the city were invited to attend. Within a week they met again, this time at Dr. Dolley's house, to approve a written constitution, constitute themselves as the medical staff, organize an advisory committee, and send a delegation to ask the city council for support. After successfully soliciting support from the

Rochester city council, the dispensary was given three rent-free rooms in a seedy, city-owned building on Front Street by the Genesee River. The council also supplied an initial budget of $100. The rest was made up from corporate donations (for example, Warner Pharmaceutical Company of Philadelphia made a "generous donation" of medicines), and from the charitable donations of laywomen on the Dispensary Advisory Board. Fifteen laywomen were recruited for the board. They were drawn from many of the local churches (and one synagogue) and acted as sponsors for the poor from their own congregations and neighborhoods. Members of the Advisory Committee were given dispensary cards to distribute that listed appointment times and the physicians' minimal fees for prescriptions. Organized at the same time as the Practitioners' Society—and by most of the same women—the dispensary was intended to fulfill the professional and charitable functions typical of other dispensaries, but with special attention to the needs of women. In January 1887 the dispensary opened for business on Wednesday and Friday afternoons, and it continued in service for a decade. Within a year the increasing patient load necessitated adding an additional office hour from 10:00 to 11:00 on both days.[52]

Rosenberg's characterization of dispensaries as "shoestring operations" certainly applies to this one. Supporters frequently supplied the extra dollars needed to cover delinquent payments to the janitor or for new supplies such as straw mattresses or linoleum. The city council never donated more than $200 in a single year. Consequently the physicians were often reminded not to prescribe costly medicines, wines, or brandies. After 1889, the poorest patients were given vouchers worth five or ten cents to use for medicine; others were to be charged that much directly.[53] The dispensary board was also glad to receive donations of service, as when orthopedic surgeon and roentgenologist Dr. Louis Weigel donated his time and labor to manufacture "orthopedic appliances" for dispensary patients. Although they were well aware of the need to keep records both of case histories and dispensary finances, lack of funding for a secretary hindered staff members from keeping careful accounts.

The summer months in most nineteenth-century cities were a time of increased diarrheal illnesses, especially among poor infants, the result of spoiled food and contaminated milk. Perhaps for this reason, during the summer of 1889 the physicians began exploring the possi-

bility of gaining hospital admitting privileges for their dispensary patients—but to no avail. They were more successful, however, in getting
two of their number, Drs. Harriet Turner and Minerva Palmer, appointed in 1890 to the city health department's staff of visiting physicians. As a result, they were paid a salary to visit patients at home and to
hold hours at the dispensary during which either dispensary patients or
nonpaying, district patients could be seen. By the end of 1890, the
dispensary abandoned its home visits, or "out practice," entirely and
placed those patients "into the hands of Drs. Palmer and Turner."[54]

For 1890, its fourth year of operation, dispensary records list a total
243 patients, down from 303 in 1889, equalling 1,439 office visits, 218
outpatient visits, and 1,386 prescriptions, for which $38.26 was re-
ceived. Of these patients, 6 were obstetrical and 2 were surgical cases.
One case was referred to a local ophthalmologist.[55] Often the practice
resembled social work more than medical care. Sometimes the physicians found employment for impoverished young patients, sometimes
they supplied shoes or crutches. They were attentive to the need to
teach mothers to prepare infant foods, such as malted milk, in bottles of
"absolute cleanliness." By 1892 they felt sufficiently proud of their work
to commission photographs of the dispensary and its staff for display at
the Columbian Exposition of 1893 in Chicago.[56]

Given the hard times in most cities during the financial panic of
1893, the Provident Dispensary was hard put to make ends meet. Although the dispensary officers hired a secretary in January to keep case
histories, by the end of the year they were forced to let her go, "owing
to lack of funds." Sarah Dolley suggested they establish a "food relief
kitchen" in response to the general hardship, but the staff decided they
could not manage the additional responsibility. In fact, each doctor had
to contribute twenty-five cents toward the cost of new dispensary cards.
The city briefly dislodged the dispensary from its quarters in January
1894 to make way for a clothing depot for relief of the poor, but by
March the physicians were able to resume operations. In June the dispensary received an extra donation of $100 from the city.

The dispensary minutes make no mention of the New York State
Board of Charities' newly stringent requirements for dispensary record
keeping passed in 1895. Few records for that year seem to have been
kept. (Penciled notes from one staff meeting were written on the back
of the table of contents from Dr. Rachel Gleason's popular book, *Talks*

to My Patients: A Practical Handbook for the Maid, Wife and Mother.) But
in February of 1896 medical staff treasurer Mary Slaight received a
stern note from the vice president of the State Board of Charities.
Noting deficiencies in the dispensary's records, he insisted she call at his
Rochester office to "arrange this matter." Failure to do so would be a
"serious matter," he reminded her, noting also that "Article 18 of Section I of Chapter 771 of the laws of 1895" made it a misdemeanor. The
following week the medical staff held an "informal discussion of the
desirability of continuing the dispensary." Among other items, they
discussed the possibility of having one of Dr. Baldwin's students write
up their case histories to satisfy the Charity Board's requirements. In
the end, however, they decided to close the dispensary for good. At a
private sale of the dispensary's effects, including textbooks, bookcases,
instruments, stove, washstand, medicine case, and other furnishings,
they netted $10.70 to pay off debts.[57] No doubt the decision to close
also acknowledged that new avenues for clinical work were opening for
women physicians. In 1895 Rochester City Hospital appointed one of
the dispensary doctors, Dr. Marion Craig Potter, as the first woman
assistant physician for the diseases of women in the Hospital's Outpatient Department. As city physicians and hospital outpatient clinics
began to take over the care of the city's marginal classes, the need for
general dispensaries seemed less pressing.

After 1895, the Practitioners' Society, parent organization to the dispensary, sought other ways to remain true to the philosophy of its
founders. For example, it strongly supported the establishment of evening dispensaries and charity wards in local hospitals. Dr. Nathan
Soble, in his presidential address of 1898 to the Pathological Society,
scornfully charged that dispensaries created "pampered pauper pets."[58]
In contrast, when Practitioners' Society president Mary Dickinson
made her annual report in 1902, she specifically addressed the "relations of physicians—especially women" to such organizations as the
Young Women's Christian Association, the Door of Hope Home for
Girls, and settlement houses.

Feminism and Professionalism

As the values of Victorianism receded into the past, a modern, scientific
conception of professionalism increasingly determined the values and

priorities of all physicians, male and female. Social reform without an explicitly professional dimension was becoming, by definition, *un*professional. For this reason, although Rochester's women physicians advocated a mode of social activism with deep roots in nineteenth-century maternalist reform, they increasingly cloaked their intentions in the rhetoric of professional as well as social duty. In this way, strict professionalism coexisted with social feminism on the Practitioners' Society's professional agenda until after World War I.

Between 1900 and 1910 the Practitioners' Society began to consolidate its achievements of the past fifteen to twenty years. Evidence of its members' increasing visibility and self-confidence in this decade is not hard to find. Three members of the society, Drs. Marion Craig Potter, Evelyn Baldwin, and Cornelia White-Thomas, for example, won positions on the junior staff of City Hospital by 1902, a great coup in the estimation of their female colleagues. Potter was promoted in January 1902, albeit grudgingly, from the Outpatient Department to the house staff, an event proudly discussed at the March meeting of the women's medical society.[59] Several months later, they began in earnest to express their newfound self-assurance. In particular, they began to seek wider recognition in the professional world beyond the city of Rochester.

The society's membership first demonstrated its ambitions by voting to confer honorary membership on a distinguished, if diverse, group of senior female colleagues. Drs. Elizabeth and Emily Blackwell, Mary Putnam Jacobi, Florence Sabin, Sarah Hackett Stevenson (the first woman seated in a state delegation to the American Medical Association), Cordelia Green (cofounder of the Castile Sanitarium), and Clara Swain (the first female medical missionary to India) represented the broad range of achievement the group wished to honor. These plans eventually led the society to reincorporate, change its name from the Practitioners' Society to the Blackwell Medical Society, introduce three levels of membership into the bylaws (active, associate, and honorary), and require that members join their county medical society. The process was complete by March 1906. The choice of Elizabeth Blackwell for the society's namesake was not unanimous; a few younger members preferred the more "professional" title of Woman's Medical Society of Rochester. The name the members settled on, however, was a testament to the continuing influence of the society's founders in maintaining both of its traditions, professionalism and social reform.[60]

Evidence for the society's strengthened self-confidence can be found in Eveline Ballintine's presidential address at the annual meeting in January 1906. Ballintine knew that revision of the society's constitution would be the major business of the evening. Nevertheless, she believed that the society first had to consider the broader implications of its reorganization. Ballintine's goal for the group was extremely ambitious. Now that the woman physician's "foothold [has been] established," she wondered, "is there not an opportunity for us to move onward?" Brushing aside concerns about the low percentage of women medical students, she posed two fundamental questions: "What is the future of the woman physician to be?" And, "In this age of differentiation of work is there not a special work for women doctors?" Dr. Ballintine offered several suggestions designed to strengthen the woman physician's position both in the profession and in the community. First she recommended expanding the organization of women physicians throughout the country, ultimately to form a national society. Next she suggested that women's medical societies affiliate with the General Federation of Women's Clubs. Finally she addressed what she believed to be the most serious obstacle to the woman practitioner, namely, the dearth of good internships. Here she pulled no punches. In Ballintine's view there was only one way simultaneously to attack the problem of internships, advance female medical careers, and fulfill the woman doctor's unique obligation to the public. She proposed that the society "use its influence" to organize and sustain a woman's hospital, staffed only by women and catering only to the woman patient and her children. As an added incentive, she suggested that such an institution might be suitable for the care of the Rochester region's unwed, parturient women currently in the care of a local shelter, the Door of Hope Home for Girls. She optimistically predicted that many philanthropic organizations would be willing to help establish such an institution.[61]

That Eveline Ballintine would suggest the founding of a woman's hospital as late as 1906, well after the closing of many all-female medical schools and hospitals and the decline of others, suggests a remarkable, if unrealistic, faith in the viability of such institutions.[62] Not surprisingly, however, no more was heard of this proposition. As Chapter 4 will suggest, the day of gender-separate medical institutions was nearly at an end. Instead, the society directed its energies and ambitions toward another of Ballintine's proposals, the expansion of women's medical societies across the state and nation. These efforts were successful. Cordial relations were well established among the women physicians of Rochester, Buffalo, and New York City by this time. They could count on the solid support of women physicians around the state in their next step: creation of a New York State women's medical society.

The women physicians from Buffalo and Rochester were, in fact, well known to each other. Their cooperation dated at least from 1901, when the all-women Physicians' League of Buffalo hosted a dinner for its Rochester counterpart at the Woman's Building of the Pan American Exposition in Buffalo. That occasion may have marked the origin of the ideas both for a regional or statewide organization and for creating honorary and associate memberships. Dr. Electra Whipple, president of the Buffalo group, was a guest at the Rochester Society's annual meeting in 1903, and the two groups seem to have stayed in touch for the next few years. Finally, early in 1907, the Rochester group—now known as the Blackwell Medical Society—invited the Buffalo Physicians' League to join them in forming a women's medical society of western New York.[63] Whipple accepted "enthusiastically" and proposed extending membership "to embrace the whole state rather than limit it to western New York." When the Women's Medical Society of New York City and the Cordelia A. Green Society of Castile also agreed to join, plans were made to launch the new society at a dinner to be held in Rochester in March 1907. In a special tribute to Sarah Dolley, the founders agreed to hold the banquet on her birthday and to name her as the first president.[64]

Beyond Separatist Organizing

Although interest in "organizing" continued to run high for another few years in Rochester, by World War I the Blackwell Society's members began to turn to other matters. Attendance at meetings after 1910 declined from the high achieved between 1904 and 1907, when the group was planning its reorganization and expansion. Although the Blackwell Society continued to function until 1926, meetings became the occasion for country club dinners followed by the presentation of a paper by an out-of-town guest; far fewer contributions were made by members themselves. While more active members, such as Potter, Ballintine, and M. May Allen, continued to give papers locally, these were usually intended for later publication in the *Woman's Medical Jour-nal.* The minutes themselves indicate declining interest or perhaps lack of available time for meetings. The last minutes preserved for the society, those of September 1913, were no more than scraps of paper inserted between the leaves of the secretary's bound leather book.[65]

The Blackwell members' presence as officeholders and committee members in the Women's Medical Society of New York State also declined from the high level seen in the society's first few years to far lower levels after 1910. Membership in the state society, in fact, provided a clear index of declining commitment to women's medical organizations by more recent medical graduates. Of the Blackwell members practicing medicine between 1907 and 1926, 100 percent of those who had graduated before 1910 became members of the state women's medical society. Of those who graduated after 1910, only 22 percent joined the state organization. Five members of the Blackwell Society became president or vice president of the state society, but all of these officeholders had graduated before 1900.[66]

Women's medical organizations, like other gender-specific groups, were by this time beginning to lose some of the feminist enthusiasm of their founders' generation. In 1906, for example, the society publicly mourned and memorialized the passing of Susan B. Anthony. By 1913, however, they merely tabled a motion by M. May Allen that the Blackwell Society be represented in an upcoming suffrage parade in Washington, D.C. Senior women such as Potter were continuing their organizational efforts, but most of the younger graduates declined to pick up where their elders left off. Instead, they tended to the business of establishing themselves in their fields.[67]

In many ways, the death of Sarah Dolley in December 1909 symbolized the passing of the older generation's fervent union of feminism, social reform, and professionalism. In the minds of her younger colleagues, Dolley was the conscience for all the women physicians fortunate enough to have known her. Dolley had encouraged female physicians to combine "intellectual assets with a purity of intention." Her personal credo, "the strength to suffer and the will to serve," expressed in the language of Victorian Christian womanhood, epitomized her personal vision of the ideal female physician.[68]

Sarah Dolley's vision of the woman physician, however, now held

only limited appeal. Although the Blackwell Medical Society continued
to exist for another eighteen years after Dolley's death, it ceased to
function effectively after 1913. The outbreak of war produced a temporary resurgence of interest in women's medical organizations.[69] Many of
the Blackwell Society's most active members, for example, gave their
time to the Medical Women's National Association (MWNA), founded
in 1915. More significantly, many of them worked tirelessly for the
American Women's Hospitals, founded in 1917 as the War Service
Committee of the MWNA.[70] Nevertheless, after the war the Rochester
women appear to have put less and less effort into their local organization, and it was dissolved between 1926 and 1927. This was not an
isolated phenomenon. Awareness of disappointing female enrollment
in medical schools and declining commitment to women's organizations by women already in practice caused consternation among the
leaders of the MWNA.[71] By 1930, the MWNA's Scholarship Committee found its only adequate source of contributions to be "the sympathetic woman of means outside the profession." Even so, it was forced
to solicit individual donations at the association's annual meeting to
liquidate a $500 deficit. That same year, the editor of the *Medical
Woman's Journal* reported and, surprisingly, concurred with the "startling" remark of Dr. Howard Kelly that "unless women were more
assiduous in interesting women students in the study of medicine,
women physicians would before a great while become extinct."[72]

In its broad outlines, at least, the forty-year history of the Blackwell
Medical Society reflected the history of women's medical societies nationwide. Moreover, the circumstances of its rise and decline help to
answer the question with which this chapter began: How did late nineteenth-century women physicians balance the ideals of feminism, social
reform, and professional advancement? Women physicians had molded
their professional identity around a core of civic duty in the form of
social housekeeping. They naturally were reluctant to give up an ideal
of professionalism so much in harmony with existing cultural norms for
women. Yet they were well aware of the changing character of modern medicine and certainly intended women's medical institutions to
facilitate their assimilation into the new professional community. But
women's medical activism was not sustained simply to satisfy a hunger
for professional success. The ambitions and professional standards of

the Practitioners' Society's members, like those of women physicians in many other cities, were tempered by genuine affection for, and belief in, the values and culture of late nineteenth-century social feminism. Although the society was meant to be a bridge between Victorian and modern notions of medical professionalism, these influences were always intended to be mutually enriching. Until the years preceding World War I, the Practitioners' Society gave its members access to both old and new professional traditions and provided a setting in which those two cultures could reach rapprochement.

⁓ YOUNGER GENERATIONS of women physicians found it difficult to maintain the delicate balance established by earlier generations. Gradually the attempted amalgamation of modern, scientific professionalism; feminism; and Victorian social reform proved untenable. By the turn of the century, local medical societies were being firmly linked to national professional networks whose importance as arbiters of professional achievement soon overshadowed the traditional centrality of the local groups. These trends affected the professional strategies of medical women no less than men. As women physicians came closer to achieving acceptance and legitimacy in a gender-integrated medical community, all-female medical institutions came to seem, perhaps prematurely, less central to their concerns.[73]

In a 1985 interview Dr. Mary Saxe, at eighty-seven the Blackwell Medical Society's last surviving member (she was its treasurer in 1926), explained the society's demise in terms that applied to many American women physicians between 1910 and the 1960s. First, she said she believed that the percentage of both women medical students and practicing women doctors in Rochester (and elsewhere in the United States) declined in the 1920s, as did the acceptance of a "woman's sphere." Women physicians felt increasingly isolated. Second, she observed matter-of-factly, "Everybody was out for themselves . . . everybody was interested in getting ahead." She candidly admitted to having thrown out the Society's last minute book many years before because, she had assumed, "It wouldn't be of any use to anyone."[74]

Thus, by World War I younger women entering the medical profession increasingly spurned all-female professional institutions. For the first two generations of women physicians, however, a gender-specific style characterized the setting in which most worked. And as the following chapter will show, some initiatives of nineteenth-century women physicians survived well beyond the turn of the century. Even after most women physicians abandoned professional separatism and feminist activism, many pursued careers that united health care, social reform, and public policy on behalf of their traditional constituency, women and children.

3

Maternalist Medicine: Women Physicians in the Progressive Era

Twenty years ago . . . a young woman who was restless and yearned to sacrifice herself, would have become a missionary or married a drinking man in order to save him. Today she studies medicine or goes into settlement work.

"*In the Social Settlements,*" Chicago Evening Post *(1908)*

SEPARATIST medical associations such as women's medical societies did not long satisfy women doctors' commitment to improving the health of their main constituency, women and children. Even while participating in organizations such as the Practitioners' Society, many women physicians also pursued more direct involvement with municipal health care initiatives. Besides, even the most successful women physicians faced substantial barriers to more competitive careers in medical schools and on hospital staffs in the first decades of the twentieth century.[1]

Many circumvented those more prestigious routes to professional advancement—by choice and by necessity—and instead pursued careers in public health or private benevolent institutions. If hospital and medical school appointments were slow to materialize, bureaus of public health and child hygiene, shelters for unwed mothers, and settlement house dispensaries offered opportunities either to supplement a fledgling private practice or to pursue an alternative career in medical benevolence.

Women Physicians and Maternalist Reform

The legitimacy of such pursuits for medical women was rooted in the rich soil of late nineteenth-century maternalist discourse. Maternalist reform, as it has been labeled, envisioned "a state in which women displayed motherly qualities and also played roles as electors, policy-makers, bureaucrats and workers, within and outside the home."[2] Drawing at first on traditional Victorian ideals of middle-class domes-ticity, Progressive women soon saw the work of "municipal housekeep-ing" a label applied by Mary R. Beard in 1915—as a bridge between the private and public moral domains: "moral motherhood" confronting politics as usual.[3]

As early as the 1830s, voluntary associations such as the Female Moral Reform Society, founded in New York City in 1834, offered nineteenth-century women the means to act on the impulses of relig-ious perfectionism, bourgeois social control, and sympathy for the poor. In addition, they bestowed a gratifying sense of feminine agency for many women, as they moved out of the household and into the commu-nity on a wave of benevolent reform.[4] But the high tide of such initia-tives occurred only at the end of the century. Not coincidentally, these years also saw the expansion of higher education for women, the found-ing of the General Federation of Women's Clubs (1890), the National Association of Colored Women (1896), the Women's Educational and Industrial Union (1893), and the revitalization of the woman suffrage movement. Particularly in the meliorist political climate of American Progressivism, college educated women found abundant opportunities for the application of their energy and idealism. All in all, women re-formers, laywomen as well as professionals, successfully extended the "domestic morality of the nineteenth century's 'separate sphere' for women into the nation's public life."[5]

This brand of maternalist politics generated reforms specifically de-signed to assist working women, impoverished and widowed mothers, and their children. During the late nineteenth and early twentieth cen-turies, when the national government's recognition of its responsibili-ties to the needy was in its formative stages, maternalist reformers helped shape government policies of assistance to poor mothers and their children through establishment of widows' pensions, campaigns to end child labor, and the establishment of pure milk stations, play-grounds, recreation centers, and parks. Women reformers, including both public health nurses and physicians, also took a particular interest in establishing child health bureaus within municipal departments of public health.[6] Their efforts culminated in the creation of the U.S. Children's Bureau in 1912.[7]

Women physicians played a large role in this process. In 1920 Dr.

Martha Tracy, dean of the Woman's Medical College of Pennsylvania and president of the Medical Women's National Association, articulated the values that underlay their participation. While conceding that women physicians should not be "separated from men in a scientific way," she insisted that the MWNA and all women physicians still had to "live to promote special work for women and children." The early twentieth century presented women physicians with ample opportunities to pursue both professional advancement and concern for women's health care. Yet by 1925 the nation's social vision would all but disappear, and with it, the reformist impulse in many women physicians.[8]

Having moved beyond an initially local concept of reform, typical of the mid-nineteenth century, Progressive women moved outward toward reform on a national scale. As middle-class women moved to combine moral suasion and political activism, "Maternalist ideologies . . . implicitly challenged the boundaries between public and private, women and men, state and civil society."[9] Progressive organizations such as the American Association for the Study and Prevention of Infant Mortality (AASPIM) typified this broad-based approach and, moreover, united men and women, physicians, nurses, and lay reformers. The AASPIM was founded in 1909 by Dr. Helen C. Putnam (a private practitioner with an interest in public health), Dr. J. H. Mason Knox (a hospital-based pediatrician associated with Johns Hopkins), Homer Folks (secretary of the New York State Charities Aid Association), C.-E. A. Winslow (a bacteriologist and public health researcher), and others.[10] As one of the leading voluntary organizations for public health education, its model of cooperation among medical practitioners, laboratory researchers, and social service leaders set an important precedent, one that was perpetuated in the coalition of groups that would support the 1921 Sheppard-Towner Act, the first national legislation to establish prenatal and child health centers.

The American Academy of Medicine (AAM), precursor of the AASPIM, was founded in 1876 with the goal of increasing the educational attainments of medical graduates. But by the turn of the century the AAM had redirected its emphasis toward the broadly conceived, quintessentially Progressive-era topic of "medical sociology." In addition to discussions of the state of the profession and the place of the specialist, for example, the annual meeting programs (especially after 1900) included talk about social conditions that were affecting the health of vulnerable populations, such as infants and school-age children. Its emphasis on preventive, environmental issues eventually made the group a haven for reform-minded physicians. The AAM also was professionally innovative in its commitment to seek out women physicians as members. Drs. Emma Culbertson of the New England Hospital for Women and Children, Alice Hamilton of Hull House, and Helen C. Putnam, for example, all participated in its meetings and were, in fact, bulwarks of the association.[11]

Dr. Putnam graduated in 1878 from Vassar College and went on to postgraduate study at the Sargent School of Physical Training at Harvard. From 1883 to 1890 she was director of physical education at Vassar. She graduated from Woman's Medical College of Pennsylvania in 1890 and followed up her medical studies with postdoctoral work at the New England Hospital from 1890 to 1891. She began her medical practice in 1892 in Providence, Rhode Island. From her work at Vassar, and later, her experience as vice president of the American Association for the Advancement of Physical Education, Putnam became a lifelong advocate of preventive public health, writing of the "oneness of mental and corporal life" and of the necessity for medicine to work hand in hand with education. She was also a firm defender of the rights of women and advocated educating both mothers and fathers in the fundamentals of environmental safety and child hygiene. For many years Putnam was chair of a standing committee of the American Academy of Medicine on the teaching of child hygiene in the schools. In 1894 and 1897 she was vice president of the AAM. In 1908, as president of the AAM, she convened a national conference on infant mortality that directly spawned the AASPIM. The new organization's first meeting was held in 1909, the year of the first White House Conference on the Care of Dependent Children. In 1911 the AASPIM board included directors of settlement houses and public health bureaus, nurses, physicians, and academicians from Chicago, New York City, Newark, Baltimore, Rochester, and several other cities. Four of the board's ten members were women physicians; also on the board were Jane Addams of Hull House, Adelaide Nutting of Columbia Teachers' College, and several pioneers in the movement for clean milk, including Dr. Henry Coit of Newark and Dr. George Goler of Rochester, New York. In 1919 the AASPIM was renamed the American Child Hygiene Association. In 1923 it merged with the Child Health Organization to become the American Child Health Association.[12]

Milk stations and "Little Mothers'" clubs, too, were initiatives begun around the turn of the century by private philanthropists, and they soon were adopted by children's bureaus and health departments in cities like Newark, Rochester, and New York to combat infant mortality and promote child welfare among the primarily immigrant poor. Dr. S. Josephine Baker, who in 1908 established the first Child Health bureau, was one of the first to utilize the power of municipal government to institutionalize and expand an initially private, philanthropic experiment. Baker, an acknowledged pioneer in the field of child health, was raised in Poughkeepsie, New York, just down the road from Vassar College. Baker was the daughter of a prosperous lawyer and a mother descended from one of the founders of Harvard College, and she had always assumed she would attend Vassar. But when her father and brother died suddenly, the need for economic self-sufficiency quickly overtook all other interests. After Baker persuaded her mother to contribute $5,000 from the family's limited savings toward her education as a physician, she moved to New York to attend the Woman's Medical College of the New York Infirmary. She graduated in 1898 and, after interning at the New England Hospital for Women and Children, started in private practice in New York with fellow intern Dr. Florence Laighton. In 1901, barely making ends meet in her fledgling practice, Baker took the municipal civil service exam and qualified to be a medical examiner for the New York City Health Department. Utilizing the political instincts that would carry her safely through thirty years of New York politics, Baker acquired her appointment from a "Tammany henchman" then serving as health commissioner. For the next fifteen years she divided her time between private practice and her "real career" with the Health Department of the City of New York. Eventually, New York's poorest children commanded all of her attention, and she gave up any interest in private practice.[13]

In 1910, two years after establishing the Bureau of Child Hygiene, Baker established the first Little Mothers' clubs in the country, to train older sisters in the sanitary care and feeding of younger siblings. Within a year New York City alone had 183 clubs. Milk stations, created through the support of New York City philanthropist Nathan Strauss in 1893 to distribute clean, unspoiled milk, became the "nuclei" for well-baby clinics, which in turn evolved into prenatal and child health stations championed by child health reformers.[14] Baker also helped found, along with public health nurse and settlement house reformer Lillian

Wald, the AASPIM. (In fact, as Baker wittily observed, the same few people, Baker, Wald, and Annie Goodrich, always seemed to belong to the same committees. One city reporter, finding them at work one evening, wanted only to know—understandably—"'What are you calling yourselves tonight?'"[15]) Baker never displayed the hostility of many physicians to the prospect of "state medicine." Rather, she insisted, "State medicine is to my mind an ideal, and the sooner it changes from an ideal to a practical reality, the better off the human race will be." She did concede the need for safeguarding American democratic ideals: "For instance, when I am ill I want to be free to choose my own doctor." Yet for much of the world she believed the choice lay between state medicine and "chaotic neglect of the vast majority."[16]

Baker's career was testimony to her ability to turn apparent disadvantages into unquestionable triumphs. For example, because she was a woman, Baker constantly was subjected to extraordinary pressures from politicians, from the public, and even from her own staff. The announcement of her appointment to head the new Bureau of Child Hygiene was met by the unanimous resignation of her entirely male staff of physicians. She persuaded them to come back and give her a one-month trial, after which they stayed on happily for years. Baker was clever enough to compromise when necessary. For example, for years she used only her initials, Dr. S. J. Baker, in her official work, as part of the Health Department's "shy evasion of the woman question." (One unknowingly grateful constituent, a barber, sent her a shaving mug engraved with her initials.) At the same time, she occasionally wished her identity were *better* known, as when—in Paris—she caused deep disappointment to those who expected to meet *Miss* Josephine Baker, world-famous exotic dancer.[17]

By the turn of the century, both middle-class and working-class mothers became the object of a movement to supplant traditional cultural norms and motherly "intuition" with the principles of "scientific motherhood." Training in the science of child care, such as the modern principles of infant feeding, came to seem the domestic equivalent of the work of medical missionaries. Women physicians played a role in directing these trends not only as public health physicians but also as medical columnists for popular women's magazines—for example, Dr. Emelyn Lincoln Coolidge's "Mother's Register" column in the *Ladies' Home Journal* and Dr. Josephine Hemenway Kenyon's child care column in *Good Housekeeping*.[18] The U.S. Children's Bureau milk stations were built on the foundations laid by the French *gouttes de lait* and the milk stations operating in some thirty American cities, including New York and Rochester. "We were the first to use milk-distribution as a way of coming into contact with mothers in order to educate them in scientific child care."[19]

Nowhere was this connection more successful than in the creation in 1912 of the Children's Bureau, a division of the U.S. Department of Commerce and Labor. The idea for the Bureau originally came from discussions among Florence Kelley, director of the National Consumers' League, and Lillian Wald, founder of the Henry Street Settlement in New York. Baker's work was another important influence. When in 1909 the first White House Conference on the Care of Dependent Children called attention to unacceptably high rates of infant and maternal mortality among the poor, a far-flung alliance was already in place, working to establish state-sponsored mothers' pensions and a federally supported program of education in the proper care of infants.[20] Settlement house workers, women's and mothers' club members, and other arms of the maternalist reform movement took the lead. Both of the Children's Bureau's first two directors, Julia Lathrop and Grace Abbott, were former residents of Hull House in Chicago and were firmly entrenched in the upper echelons of Progressive reform.

The Bureau was mandated chiefly to conduct educational campaigns to improve maternal and child health. Its governing philosophy of child welfare presumed a comprehensive approach to the subject, including social, economic, cultural and medical aspects of child health. With widespread support of women's groups around the country and Lathrop's astute leadership, the Bureau persuaded Congress to increase its initially low budget of $25,640 in 1912 to the generous appropriation of $139,000 by 1914. Women of all social classes were united in their support for and appreciation of the Children's Bureau's initiatives. Women's clubs organized child health fairs designed to bring in mothers and babies so that the babies could be weighed and measured. Mothers received Children's Bureau pamphlets entitled *Infant Care* and *Prenatal Care*. A "baby-saving" campaign organized by Lathrop had enrolled 11 million women by 1918. In 1913 the Bureau began one of its most advanced programs, a survey of infant and maternal mortality authored by Dr. Grace L. Meigs. The survey documented that maternal and infant mortality rates were nearly twice as great in low-income families as in prosperous ones, thus correlating poverty and poor education with high death rates for babies and mothers. In 1915, the survey found that among African American women, 11 mothers and 181 infants died per 1,000 live births; for white women the mortality rate was 6 mothers and 100 infants per 1,000. Lathrop concluded that poor families could not by themselves avoid or control such "hazards to the life of the offspring . . . because they must be remedied by community action."[21]

Yet by claiming to assist working-class, immigrant, and African American women, these predominantly middle-class, white, reform-minded women implicitly presumed the moral superiority of their maternalist perspective over the often different social values of their intended beneficiaries: women from other sociocultural domains. The history of mother's pensions illustrates this point. According to Theda Skocpol, a coalition of women's groups, including the General Federation of Women's Clubs, the National Congress of Mothers, the Women's Educational and Industrial Union, the Women's Trade Union League, the National Consumers' League, settlement house leaders, and the Children's Bureau, gave crucial support to maternalist initiatives such as the establishment of juvenile courts, mothers' pensions, the abolition of child labor, and laws regulating the wages and hours of women workers. They justified these reforms on the grounds that they would enable mothers to spend adequate time caring for their children and, thus, contribute to the stability of the family. Yet middle-class women's groups mustered far less support for initiatives promoting women's fuller participation in the workforce. Indeed mothers' pensions, which were instituted by forty states between 1911 and 1920, were structured to supply family income supplements rather than the equivalent of a working wage.[22]

It would be unfair to conflate maternalist policy objectives—the health and well-being of poor women and children—with legislative outcomes dictated by mainstream (and mostly male) politicians. Maternalist reformers appear to have known and privately acknowledged the depth of the compromises they felt were necessary to accomplish anything at all. Nevertheless the origin of many of today's deepest divisions within American feminism can be traced to these decades. Black mothers, for example, received far fewer benefits from mothers' pension legislation than did whites. A Children's Bureau study of 1931 showed that only 3 percent of the recipients were black; other women of color received only 1 percent of benefits. Moreover, some groups, notably the General Federation of Women's Clubs, refused to accept African American women's clubs in the Federation, spurring black women to form the National Association of Colored Women in 1896.[23] Thus, despite the possibilities for feminine agency created by maternalist discourse, as a political strategy it harbored many dangers. Although genuine gains resulted from maternalist reform initiatives during the Progressive era, both for women reformers and for the beneficiaries of their efforts, as a discourse of gender difference, it contained many drawbacks. Not only did middle-class women frequently misunderstand the goals of working-class women but they also risked marginalization for themselves by their devotion to the rhetoric of gender difference. Maintaining the claim to women's distinctiveness, writes historian Nancy Dye, may even have served to "codify a limited domain for women . . . in the workplace."[24]

Where did women physicians stand on these issues and what roles did they play in the institutions of Progressive reform? Many women in medicine saw no contradiction between their professional and scientific values and the goals of benevolence and maternalist social reform. These were, after all, among the values that had legitimated the cause of women physicians in the first place. Atina Grossmann writes, "Women physicians stood at the forefront of bureaucratized health systems not only because they had been relegated to subordinate employee positions within the medical profession but also because of the motherly and nurturing qualities they both claimed and were assigned."[25] Reform-minded physicians held themselves accountable to rigorous professional standards. Nevertheless, it was not always possible to satisfy modern medical standards in volunteer-run institutions such as maternity homes and settlement houses, where professionalism often ran counter to the traditionally evangelical and benevolent values of lay boards of directors. In government-run public health bureaus, in contrast, professional standards sometimes outweighed traditional values. Yet on the whole, agencies such as the U.S. Children's Bureau and its local counterpart in New York City did a fine job, in part because of their ability to reconcile maternalist and professional accountability.

Public Health

The association of women physicians with public health and preventive medicine was well established. Regina Morantz-Sanchez has shown how maternal responsibility for maintaining the health of the family flowed directly into women's earliest campaigns for higher education, especially admission to medical schools. Soon after its founding in 1850, the Female Medical College of Pennsylvania (later, WMCP) established the first professorship of hygiene in the country, thus acknowledging the woman physician's special responsibility for preventive medicine. In 1855, future dean Ann Preston became professor of physiology and hygiene. In her introductory lecture she wrote, "The frightful amount of ill health in this country, especially among our women, is arousing attention and calling for some change in our regulations. Physicians abound, but health comes not!"[26]

Over the next fifty years WMCP continued to inculcate a belief in the importance of public health and preventive medicine. Preston and her nineteenth-century successors in the Chair of Physiology and Hygiene understood that hygiene had to be grounded in physiological science. Yet they increasingly emphasized the moral necessity of teaching hygiene rather than expanding the bacteriological study of infectious agents. In the view of Martha Tracy, dean of WMCP, the job of public health officer (typically involving laboratory studies, infection control, and the like) was better suited to male physicians.[27] Women physicians who trained at Johns Hopkins, such as Anna Wessel Williams (codiscoverer of diphtheria antitoxin) and Dorothy Reed Mendenhall (discoverer of the Reed cell), would have rejected this gender-based division of labor.

But at the turn of the century, it was widely believed that public health education was an appropriate field of activity for women physicians. In 1909 even the American Medical Association, responding to a suggestion made by surgeon and WMCP graduate Dr. Rosalie Slaughter Morton, established a Public Health Education Committee in which women physicians were explicitly urged to take the initiative. They were instructed to educate the public, by working with women's clubs, mother's associations, girls' schools and colleges, and "other similar bodies," in "the nature and prevention of disease and . . . the general hygienic welfare of the people." Morton was named to chair the committee, with Dr. Sarah Adamson Dolley (then nearly eighty) as honorary chair. Subjects suggested for public lectures ranged from the most general (the cause and prevention of ordinary colds; the value of pure foods; the relation of pure water to the public's health; the value of rest and exercise; the relation of flies to public health) to topics of more specifically feminine interest (care of the sick at home; pure milk and infant hygiene; care of health during menstrual periods; pregnancy and menopause; the prevalence and prevention of venereal diseases; the value of early diagnosis of cancer in women). Other subjects were clearly designed to win over public support for some of organized medicine's most cherished projects. These included the need for medical inspections in the public schools, the benefits of early diagnosis and treatment of infected tonsils and adenoids, the value of animal experimentation for surgery, nutrition, diabetes, nervous diseases, tuberculosis, and other infectious diseases.[28]

Although male physicians participated in the AMA campaign for public health lectures throughout the country, by 1913 twice as many women as men had signed up to lecture.[29] Eventually, control of the program was taken out of the hands of Morton's committee, yet women physicians continued to play a large role in carrying out its activities over the years. The AMA also promoted education of mothers in child health, nutrition, and hygiene.[30] In a review of women's work in public health in the *Woman's Medical Journal* in 1914, Dr. Kate C. Mead argued the "duty" of women physicians to teach public health. She noted the contributions of Dr. Josephine Baker, Rachelle Yarros's work in obstetrics in Chicago, Marion Craig Potter's work in New York State, and several others. Mead urged more women to enter the field and noted that a physician with an additional degree in public health could earn a salary ranging from $500 to $7,000.[31]

Perhaps the most prominent medical woman to combine devotion to medical science with maternalist reform was Hull House resident and nationally known industrial toxicologist Dr. Alice Hamilton. According to Hamilton's biographer, she "refused to put purely professional concerns . . . ahead of the obligations of citizenship as she saw them." As Hamilton herself once wrote, "It would be quite impossible for me to enter into a relation with any institution if by doing so it was necessary for me to detach myself from purely human problems."[32] Thus, she refused a full-time position at Harvard so that she might be free to live half the year at Hull House and continue her freelance work in toxicological investigation. Like many other reform-minded women professionals, Hamilton, who had graduated from medical school in the 1890s, saw her commitment to social progress as perfectly congruent with the Victorian gender norms with which she had grown up. Indeed, her strong commitment to "municipal housekeeping" may have been essential to her sense of calling as a woman physician. The many women physicians who worked on the health committees of various laywomen's organizations or volunteered their services to maternity homes and settlement houses likely saw their participation as necessary to their sense of themselves as women, too.

But full-time professional positions in public health also were opening up for women physicians. During World War I the Public Health Service established a Women's Division for public education on hygiene and social hygiene (prevention of the spread of venereal disease). A year later Martha Tracy of WMCP created a new graduate medical course in social hygiene education for women physicians who wanted to specialize in teaching public and personal hygiene, nutrition, dietetics, child hygiene, medical inspection of schoolchildren and anything else pertaining to the work of "the woman physician in educational and municipal health work."[33] Some women physicians, such as Dr. Dorothy Reed Mendenhall, who chose to combine marriage, motherhood, and medicine, and who moved from New York City in 1905 to the less professionally active town of Madison, Wisconsin, went into public health work in health education and hygiene because it was the only viable career choice open to them. (Years later, when Mendenhall and her husband moved to Washington, D.C., she began working for the U.S. Children's Bureau.)[34] Thus women physicians approached their work in social medicine both idealistically and pragmatically, as reformers and as professionals seeking career opportunities.

The increasing vitality of urban boards of health in the late nineteenth century made them, initially, a strong ally of maternalist reform, particularly in the fight to improve health in the increasingly overpopulated cities. Prior to the development and dissemination of bacteriological science during this period, physicians had participated in the public health movement, in Dorothy Porter's phrase, "more as philanthropists than as specialists in disease control."[35] But with the advent of an understanding of etiological specificity, public health officers from Providence to Seattle were winning enough political power to begin effective enforcement of municipal sanitary legislation aimed at cleaner water, milk, and sewage disposal.[36] While voluntary associations such as the American Association for Study and Prevention of Infant Mortality investigated the social and environmental causes of disease, public health officers such as William T. Sedgwick in Massachusetts; Charles V. Chapin in Rhode Island; Hermann M. Biggs, Dr. William H. Park, and Dr. Anna Wessel Williams in New York City; and Dr. George W. Goler in Rochester, New York, launched a multipronged attack on diseases such as smallpox, diphtheria, cholera, typhoid, tuberculosis, and the deadly infant summer diarrheas. Bacteriological research on smallpox vaccine and diphtheria antitoxin was widely heralded, but the equally crucial passage of ordinances against unclean milk, public nuisances (such as a noxious privy or stable), and contamination of drinking water through poor sewage disposal was far more difficult to accomplish.[37]

During the Progressive era public health training began to combine the special expertise of medical science with the environmental concerns of previous decades. Yet because of the etiologic and professional eclecticism of the field, relatively few male physicians made a career of it. Also, many private practitioners viewed public health work as unfair competition. This left the field more open to incursions by women physicians than other modernizing institutions such as hospitals and medical schools. As illustrated by the relatively brief ascendancy of the Children's Bureau, however, the period during which public health physicians accrued a degree of prestige equal to that of those in private practice, did not last long.[38]

For women physicians public health work represented both a professional opportunity and a professional risk. Prior to World War I, social housekeeping initiatives in medicine seemed full of promise for professional as well as personal fulfillment. Yet these were, simultaneously, the years in which most of professional medicine began moving decisively in the opposite direction—toward specialization, biomedical reductionism, and the rejection of holistic, environmental, culturally contextualized conceptions of illness. The marginalization of public health efforts and medical school programs in preventive medicine and community health began in the early decades of the twentieth century. By putting their weight behind social medicine, women physicians may have remained true to their professional traditions of generalist medicine and maternalist social reform, but they also were following a path few others wished to pursue.

The history of the Board of Health in Rochester, New York, and its early alliance with maternalist Progressive initiatives illustrates the opportunities and drawbacks public health work could provide to women physicians. Rochester's first city health officer held a part-time appointment beginning in 1876; he had a staff of four other part-time city physicians who volunteered their services. By 1890, two of these volunteers were women, both members of the Practitioners' Society. The Health Board was nominally governed by the city's Common Council but was in fact controlled by the local political boss, George Aldridge. In 1892, under threat of an impending outbreak of cholera, the city doubled the board's funding, hired Dr. George W. Goler as a full-time assistant city physician, and attempted to at least reduce political interference. Still, when Dr. Harriet Turner's reappointment as city physician seemed in doubt in 1895, her savvy colleague Dr. Sarah Dolley went directly to the Republican political caucus to ensure her reappointment. And when Goler was proposed as the new chief city physician, one of his supporters was Susan B. Anthony, who wrote him that she would send a letter of support to the governor and hoped that the latter would "do his duty."[39]

In 1900, when the Health Board was renamed the Bureau of Health, the seven-physician staff included two women. Throughout most of Goler's career as head of the Rochester Health Bureau, his staff included two or three women physicians, usually serving as medical school inspectors and child health experts in city clinics. Six of the seven women physicians who worked for the city over the years remained members of the Practitioners' Blackwell Medical Society until its dissolution in 1927. Three of the six women physicians active in Rochester's two settlement house dispensaries also were members, as were three of the seven women on the medical staff of the Door of Hope maternity shelter.[40] In 1897, Goler's rapport with the Rochester branch of the Women's Educational and Industrial Union (WEIU), founded by Sarah Dolley, Susan B. Anthony, and others in 1893, made possible one of Goler's earliest innovations, the first publicly supported milk station in the country.[41] Doubling as a child health clinic, the station was located at a settlement house and was staffed by a public health nurse whose salary was at first paid by the WEIU. By 1900 the Health Bureau's outlay for milk stations was a generous $900 per year.[42]

Between 1903 and 1906 Goler oversaw the establishment of school health clinics; by 1911 a staff of twelve city physicians combined home visits to indigent patients with attendance at weekly clinics in downtown public schools. Visiting nurses attended a total of 934 families in 1913, treating 2,547 children. Seventy-one of these infants were sent out to Lake Ontario, to the Infants' Summer Hospital established by Goler and other local physicians in 1893. In 1916 Dr. William Brown, an obstetrician with appointments both at Rochester General Hospital and the Bureau of Health as well as an interest in research, volunteered to open the first prenatal clinic in the city. It opened at School No. 27, located in a neighborhood of high birth rates and low rates of obstetrical service. By 1924 prenatal clinics had been established at four public schools, each staffed by a bureau physician and nurse. These prenatal clinics, one might reasonably speculate, served the same dual purpose for doctors as the nineteenth-century dispensary: community service and professional opportunity. Rarely was one staffed by a woman physician. Usually women were employed at child health clinics established at the Lewis Street and Baden Street settlements or as medical inspectors for schoolchildren. In contrast, medical care at maternity shelters was often the exclusive province of women physicians. (Such shelters, however, sometimes were targets of complaints from the municipal health officer. George Goler's records reveal numerous complaints lodged against one maternity shelter, called the Wilhelmine Hospital.)[43]

Such shelters, part of a national movement to establish maternity homes as rescue missions for "ruined girls" and "fallen women," proved of special interest to some of Rochester's women physicians. In 1894 Dr. Marcena Sherman-Ricker, physician to Susan B. Anthony and a well-known Rochester homeopath, helped found the Door of Hope Home and worked there both as a physician and as a member of the board. Drs. Mary Dickinson, Evelyn Baldwin, Katherine Daly, Charlotte MacArthur, Ida Porter, Mary Stark, and Cornelia White-Thomas, all Practitioners' Society members, also served on the home's medical staff and as board members.

The earliest such homes were founded in the 1880s during a time of increasing anxiety over the evil influences of the modern city, especially prostitution and the "white slave trade." Both the Salvation Army and the Women's Christian Association (known as the YWCA after 1893) established homes for single women workers in various American cities, but pregnant unmarried women posed a more difficult problem. Previously such women would have been consigned to a municipal almshouse. But by the 1880s, in an effort to rationalize and economize municipal charity, the medical functions of such institutions were being drawn off into newly established city hospitals. Around the same time, privately supported maternity shelters were being established by evangelical reformers to provide a setting in which unmarried women could receive both prenatal care and the moral and vocational training to make a better life for themselves after the birth of their baby. In 1882, evangelical philanthropist Charles Crittenton founded the first Florence Crittenton Home in an organization that became known as the National Florence Crittenton Mission. A decade after founding the first home, Crittenton was joined by Dr. Kate Waller Barrett, who became vice president of the mission. Waller, a wife and the mother of six, had sought a medical degree specifically to further her capabilities in rescue work. A believer in the power of maternity as a "means of regeneration," Waller insisted that the "instinct of motherhood [was] one of the strongest incentives to right living."[44] By 1918 Florence Crittenton Homes were operating in many cities and had received a national charter. The Salvation Army also organized maternity homes for unmarried women, beginning in 1887 with one such home in Brooklyn and reaching a total of fourteen nationwide by 1894. The YWCA began operating maternity homes in Cleveland, Pittsburgh, and St. Louis. Another organization, the Door of Hope Homes, was founded during the 1880s by Emma Whittmore and eventually deeded over to the Salvation Army. In these maternity shelters, with the exception of homes run by the Salvation Army and at least some of those established by the Roman Catholic church, women physicians were always preferred over men as medical attendants. In the words of the Florence Crittenton Mission guidelines for 1907, "This is pre-eminently women's work to care for and succor her unfortunate sisters." It may well have been, in addition, uncongenial work to many male physicians.[45]

The vast majority of rescue homes accepted only white residents.

The years of the greatest success in the United States for progressive-maternalist institutions, the half century following the end of the Civil War, were precisely those years in which racism and segregation reemerged both in the North and South. Little of the national effort to care for or educate poor children, or improve public health, was directed to the needs of African Americans.[46] In 1883, for example, only 9 of 353 child-care institutions had been built for black children; 60 or 70 others claimed to accept white and black children, but the numbers of nonwhite child residents were very low. The Salvation Army did operate several segregated homes for black working women and for unwed mothers, but in general the prevailing prejudice that illegitimacy was less stigmatizing for blacks than for whites made it easy to shortchange black institutions.[47] As a result, black women's groups, such as the National Association of Colored Women (NACW), organized philanthropies of their own. Mary Church Terrell became the first president of the NACW in 1896. The middle-class and elite women who led the group were quite conscious of the need to counter prevailing stereotypes of black women's promiscuity. They infused the ethos of rescue and reform with a doctrine of racial self-help: Lifting As We Climb was the formal motto of the NACW and a personal motto for many.[48]

The Door of Hope Home in Rochester was not formally allied with any of these networks but chose the name Door of Hope to signal its agreement with the goals of the better known establishments and to facilitate future exchanges of information with the other Door of Hope Homes. Like other such institutions, its stated purpose was evangelical: redemption for the "fallen girl." Minutes of staff meetings are studded with references to residents as members of a "family." Both in its ideals and its rhetoric, the Door of Hope in Rochester declared itself at one with the moralizing discourse of evangelical religion. For another two decades this worldview blended easily with the professional ideals of medicine.[49]

On November 1, 1894, at the original planning session for the home, the chief organizers included two women physicians affiliated with Baptist congregations, Dr. Marcena Sherman-Ricker and Dr. Catherine Walker. Subsequently, the board resolved to admit men to the governing body of the home but decided that, "given the difficulty of persuading men to join this Board on account of their fear of unfortunate publicity . . . their names would be withheld if desired." For the most part, charity-minded laywomen drawn from local Protestant congregations held sway over the board's policymaking—with the exception of financial decisions. Like dispensaries, these homes operated on a straitened budget. When, for example, they were offered a donation of $500 by one of the city's brewers, the board wrestled with its conscience. Finally the Reverend Mr. Gannett, a Unitarian, suggested accepting the money, "but with the understanding that they should use it in helping to subdue the evil that the making of this money had produced."[50]

By the spring of 1895 organizers had furnished a modest, steeply gabled clapboard house at 293 Troup Street in a downtown district. The house could accommodate about ten women per night, mostly unwed mothers, plus their infants and children. During 1900 the Troup Street house took in a total of forty-five adult residents, at a rate of five to ten each month. The first entries in the residents' ledger, dated January 30, 1895, read, "Edith C.—anaemia—babe 4 weeks old; Ida T.—incontinent." At the end of 1904 the board opened a second house, on Fitzhugh Street, for "rescue work" with female prostitutes and alcoholics. But by 1905 the budget would not stretch far enough to maintain the second house, and it was closed. In the year from 1905 to 1906, the minutes record the receipt of two donations of $2,660 from Henry Alvah Strong of the Eastman Kodak Company, and two bequests totaling $14,000 to be used in part to pay outstanding debts for a new furnace and coal. That year during the months of July through December alone, the home had forty-six adult women residents, about seven or eight each month. By 1911 the residents' census was down to twenty-seven adults and a total of twenty-eight infants, two of whom died. Another eighty women and fourteen children were classified as transients.[51]

As in other homes of this type, the Door of Hope staff imposed an environment that combined evangelical moralizing and compulsory domesticity with provision of social welfare services, such as finding jobs for former clients. In its annual report for 1897–98, the board proudly declared that "the Door of Hope Association . . . now has in its possession a large and convenient Home . . . where it gives shelter to erring women, who manifest a desire to reform. Here they are taught housework, sewing, and nursing; are tenderly cared for when ill, and are given valuable spiritual instruction." After a few years, a laundry was established on site to provide a form of vocational training and a source of wages for the residents. Just as important, the home provided residents with prenatal, obstetrical, and general medical care. Initially its staff recruited residents directly from the downtown streets. Eventually, prostitutes, industrial-school inmates, and homeless mothers-to-be found their way to the Door of Hope or were referred there by the police; unmarried young women were also brought there by their own desperate parents. In 1898 the board moved to get a license from the State Board of Charities to enable it to place unwanted infants with adoptive parents. By the 1920s most of the home's work centered on unmarried mothers rather than prostitutes or juvenile delinquents.[52]

Historians have examined the evolution of such institutions from evangelical missions run by middle-class reformers (usually female) to bureaucratized welfare agencies structured according to the discursive constraints of professional social work.[53] Little has been written, however, about the role of female physicians in this transition. Between the 1890s and 1912, the Board of Managers of the Rochester Door of Hope Home insisted on the exclusive employment of women physicians.[54] Perhaps because of the home's evangelical underpinnings, it was the one health care venue in Rochester in which women doctors of both the regular and homeopathic varieties joined forces. The physicians participated actively in caregiving and policymaking.

From the first year of operation, members of the medical staff demonstrated a perspective somewhat distinct from that of their lay colleagues. Character reform was deemed essential to the success of these missions by the laywomen who organized the home in its first two decades, while the medical staff emphasized their clients' need for sound information as well as moralizing. Dr. White-Thomas, for example, spoke at the board meeting of May 27, 1897, on "the duties of mothers towards their daughters, to the end that they should be so instructed and cared for that there should be no need for Doors of Hope." The minutes continue: "Dr. Thomas holds that it is mostly through ignorance that these unfortunate girls are where they are." Providing good obstetrical and general medical care was a top priority for the physicians who allied themselves with this and other such institutions. Maintaining respectably low rates of mortality and morbidity among residents and their newborn children contributed to the home's success in gaining referrals and in fund-raising. It was also essential to women physicians' willingness to risk their professional reputation by donating their services. At a time when home births were still over-whelmingly preferred by doctors and parturient mothers, physicians were asked to provide both obstetrical and general medical care for the home's residents. In cases of emergency, the Ladies' Board of Managers of City Hospital pledged its "sympathy and cooperation" in receiving residents of the home as inpatients. Over the years City Hospital, an allopathic institution, and the homeopathic Hahnemann Hospital accepted patients from the home in about equal numbers.[55]

The question of whom to hospitalize, and for what reasons, weaves through the minutes of the Door of Hope, marking the boundary between the medical and lay perspectives. At Rochester's Door of Hope, medical authority initially was subsumed under the authority of the matron, a position held by a woman without formal nursing or medical training. For example, in August 1900 Mrs. Helen St. John was hired at $25 per month as matron of the home, with Dr. Dickinson named as her "assistant." One month later, the medical staff attempted to gain more control over medical policy. That year, according to Door of Hope records, the home served forty-five residents and fourteen babies; five were born at the home and thirteen at the hospital, leaving an undeclared total of four infant deaths. Since the minutes also state that no infant deaths occurred at the home, those babies must have died while at a hospital. Nevertheless, the physicians believed that some of the deaths were the result of unnecessarily unsanitary conditions at the home. Under the leadership of Dr. Evelyn Baldwin they requested that if maternity cases were to be kept in the home, a trained nurse be employed as matron. Otherwise, the doctors maintained, maternity cases should be sent to the hospital. They also requested the services of a laundress. The board of directors instead decided to bring in trained nurses for three or four days at a time whenever needed. In emergencies, the "girls" would be sent to a hospital. The board did agree to hire a laundress, but only at the discretion of the matron.[56]

The board of directors never addressed adequately the medical staff's dissatisfaction with sanitary conditions at the Door of Hope. In 1903 the physicians called attention to the "unsanitary conditions of the cellar," and as a result, the matron was authorized to purchase lime and charcoal for disinfection. Yet, given the widespread risk of puerperal infection in those years, hospital conditions were not necessarily any safer. The following month two of the home's residents gave birth at the hospital, where one of the two babies died. Later that year another mother insisted on leaving the hospital with her baby because of unsatisfactory care. Over the next eight years board policy shifted back and forth between retaining confinement cases at the home and sending them to one of the local hospitals. The medical staff insisted that at least one room at the house be set aside especially for confinements to try to minimize cross-infection. The medical staff's annual report of May 1911, however, reported "several cases of severe illness, as well as many minor complaints." In their report to the board in July, the staff requested that "the Board procure an assistant who will give her personal attention to the health and welfare of the girls as a nurse and see that the doctors' orders are carried out." Finally, between January and April 1912, the medical staff decided that their patients' best interests were not being served "under the present rule" at the home. In May the entire medical staff resigned. Thenceforth, medical patients from the Door of Hope were sent to the Outdoor Department of Rochester General Hospital (formerly City Hospital); confinements were attended on the inpatient wards.[57]

Settlement House Health Centers

Settlement houses provided yet another site where women physicians could fulfill their ideal of maternalist medicine within a professionally acceptable setting. The American settlement movement began with the opening of the Neighborhood Guild by Stanton Coit, Charles B. Stover, and Edward King on New York's Lower East Side. Several years later, in 1889, several women college graduates, including Vida Scudder, founded the College Settlement in New York, and at the same time, Jane Addams and Ellen Starr founded Hull House in Chicago. Lillian Wald's Henry Street Settlement opened in 1893. Founders of the earlier settlement movement in England had been imbued with Christian socialism, romanticism, and a dislike of industrialization. In contrast, the more pragmatic American settlement founders soon discovered the necessity of combining social and moral reform with programs directed toward individual self-help.[58] Most American settlement houses offered classes in housekeeping, sewing, music, reading, arithmetic, and other subjects, as well as social clubs for children, and gymnasiums. However, many of the leading settlements also engaged in more political efforts, such as surveying local economic conditions and lobbying for improved garbage collection, street paving, and hours and wages legislation.

Medical and public health clinics were intended as part of the settlements' general effort to reach out to immigrants and other impoverished populations in their neighborhood. Believing that immigrants were in particular need of education in infant care, child hygiene, and prenatal care, nurses and doctors at the clinics emphasized the educational and preventive aspects of their work. Not surprisingly, a disproportionate number of women physicians participated in these clinics, although probably very few followed Alice Hamilton and chose to live in or near the settlements themselves. The first settlement house clinics opened in the 1890s, and the number increased dramatically after 1900. According to the *Handbook of Settlements*, 74 settlements were established by 1897; by 1911 they numbered 413. Of these, at least 15 percent (63) included a clinic. Of those with a clinic, at least 20 percent (13) listed women physicians as members of the clinic's staff—at a time when women constituted, at most, 6 percent of all physicians in practice.[59]

In Rochester, the first settlement to establish a free clinic was the Baden Street Settlement, located in the heart of a downtown neighborhood of German and Polish Jewish immigrants. The settlement was opened in 1901 at the instigation of women active in Temple B'rith Kodesh, the center of reform Judaism in the city. By 1904, concern for the health of the neighborhood prompted the Baden Street directors to petition Dr. Goler at the Health Bureau to locate a milk station at the settlement. Soon after, the settlement hired two nurses, one to operate a baby weighing station and another to make house calls. By 1908 the operation was expanded to include both a general medical clinic and a special clinic for ear, nose, and throat diseases. In 1917 an all-new dispensary building was opened, which made the nurse in charge of Rochester's other dispensary, the Lewis Street Center, "green with envy." The clinic staff included a resident nurse and fourteen physician volunteers.[60]

Most medical records for the Baden Street Settlement were destroyed, and scattered surviving evidence identifies only one woman physician on the staff, Dr. Dorothy Worthington, a 1921 graduate of the medical school at the University of Pennsylvania.[61] Records from the Lewis Street Center dispensary, however, located in a nearby neighborhood of Italian immigrants, reveal women physicians to have been an integral and essential component of the clinic's success.

The Lewis Street Center was founded in 1907 as the Practical Housekeeping Center of Rochester, named after the Practical Housekeeping Centers of New York City. The center was founded in the heart of Rochester's Italian immigrant community at a time of dramatically increasing Italian immigration. Located in a ward with the highest population density in Rochester, its initial purpose was to teach the neighborhood's immigrant women the "proper" techniques of housekeeping and child care. After outgrowing its first location on Davis Street, "a little double-decker house . . . precisely like dozens of other homes in the neighborhood," in 1911 the center moved to 57 Lewis Street and incorporated under New York State law. From an initial attendance at the settlement of only 6,000 persons in 1907–1908, approximately 46,000 attended at least one program in 1923.[62]

The clinic got its start on Davis Street when the Health Bureau, under Dr. George Goler, established a milk station there and assigned a nurse to weigh babies and give advice on child hygiene and feeding. In 1916 a full medical clinic was established under the direction of Dr. Alvah Miller. The following year, a gift from George Eastman permitted the purchase of a second building on Lewis Street, and this became the site of the clinic. By 1917 the Health Bureau was sponsoring three different clinics at the center: Dr. Miller directed the children's clinic and Drs. Elsa Will (Leveque) and Lucy Baker shared the mothers' and prenatal clinics; the nurse-in-charge, Agnes Kahaly, attended at all clinics and conducted home visits when necessary. The doctors charged ten cents per visit but would waive this fee on occasion.

The clinic's application to the New York State Board of Charities for a dispensary license declared that it would not make use of "clinical material" for instruction of medical students. Nevertheless, such clinic appointments may have seemed a good way for young physicians, especially women with lesser access to hospital appointments, to gain clinical experience. Just as important, Miss Kahaly, who exercised the greatest authority in hiring physicians, was convinced of the value of having women doctors. When at the outset she began planning for the prenatal clinic, she told the board, "I haven't as yet decided on the doctor, although I think a woman doctor would be much more preferable than a man." Dr. Will (who took charge of the prenatal clinic) soon received a letter from a woman physician who had recently opened an office nearby. She requested that Dr. Will divide her clinic patient list between them. The board replied that "a mechanical division of patients was not thought wise, that we wished all patients to choose their own doctors and that we would be glad to place her name on our list of available doctors with established reputations, etc." Seven years later, Kahaly's successor Miss Mary Harriman, told the board how fortunate she felt in having gained the services of Dr. Dorothy Worthington. "She is a most charming woman and a most capable physician. The children just love her and the mothers have a great deal of confidence in her." Of another of the center's women doctors, Dr. Gertrude McCann, she noted: her "clinic is very popular. Her practical experience of [having had] children helps greatly in dealing with our mothers." Between 1917 and the clinic's closing in 1931, a total of eight different women physicians worked for at least part of a year at the clinic. Two of them, Dr. Elsa Will Leveque and Dr. McCann, also served six-year terms as manager of the center.[63]

In those settlement houses where nurses managed the clinics, clinic personnel went to great lengths to establish good relations and a strong rapport with patients. Given the ethnic and educational distance that must have separated them from their clients, mutual understanding was not so easy to accomplish. In her first year at Lewis Street, for example, Agnes Kahaly told the following story in her monthly report to the board: "Some of the Italians have a perfect horror of going to the hospital, although this is not true of all. One woman took her baby to Rochester General to the clinic, and the baby had a scalp wound that was very bad and infected. The doctor begged her to leave the baby, but she refused. All she would say when I asked her was, 'Do you want my baby to die?'" Six months later, however, Kahaly was beginning to understand her neighbors a little better. She reported on her amusement when a doctor at the rival Baden Street clinic prescribed for a baby with rickets "milk, eggs, orange juice, beef broth, chicken broth, oatmeal, rice, bread and milk, and also let it play in a sand pile." She said, "The father has been on strike for four weeks, and [the baby] will consider itself lucky if it even gets milk." By 1924 Kahaly's successor displayed genuine solidarity with the mothers in her care. For example, after one woman was sent to the hospital for an operation (taking her infant child with her), she made the following report: "I had been telling the husband that his wife works too hard, but he could never see it. The first few days she was in the hospital . . . the man had to take care of the . . . children as he was not working. Every day he would come over and beg me to do something as he couldn't stand it. I think now he realizes what his wife had to do with four children of three years and under."[64]

Clinic attendance records suggest that the clinic made a major impact on its neighborhood. In 1916, its first year of operation, about 2,800 patient visits were recorded; two years later the number had increased by nearly two thousand visits. After 1920, the numbers fluctuated between about 2,500 and 3,900 patient visits per year, with children accounting for most of the increase. But by 1931 when the clinic closed, attendance figures were low. Settlement leaders offered various explanations for this decline. In the words of one longtime supporter, "By then it had fulfilled its purpose. More clinics had been developed in all the hospitals. A new generation had been trained to venture out of the immediate community to attend them." In addition, the competition from the Health Bureau's school and prenatal clinics, which charged no fees at all, caused a large percentage of the decline. Lastly, as Harriman admitted, "we seem to have been deserted by the doctors." Many of the center's physicians found more lucrative or respectable work as full-time physicians for the Health Bureau, as school or college physicians, or by joining the staff of one of the local hospitals—positions that left them neither the time nor the professional incentive to volunteer at Lewis Street. Both the Health Bureau and the local Medical Society turned down the center's requests for volunteer city or private practitioners. By 1930, when only one physician was left, Harriman confided that "she was doing her best not to let [him] know that he was the only one coming." Finally she, too, left, taking a job as a nurse for the Rochester Health Bureau in 1931, when it was clear that the clinic could not long survive.[65]

4

Redefining the Margins: Women Physicians and American Hospitals, 1900–1939

> Hospital appointments are crucial for successful medical practice. The more important posts are associated with the highly specialized practices . . . The two form an interrelated system.
>
> *Sociologist Oswald Hall, 1948*

> One of our most important problems [at this hospital] is picking interns. The main qualification as far as I can see is "personality."
>
> *Hospital department chair, quoted by Hall, 1948*

IN JANUARY 1900, Superintendent Sophia Palmer of Rochester City Hospital sent a long letter to the hospital board's executive committee. In an attempt to attract more patients to the hospital, she urged that Dr. Marion Craig Potter, a leading physician in the city and the only woman on the hospital's outpatient staff, be "granted the privilege of bringing [women patients of the public ward class] into the hospital," adding: "Dr. Potter brings us quite a large number of private patients during the year." Although the Board of Directors readily agreed to the proposal, the medical board flatly rejected it. Rather than promoting Dr. Potter to the regular inpatient staff and giving her charge of a women's ward, they grudgingly agreed to name her assistant to the staff, a position with little authority. In "all cases," they insisted, patients would be assigned to her by the attending physician or surgeon. At the time, Potter was a successful practitioner with fourteen years' experience. She continued her prosperous private practice until her retirement in 1938. But neither she nor any other woman physician was appointed to the regular inpatient staff of City Hospital for another quarter century.[1] Marion Craig Potter's experience provides a rare window on the means by which even successful women doctors were deflected from full participation in hospitals, the flagships of American medicine in the twentieth century. Indeed, the growing centrality of the modern hospital was a critical factor in the development (or retardation) of medical careers, especially after World War I.[2] Womens' underrepresentation on most hospital inpatient staffs, on hospital staffs overseas during the war, in most internships, and in most residency programs, significantly slowed their move into the core institutions of twentieth-century medicine—hospitals, medical specialties, and medical school faculties.[3]

Sidelined: Women Physicians ca. 1900–1920

Prior to World War I, most doctors, whether male or female, would have identified themselves as general practitioners; GPs made up three-fourths of the profession in 1928. But by 1942 fewer than half of all physicians in practice described themselves as generalists, and this declining trend would accelerate in succeeding decades.[4] By the 1920s general practitioners were vulnerable to a concerted campaign against "unqualified" practitioners, defined as anyone lacking either the imprimatur of lengthy specialty training or membership in a specialty society. The added factor of gender discrimination ensured that the effect on women was even more enduring. The American Gynecological Society (established 1876) elected its first woman member, Dr. Lilian K. P. Farrar, in 1921; the second was elected in 1970.[5] Moreover, since many women, like Dr. Potter, were nevertheless de facto specialists in obstetrics and gynecology, they were often caught in the cross fire between surgeons and gynecologists in the battle to control the specialty.[6] For decades, a cascade of discriminatory practices kept most twentieth-century women physicians from participation in hospital-based and specialty practices.[7]

Previously, although health care institutions had deliberately instituted discriminatory policies that mirrored social distinctions of class, gender, religion, and race, the first generations of medical women had achieved visible success in the protected niche of gender-specific health care settings.[8] Subsequent ranks of fledgling women graduates had looked to such women for inspiration and for help in launching a career.[9] The erosion of a "woman's sphere" in medicine diminished the supply of visible senior women to recruit and mentor promising younger women, contributing to a scarcity of younger women to follow in the older generation's footsteps.[10] When these separatist health care settings succumbed, young women graduates found few allies when they applied for positions as hospital interns or, after 1910, residents.

Some studies trace the slowed progress of women physicians in part to the declining percentage of women medical students between 1904 and 1915, emphasizing external factors that repelled potential female medical students.[11] A wide array of factors discouraged women from applying to medical school. These included the decline of homosocial institutions and discourse in the culture at large and, in medicine, the disappearance of almost all of the women's medical schools; the availability of new fields for educated women, such as social work and academic science; the overall decline in numbers of openings in medical schools; the increased expense in time and money of obtaining a medical degree; and, in some schools, deliberate discrimination against, and reduction in the numbers of, women students.[12] Many financially or educationally weak medical schools (notably sectarian, women's, and African American schools) succumbed to external pressures to close

Table 4.1. American medical graduates who were female, 1904–1950

Year	Percent female
1904	3.4
1910	2.6
1915	2.6
1920	4.0
1925	5.1
1930	4.5
1935	4.1
1940	5.0
1945	5.1
1950	10.7

Sources: "Medical Education," *JAMA* 105 (1935): 685; Kenneth M. Ludmerer, *Learning to Heal: The Development of American Medical Education* (New York: Basic Books, 1985), p. 248; Carol Lopate, *Women in Medicine* (Baltimore: Johns Hopkins University Press, 1968), Appendix 1, p. 193. In earlier years, the AMA reported slightly higher percentages: 4.0% for 1904 and 3.7% for 1915. "Medical Education Statistics," *JAMA* 65 (1915): 691.

Table 4.2. American physicians in practice who were female, 1900–1950

Year	Percent female
1900	2.5
1915	3.6
1920	5.0
1930	4.4
1940	4.6
1950	6.1

Sources: For 1900–1920, Thomas Neville Bonner, *To the Ends of the Earth: Women's Search for Education in Medicine* (Cambridge, Mass.: Harvard University Press, 1992); Mary Sutton Macy, "The Field for Women of Today in Medicine," *Woman's Medical Journal* 27 (1917): 52–56; Marion Craig Potter, *Census of Women Physicians*, Marion Craig Potter Collection, Edward G. Miner Library, University of Rochester School of Medicine and Dentistry. For 1920–1950, cf. Mary Roth Walsh, *Doctors Wanted: No Women Need Apply* (New Haven, Conn.: Yale University Press, 1976), p. 186.

their doors. Surviving schools drastically curtailed enrollments. In 1906 the AMA Council on Medical Education began ranking medical schools in three categories, A through C. By 1914 thirty-one states barred recognition to Class C schools. The disappearance of these schools was almost certainly linked to the disproportionate decline in the number of female and African American medical students.[13] Between 1904 and 1915 the total number of medical graduates fell from 5,574 to 3,536, a decline of 37 percent. Male students declined by 36 percent. But in the same period the number of female graduates decreased from 198 to 92, a decline of 54 percent, reducing the proportion of women medical graduates from 3.4 percent in 1904 to 2.6 percent in 1915 (see Table 4.1).[14] Following this twelve-year decline, the percentage of women graduates increased modestly to about 5 percent by 1924.[15] The percentage of women in practice fluctuated between 4.4 percent and 5 percent from 1920 until World War II (see Table 4.2 and graph).[16]

Just as significant, those women who did become physicians faced extremely limited opportunities for internships, residencies, hospital staff positions, medical school faculty appointments, specialty society memberships—that is, for entrée to the profession's upper tiers. Many women physicians continued to work as generalists in private practice or in community health long after male physicians began joining the ranks of hospitals, medical schools, and specialty societies.[17] What kept

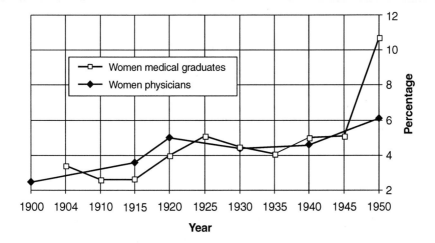

women from gaining access to the newer professional pathways? What kept women physicians from continuing to build on their late nineteenth-century successes?[18]

The career of Rochester physician Marion Craig Potter illustrates the obstacles and constrained career choices faced by second- and third-generation women physicians. Potter, a protégée of Sarah Dolley, was a successful private practitioner who limited her practice to obstetrics and gynecology. For many of her professional contemporaries—male or female—her career would not have seemed marginal at all. She was admired by her female colleagues as a model of professional fulfillment. But viewed from the perspective of her own ambitions and (admittedly) from the vantage of late twentieth-century hindsight, her failure to rise in the Rochester City Hospital hierarchy signaled an important professional defeat. Potter may not have been the typical woman physician, but she is a fitting exemplar. Her successes and failures illustrate the impermanence of women physicians' place in the hierarchy of modern medical institutions. Potter's experience was no anomaly. It typified the relationship between the restructuring of medical institutions and the static careers of many women physicians.[19]

Marion Craig Potter: Background and Training

Marion Craig was born on September 14, 1863, in Churchville, New York, a village fourteen miles south of Rochester and the hometown of temperance leader Frances W. Willard. One of six daughters of Dr.

James and Sarah Craig, she and two elder sisters, Anna and Sarah, all eventually became physicians. But Marion took the lead. At the age of fourteen she regularly accompanied her father on his rounds, once gamely assisting him during surgery after the patient's family all had fainted. Two years later, following her graduation from Geneseo Normal School, Craig announced she was going to study medicine. Her parents reluctantly agreed—if she would attend the nearby University of Buffalo. Instead, she consulted Dr. Sarah Dolley in Rochester, who always advised women to study medicine at the coeducational University of Michigan, which was at the time one of the best medical schools in the country.[20] But Craig was only sixteen. Her parents agreed to send her to Michigan on condition that her sister Sarah would accompany her and act as chaperone. At the end of the three-year medical curriculum, they both graduated. Their eldest sister, Anna, graduated from Michigan in 1893 after first working for twelve years as a teacher.[21]

For women studying medicine—especially those like Dolley and Craig, who valued coeducation as the way of the future—the University of Michigan Medical School was arguably the very best. Until 1877 teaching there was conducted according to the old style, through lectures and demonstrations. By 1880, however, laboratory instruction was given in all scientific subjects—thirteen years before the Johns Hopkins medical faculty made this the standard of excellence. Prior to Michigan, only Harvard had undertaken such reform. Even after women were admitted to the University of Michigan Medical School in 1870, many of their classes were held separately from those of the men. Not until 1883 was integration of the sexes completed in most classes (except anatomy), and even then women were required to sit on one side of a red line. Women comprised 19 percent of the graduating class in 1880, more than 25 percent in 1890, and 24 percent in 1900, a consistently higher proportion than in the rest of the university. The Craig sisters' graduating class of 1884 contained eighteen women. The first African American woman at the medical school, Virginia Jane Watts, graduated a year later. Attrition rates were about even for men and women. On average, women's grades were just slightly higher, except in Dr. Dunster's gynecology lectures, where the grades were about even. (He taught the discouraging doctrine that menstruation controls the female's destiny, a subject of intense controversy among educated women.) But women held no positions on the faculty, nor were they admitted as house officers to University Hospital, until the 1920s. Moreover, after 1910 the number of women students never reached more than 10 percent of the class. (In 1910 women made up only 2 percent.)[22]

Following her graduation, Marion Craig returned to Churchville to join her father's practice. At the time, most women were accepted into internships only at all-women's hospitals. Cook County Hospital in Chicago first accepted a woman intern and a woman attending physician in 1881; Blockley, where Sarah Dolley had spent her postgraduate year in 1851, began regularly accepting women interns in 1882. But these were the exceptions. And besides, Craig was only twenty years old. In fact, Michigan refused to send her a diploma until she reached the age of twenty-one.[23]

Marion Craig described work with her father as "a large country practice which spilled over into Rochester . . . I had wonderful training and experience in obstetrics and gynecology and considerable surgery, including extracting teeth."[24] Her casebook from August 1884 to November 1885 illustrates the variety of problems encountered during a typical year by rural and small-town physicians in New York State in the 1880s. Of the thirty-three different patients recorded, twenty-one were female, including four children (nearly 65 percent); twelve were male, including two children. The evidence that 35 percent of her patients were male seems to challenge the stereotype of the woman physician's typical practice. More likely, however, because Marion Craig was her father's partner and because the population as a whole was sparse, de facto specialization by sex would have been far less feasible than in an urban practice. Out of thirty patients whose case dispositions are known, nine died during the sixteen-month period covered in her casebook, a mortality rate of 30 percent.[25]

Her notes on an early case betray the inexperience of a fledgling physician: The patient, Mrs. W. (no age given), "enjoyed good health until . . . an attack of uterine hemorrhage following menstruation attended w/ dark, bloody, semi-solid pieces appearing more like retained placenta than bl. clots. It could not be disorganized by water. Had no pain except slight pain in back not much different from those she had when tired. These pieces have come away every 2 or 3 days for three weeks, having slight hem. all the time which is worse when these pieces are thrown off." Four days after the initial entry, Dr. Marion Craig wrote: "Has had no hem. in 2 days. Dig[ital] examination shows the os contracted where it was previously dilated. Query what's the trouble."[26] This likely was a case of miscarriage or possibly an incomplete abortion. But both doctor and patient were lucky: the patient survived—as did her physician's reputation.

Other cases included fractured limbs, Bright's disease, tuberculosis (this patient died after a year), eclampsia of pregnancy, an infected foot ("rapid recovery"), and a case of scurvy, for which she prescribed bananas, lemons, and "very fresh meats." On December 25, 1884, Craig was consulted by a twenty-six-year-old Irish seamstress whose condition she diagnosed as "hysteria." Her symptoms included exhaustion from "sewing early and late," amenorrhea, abdominal pain, transient deafness, and mental confusion: "Patient nervous and excitable." A course of morphine, chloral (a sedative), and salicylate of soda over the course of three days finally quieted her and eased her pain so that she could sleep. Craig aimed her treatment at the underlying cause of her patient's symptoms which, in Craig's view, was exhaustion. By February, she was "improving."[27]

In 1887, after practicing medicine in Churchville for almost two years, Craig opened her own office in Rochester. Soon after, she moved to South Clinton Avenue, a neighborhood of many doctors' offices, closer to the central business district and to City Hospital.[28] Craig arrived in Rochester just weeks after the formation of the Practitioners' Society, the first women's medical society in the city, and the Provident Dispensary for Women and Children. Within months, she became a fixture in both organizations.[29]

In 1891 Craig was diagnosed with an irregular heartbeat—seemingly a serious threat to her health. Instead of entering into a passive convalescence, however, she decided to travel around the world. In the course of nearly a year she toured England, visited the Pasteur Institute in France (Pasteur was her lifelong hero), enrolled in postgraduate study in Vienna, and visited her sister Sarah, who was now married and practicing medicine in Kyoto. While in Europe she met an old friend, Dr. Ezra B. Potter, assistant superintendent of Rochester State Hospital, a psychiatric hospital. They were married in January 1893, shortly after her return home.[30] The couple first moved into Ezra Potter's quarters at the State Hospital, although Marion Potter maintained her own private office downtown. Five years later, after the birth of their first son, they built a new house directly across from the hospital campus. The Potters had two children, James Craig (called Craig), and Ezra Barker, Jr. (called Pinkey), born in 1898 and 1901, respectively. Craig Potter recalled a household full of servants—many of them asylum inmates—including a nurse for the children, a cook, a gardener, maids, and a chauffeur.[31] Young Ezra died at the age of nine of pneumonia. Craig became a physician, at first practicing with his mother. Bright, self-confident, and authoritative, Marion Potter was the foremost influence on her son's career, something he readily acknowledged. She urged him to specialize in medical research, an opportunity she had never had, and so he spent two years at the Mayo Clinic in Rochester, Minnesota, earning a master's degree in pathology. She also insisted he learn to knit so that his rather large hands would stay supple enough for surgery. Eventually he headed the department of Obstetrics and Gynecology at Rochester General Hospital.[32]

After the death of her husband in 1921, Potter bought a capacious and elegant house on University Avenue where she, her son, his wife, and their children all lived. Mother and son both had offices there. One of Potter's granddaughters remembers with what little regret the two doctors reorganized their "beautiful dining room" to install X-ray equipment. The third floor housed Marion Potter's private quarters as well as a roomful of caged white rats, for research.[33] Potter never sacrificed her love of fine clothes and fine furniture, however, nor her decidedly feminine manner with patients and colleagues. She made her office into a place of comfort and warmth and often kept a fire going throughout the day. Her patients included women and children from both the highest and the lowest classes of Rochester society. Potter always began her appointments with conversation. Talk was followed by a complete physical exam and, finally, some straightforward advice about marriage, sexuality, pregnancy, menopause—whatever concerned the patient. She would not do abortions, according to her granddaughter, but she often cared for young women suffering from a botched one. Often, too, she acted as an unofficial adoption agency, not an uncommon practice for doctors of the early twentieth century. Confidentiality was supposed to be absolute: her granddaughters, whose room was directly above her office entrance, were forbidden even to look out their windows during office hours.[34]

An office ledger for the years 1890 to 1939 reveals Potter had as sound a grasp of the financial side of medicine as of the interpersonal. Given the oversupply of physicians in the late nineteenth century, both men and women faced a rocky financial road during their first years in practice. Contemporaries estimated it took anywhere from two to five years before an office practice became self-supporting. Some locales were more hospitable than others. In California, a combination of opportunity and open-mindedness launched many women doctors directly into successful careers. But it took Bertha Van Hoosen, first president of the Medical Women's National Association, years to build up a thriving practice in Chicago.[35]

Marion Craig had enjoyed many advantages when she first arrived in Rochester: her father had been well known, she had been accepted quickly into Sarah Dolley's circle of women physicians, and she had been able to join the newly founded Provident Dispensary and Practitioners' Society. Later, her marriage brought her into social contact with many leading physicians of the city, including presidents of the Monroe County Medical Society, founding members of the Rochester Academy of Medicine, and at least one leading surgeon at Rochester City Hospital.[36] The climate for women physicians had been relatively welcoming, and she had soon prospered. Although she earned only $718 in her first year in the city, between 1890 and 1899 she recorded charges totaling $54,523, for an average of $5,452 per year—well above the average physicians' income.[37] Her actual collections totaled $45,689 for the decade, an office income averaging $4,569 per year, even though she earned nothing at all during her ten-month trip around the world. During the decade from 1916 to 1926, her yearly receipts averaged a substantial $7,232; if she had collected all she was owed, her average yearly office income would have been $8,540 per year. In comparison, a large survey of relatively prosperous women workers in 1927 found that women physicians in private practice earned about $3,000 on average. Potter's peak earning years were 1926 and 1927. After 1932, her earnings slowly declined until her death in 1943. Excluding the value of her house, her net estate totaled around $23,000.[38]

All of Potter's substantial success in private practice, however, did not help her gain access to the inner circles of hospital medicine in the city. Although she was named in 1898 to the position of assistant physician in the City Hospital Outpatient Department for Diseases of Women, her appointment resulted from pressure by the hospital's Ladies' Board of Managers. Three years later, when the Ladies' Board, Nursing Superintendent Sophia Palmer, and the Board of Directors attempted to establish an inpatient women's ward with Potter as the assistant physician in charge, the medical staff refused to approve it.

Hospitals and the Original "Glass Ceiling"

The resistance Potter encountered was not unusual for her generation. Chicago obstetrician-gynecologist Bertha Van Hoosen spent years after her 1893 graduation from the University of Michigan Medical School seeking a secure affiliation with a coeducational hospital or medical school faculty. Van Hoosen's autobiography describes her many setbacks as she attempted to gain hospital and teaching privileges in one setting after another. Van Hoosen completed an internship, became an instructor in anatomy at the Chicago Woman's Medical College, and studied privately with a general surgeon. Yet she was at first limited to privileges at the all-women's Mary Thompson Hospital and the Northwestern University School of Medicine. Eventually, after pursuing innovative clinical research in the use of scopolamine-morphine anesthesia in childbirth (published primarily in the *Woman's Medical Journal*), she was named clinical professor of gynecology at the College of Physicians and Surgeons in Chicago. Her hospital privileges were soon revoked, however, as a result of male faculty resentment, and she was forced to found her own private gynecological hospital to provide a teaching theater for her students. Without a hospital affiliation, she would have had to resign her faculty appointment.[39]

The resistance encountered by Potter and Van Hoosen must be understood in the context of hospitals' rising professional importance between the 1880s and the end of World War I.[40] In early nineteenth-century America hospitals served as repositories for impoverished, chronically ill men and women who lacked the means to receive care in their own home. With few exceptions, they functioned as extensions of almshouses. Until the 1870s, only about 170 hospitals were in existence in America. Funded by a combination of municipal and private charity, they provided care but little cure. Prior to the founding of nurse training schools in the 1870s, attendants were drawn from the ranks of the almshouse inmates. Fortunately the care was often sufficient for nature to take its own restorative course; patients frequently survived their stay.[41]

In the decades following, although a few hospitals remained attached to city almshouses, most became independent institutions, either privately owned general hospitals established by urban elites or religious foundations, or proprietary hospitals owned by the physicians who ran them. Improvements in hospital organization, nurse training, asepsis, anesthesiology, and surgical technique helped to alleviate the public's distrust of hospitalization. In 1904 the number of hospitals in the United States had grown to about 1,500—a mix of public and private, nonsectarian and religious, charitable and proprietary institutions. Revitalized medical school curricula further reshaped the culture of the hospital. Especially in urban centers, medical schools eagerly sought permanent hospital affiliations to enhance research and teaching. By 1925 nearly 7,000 hospitals were in operation in cities large and small.[42]

Hospitals thus were slowly reshaped into "medical workshops," and this had major implications for their structures of governance. Formerly they had been operated by local philanthropic elites, often through the partnership of a board of directors (all men) that handled financial management and a "ladies' board" that oversaw day-to-day supervision and small-scale fund-raising. But as the professional sophistication and ambition of physicians expanded and the costs and complexity of hospital management increased, hospitals became the site of a struggle for authority among lay boards, hospital superintendents, nursing staffs, and, especially, physicians.[43] At stake was the authority to make decisions regarding patient admissions, medical admitting privileges, the organization and staffing of the wards, the hiring of supervisory and nursing personnel, and the purchase of equipment and supplies. In the case of municipal hospitals, challenges to the authority of political patrons by the medical staff were also a pressing concern.[44]

Staffing patterns reflected these tensions. Part-time appointments, approved by lay governing boards, were the norm for the nineteenth-century hospital. By World War I, senior medical boards selected the attending physicians, who were still predominantly part-time appointees. They were divided into senior and junior staffs, while a staff of assistant physicians attended the increasingly important outpatient clinics, also on a part-time basis. Finally, as the number of patients, the acuteness of their conditions, and the frequency of surgical admissions all increased, the role of a full-time roster of interns became much more important, to assure twenty-four-hour oversight. In 1904 hospital internships were available for only about half of all graduates, although many more sought to find one. Between 1914 and 1924 the number of approved internships grew from 2,667 to 3,866. By 1924 the AMA calculated that hospitals' demand for interns far outstripped the supply.[45]

The creation of hospital residencies in the major specialties was the last in this series of changes. Few hospitals had followed the early lead of Johns Hopkins in establishing formal residencies; many American physicians, like Potter, traveled to Europe for some postgraduate training, usually to advance their technical rather than their research skills. In addition, from the 1880s through the early twentieth century, postgraduate or "polyclinic" medical schools in the United States offered specialized courses for physicians wishing to hone their skills in surgery, gynecology, neurology, pediatrics, and other emerging specialties. Some of the leading polyclinics, such as the Post-Graduate Medical School in New York City and the Philadelphia Polyclinic, enrolled sizable numbers of women physicians.[46] But when World War I closed off postgraduate training in Europe to American physicians, many leading doctors insisted on the need for more rigorous specialty training in the United States. The AMA Council on Medical Education recommended that two or more years of hospital-based specialty training, following a general internship, be required for specialty practice.[47]

Hospital boards could not keep up with the rising cost of hospital modernization by relying on traditional philanthropy. To increase the number of paying patients, hospitals had to expand the network of physicians who would consent to admit patients. Commonly, physicians began to admit their patients—and patients became less reluctant about going to the hospital—when they themselves retained the right to continue patient supervision and the collection of fees. In return, the physicians donated some of their hours to the care of the hospital's indigent inpatients. By broadening community physicians' access to admitting privileges while allowing the medical board to control medical staff appointments, hospitals attracted paying patients and sustained the elite quality of their senior medical staff.[48] Doctors' economic self-interest also explains senior physicians' insistence on controlling inpatient staff appointments. Unlike those working in the outpatient department, which provided free or low-cost care, physicians on a hospital's inpatient staff could collect fees from private patients, a source of additional income. Imposing some limits on the competition made good economic sense.[49] Women physicians, in particular, represented considerable financial competition, because many women patients explicitly requested treatment by a woman, especially for gynecological problems. Moreover, gynecological procedures such as the removal of ovarian cysts and uterine fibroids, and the repair of cervical or perineal lacerations, were among the most expensive of the typical medical fee schedule.[50] As Marion Potter's office practice and earnings reveal, she was capable of drawing considerable numbers of patients into her economic orbit—and away from her competitors.

To accomplish their diverse goals, hospitals eventually granted admitting privileges to a much wider range of physicians, but appointments to inpatient services, internships, and residencies, were controlled by the senior medical staff.[51] Hospitals became important both as centers of specialty training and for the solidification of clinical reputations. Where once an appointment to a hospital staff was a relatively exclusive privilege, by 1930 the AMA estimated that nearly 98,500 physicians (more than two-thirds of all doctors) had a hospital affiliation of some sort.[52]

Women's access to hospital privileges, however, remained limited. In the nineteenth-century, hospitals, except those managed by (and for) women, were loathe to take on women medical graduates either as members of the house staff or as attending physicians. The prejudicial belief that women were less rigorous, less objective, less "professional" dogged even the best women physicians throughout their careers. Even the University of Michigan Medical School refused to give a faculty position to Bertha Van Hoosen, one of its star graduates; as Dean Victor Vaughan admitted, he just couldn't hire a woman.[53] By 1915 Harvard physician Richard C. Cabot still advised women to avoid taking up the more "competitive" branches of the profession, chivalrously suggesting they confine themselves to "noncompetitive" specialties such as public health or "social medicine."[54]

Such prejudices were reinforced by senior women doctors' virtual absence from the medical staffs making decisions about whom to admit. In 1914 the first AMA study of internships found that a total of 508 general hospitals of varying levels of professional competence offered 2,667 internships, more than enough to supply the number of graduates.[55] Three years later, a survey of American hospitals by the MWNA revealed that of the 688 general and specialized hospitals for which information was obtained (including psychiatric hospitals), only 113 either accepted women for internships or claimed to be willing to do so. Cook County, Philadelphia General, and Bellevue Hospitals were among those general hospitals—usually situated in large urban settings—they did accept a number of women. Johns Hopkins accepted a small number by virtue of having a number of women medical school graduates who moved on to internships at the university hospital. State psychiatric hospitals, too, were an exception; approximately eleven states, most on the East Coast, required psychiatric hospitals to include at least one woman on the medical staff.[56]

But acceptance of women as interns was not the norm for most hospitals. Since only 92 women graduated from medical schools in 1915 (2.6 percent of graduates), the positions available would have been adequate in number. The important issue was the lower standing of the hospitals that did accept women.[57] As more and more young physicians began competing for internships at "teaching hospitals," letters of reference from well-connected senior physicians often eased their way. Such an internship could become the stepping stone to a competitive residency. The scarcity of women faculty members (always excepting the all-women's institutions) made it that much more difficult for women graduates to gain access to desirable internships. As late as 1925, when most medical graduates found internships, of the 524 hospitals with AMA-approved internships only 128 would consider accepting women.[58] In 1935, 6 percent of all medical graduates failed to enter an approved internship but 8.4 percent of women graduates failed to do so (out of only 226 women graduates).[59]

Of course some women physicians, like Emily Dunning Barringer, who interned at Bellevue in New York, managed to train at well-respected coeducational hospitals. But, as she was the protégée of the well-known Mary Putnam Jacobi, hers was an exceptional case. Likewise, although Rochester State Hospital (a psychiatric hospital) had hired women physicians for more than twenty years, by 1917 neither of the more prestigious general hospitals in Rochester had ever appointed a woman intern. The superintendent of Rochester General Hospital (called City Hospital until 1911) sent the following reply to a MWNA survey in 1917: "I would say that we have never yet received medical women as interns in our hospital. I have, however, filed your communication in order that I may take the matter up with you again in the event of a future change of policy."[60]

For African American women medical graduates of the time, prestige was not the first consideration. Prior to the 1960s, about 80 percent of all black physicians had graduated from either Howard University or Meharry medical school; in 1920, more than half of all black women medical graduates were Meharry alumnae. Few could obtain internships; even fewer could secure one at an integrated hospital. Dr. Virginia Alexander, a 1920 graduate of Woman's Medical College of Pennsylvania, was told she would not be considered for an internship at Philadelphia General if she were "first in a thousand applicants!"[61] Consider the situation of Dr. Isabella Vandervall, a 1915 graduate of the homeopathic New York Woman's Medical College. Vandervall was unable to gain a license to practice in her home state of Pennsylvania because of that state's 1914 legislation requiring new licensees to have completed an internship. Vandervall was turned down for every hospital internship for which she applied—not so much on account of her gender, which assuredly didn't help, but explicitly on account of her race. Adding insult to injury, when she protested to the *Woman's Medical Journal* in 1917, she received only a lukewarm response. Dr. Emma Wheat Gillmore of Chicago was sympathetic but reminded her that, after all, Pennsylvania wasn't discriminating against her *race*, only her lack of qualification. "For centuries," Gillmore pontificated, "the general public have so discriminated, and with fearless veneration for the truth [Vandervall] should recognize that the prejudice was originally founded upon a very real objection." Vandervall, although ranked first in her class, had graduated from one of the less prestigious medical schools. Yet women such as Dr. May Chinn, a 1926 graduate of Bellevue Medical College who became a leading member of the black community in New York City, and Dr. Margaret Lawrence, a graduate of Columbia's College of Physicians and Surgeons who became the first black trainee at the Columbia Psychoanalytic Clinic, had almost no choice but to intern at predominantly African American institutions such as Harlem Hospital. In fact, in 1926 Chinn became the first woman to intern there. As late as the 1940s black medical graduates encountered strong opposition to their interning at any but all-black hospitals. Yet in 1939, only 168 accredited internships were available at black hospitals. In addition black women physicians also battled against sexism, sometimes even at all-black institutions. In 1929 the dean of one such college admitted, "I think the general attitude of the male student [here] is, so far as work is concerned, that he would rather the women students would go to a school for women."[62]

The system of residency training for specialists was formalized during the 1920s and 1930s. Women interns of whatever race or ethnic category rarely enjoyed access to the mentoring and recommendations needed for acceptance in a competitive residency program. The number of residencies in 1925 and 1930 was much lower than the number of internships, and competition was keen; in 1930, 338 hospitals offered a total of only 1,921 residencies approved by the AMA Council on Medical Education. By 1940, when the number of residencies had more than doubled, only 6.5 percent of hospitals with residency programs enrolled women residents. By then, 97.3 percent of board-certified specialists were men.[63] Some women were remarkably persistent, however. Dr. Lena Edwards, a 1924 graduate of Howard University Medical School who completed an internship at Freedmen's Hospital in 1925, was hired in 1931 by the Margaret Hague Maternity Hospital in Jersey City, New Jersey, as an assistant attending physician. Five years later, when she applied for a residency there so that she could apply for board certification in obstetrics-gynecology, the hospital refused. She applied—and was refused—every year until 1945, when the hospital relented. She then completed her residency, passed her boards, and became a member of the gynecology service at the hospital, two decades after she completed her internship.[64] Bertha Van Hoosen found that all but one of the fifty-one women in the American College of Surgeons (ACS) in 1926 had graduated *before* 1915; 75 percent were either graduates of women's medical colleges or had interned or were affiliated with an all-women's hospital. By 1939, the number of women in the ACS had risen only to 81, or .6 percent.[65]

Prior to the 1930s, some women did specialize in fields such as gynecology and pathology, but they remained absent from most specialty societies. A MWNA study of 1917 found that 258 women physicians, or about 5 percent, restricted their practice to one specialty; of this number, more than half specialized in gynecology, ophthalmology, psychiatry, and neurology. By 1939 the percentage of women physicians who specialized had not changed, although the percentage for all physicians was five times higher. What had changed was the ranking for gynecology: the percentage of women choosing that specialty had declined by half.[66] As late as 1928 many specialty societies neither accepted women physicians as members nor included women on the program at their national meetings. To take a field in which women had achieved relative visibility, the American Pediatric Society (APS), established in 1888, admitted no women until 1928.[67] Those few women who did gain access to training either in a clinical specialty or in bench research, women such as Johns Hopkins medical graduates Florence Sabin and Dorothy Reed (Mendenhall), were expected to remain research fellows or technical staff. In 1902, for example, Dr. Reed had completed a year as a pathology fellow for Dr. William Welch, dean and "director of the most famous teaching laboratory for pathology in America." Her research into the "Dorothy Reed" cell had shown the unique cellular properties of Hodgkin's disease. Yet when she asked about her prospects of getting, in time, a faculty position, Dr. Welch only looked "puzzled and embarrassed." He did make special efforts to find her a well paying position as a staff pathologist or in a choice residency, a degree of consideration that few women of the time received, but a faculty position was out of the question. Dr. Clelia Mosher, who eventually became a professor of physical hygiene at Stanford University, wished to train as a gynecological surgeon with Dr. Howard Kelley at Johns Hopkins but was similarly rebuffed.[68] A 1917 survey of women in the profession discovered only 12 women holding the rank of assistant, associate, or full professor in any American coeducational medical school.[69] By 1927 the number of women faculty had jumped to 198 at the forty-three "class A" medical schools, but three-fourths (148) held positions at the level of clinical assistant professor or instructor. Of the rest, 13 were on the faculty of Woman's Medical College of Pennsylvania.[70]

Women Physicians at Rochester City Hospital

The situation encountered by Marion Craig Potter in Rochester thus was quite typical. During the 1890s the hospital's governance lay with the all-male Board of Directors; the Ladies' Board of Managers; and a roster of visiting or consulting physicians (all men) who recommended, but did not appoint, the physicians and house staff. For members of the hospital's boards, male and female, professional advancement through the use of patients as "clinical material" took second place to moral stewardship. Within the hospital universe, an analogue to the idealized

Victorian household, the discourse of gender loomed large.[71] Agents of feminine governance included the Ladies' Board of Managers, consisting mostly of the wives and sisters of the city's powerful elite, and the hospital superintendent, Miss Sophia Palmer. The introduction of women physicians to the hospital can be credited directly to an alliance between the Ladies' Board and Sophia Palmer. Still, women physicians did not have an easy time during the five years of Palmer's tenure. She came to City Hospital from Washington, D.C., in 1896 after twenty years in nursing. Although she was not the first woman to combine the roles of director of nursing and superintendent of the hospital (Miss Helen Gamwell had held the two posts for a year before marrying a member of the medical staff and retiring), Palmer was the first to function as a forceful manager. At her retirement, she was credited with efficient management, increasing public trust in the nursing and house staff, and completing the nurse's home. At her retirement in 1901, Palmer became the editor of the *American Journal of Nursing*, of which she had been a founder, and a leader in the campaign to pass New York State's nurse registration law.[72]

During Palmer's tenure City Hospital faced several major crises. The hospital had begun a deliberate move toward modernized management practices around 1880. The Board of Directors made a conscious decision to attract more patients—especially more middle-class, paying patients—by adding a nurse training school and private rooms furnished in a style consistent with middle-class tastes. In 1888 the hospital built a "clean" room specifically for surgery, and this, along with private-duty graduate nurses and nurse trainees for the wards, contributed to a rising number of private patients. The same year, construction was completed on the Magne-Jewell Memorial Dispensary for "outdoor patients" and an ophthalmic clinic. By 1890, income from private patients accounted for one-fourth of the hospital's total income for the year, but overall patient numbers and revenues were failing to keep up with expenses. The board hoped that, by adding an evening clinic, the community's broader familiarity with the hospital would increase inpatients, too. Hospital surgeons, too, were anxious to increase their service and shorten the lag time between the arrival of an emergency case and the availability of an operating room.[73]

The directors hired Palmer in 1896 only after failing to find a male superintendent who could oversee both the hospital and the School of

Nursing.[74] She had charge of hiring junior house officers (interns) as well as approving (or disapproving) their promotion to senior level. Both boards expressed "excellent satisfaction" with her "splendid work." But within a year the medical staff expressed quite a different sentiment. After Palmer disapproved the promotion of a junior house officer for insubordination, the medical staff objected to the Board of Directors that she ought merely to give them "advice" rather than "embarrass them, as in the case [at] present." Soon after, the medical staff persuaded the board's executive committee to take over responsibility for house staff promotions. They also created a physician's executive committee to advise the board.[75] The superintendent's authority thus was directly and successfully challenged. Nor was the authority of the Ladies' Board immune to attack. Neither would be allowed to challenge the physicians' increasing authority.[76]

Potter's appointment to the hospital's outpatient department occurred on the heels of this contretemps. Her difficulties, and those of other women subsequently appointed to the outpatient staff, can best be understood in this context. In 1898, after the Provident Dispensary for Women and Children closed, City Hospital arranged to incorporate it into its outpatient dispensary as a Women's Clinic, staffed by women physicians. Two motives underlay this innovation: a desire to broaden the scope of the hospital's still struggling clinic, and a concern, shared by Palmer and the Ladies' Board, that "as the Dispensary for Women and Children connected with the City Poor Department has been closed, it would seem proper that a woman be in attendance for that class of patients coming for treatment at the Outdoor Department of the Hospital." They specifically requested the appointment of Dr. Marion Craig Potter. Neither the minutes of the medical staff nor of the Board of Directors record any discussion of Dr. Potter's appointment. But the outpatient department staff listings for 1898 do record her among the seventeen physicians named. The following year Dr. Evelyn Baldwin's name was added, again without comment.[77] The Board of Directors' annual report, published in the *Hospital Review*, lavishly praised the work of Dr. Potter. Privately, some of the physicians on the staff were not so enthusiastic.[78]

Potter's uneasy relations with her male colleagues became apparent two years later, in 1900, when Superintendent Palmer attempted another bold step. In a long letter to the Board of Directors' executive committee she recommended that a suite of rooms currently unused be turned into a public ward for women, under Dr. Potter's direction. She justified the move on several grounds:

> Dr. Potter . . . frequently has patients of the public ward class who desire to be treated by a woman, and who are unwilling to come to the Hospital because, with the present arrangement, they are unable to have the services of a woman physician. I believe that if Dr. Potter were granted the privilege of bringing this class of patients into the Hospital under the same conditions upon which patients are received in the public ward, that we should broaden the scope of the hospital, and have the support of the leading women physicians of the city. By using this second suite of rooms for this purpose the regular Staff service would not in any way be interfered with. Dr. Potter brings us quite a large number of private patients during the year.[79]

In the same letter, Palmer also recommended that deliberate action be taken to increase the use of the newly renovated maternity wards, especially in light of the modern ward just opened by their chief competitor, Rochester Homeopathic Hospital. The Executive Board moved that the medical staff "be advised" of their approval of both suggestions, but added that they "would be glad to receive recommendations of the Staff in regard thereto."[80]

Four months later, after the medical staff failed to respond, the directors sent another note requesting their advice and noting that three rooms on the second floor of the outpatient building already had been set aside for a women's ward under the charge of Dr. Potter. Julius Wile, a leading Rochester banker and a member of the board's executive committee, wrote again to the secretary of the medical staff, Dr. Charles Dewey: "Up to the present time no word has come from the staff indicating their willingness to have certain rooms set aside for that one of the female physicians of the city who has done most for the Hospital."[81]

The medical staff and the board held a series of private meetings in May, following which the physicians put their objections in writing. As they wrote, "The Staff did not approve so radical a departure in the policy of the Hospital . . . at least without essential modification. It was felt that no such discrimination should be made in the case of members of the Junior Staff—that it would be unjust and unwise to give to certain ones privileges not granted to all alike."[82] An angry Dr. Potter met privately with Julius Wile to offer her resignation as a result of her treatment by the senior medical staff, especially the surgeon Edward W. Mulligan. But Wile assured her that the real question was not whether she was qualified, but whether the trustees or the surgeons—particularly Dr. Mulligan—would control the hospital.[83]

Nevertheless, after more than a month of wrangling, the trustees rescinded their order "establishing a Woman's Ward under the care of Dr. Marion C. Potter." In return, the medical staff agreed to make her an assistant physician on the inpatient service, but without control of a ward. A letter from Dr. Dewey conveyed just what her new title entailed: "This means that you will attend such ward patients as the Staff may assign to you, specially those in the regular Female Ward who may prefer the services of a woman physician." In other words, as she was pointedly reminded in 1902, she must "get the consent of the attending physician or surgeon in all cases."[84] Further cementing their victory, the senior staff won an agreement from the board that the two groups' executive committees would meet together at least once a month. The senior staff, however, was still dissatisfied with the hospital's medical administration; in continued discussions with the board, the physicians now focused on the "future status of the Superintendent." About six months later Sophia Palmer retired, a clear indication that board member Wile's question had been answered definitively—in favor of the surgeons.[85]

Potter's position within the hospital hierarchy never recovered its upward movement after this setback. In 1902, during an investigation of sloppy management in the outpatient department, for example, Potter was targeted for special criticism by Dr. L. W. Rose, secretary of the department. "The records of the 'Female Clinic,' he wrote, "are not worthy the name." He also accused Potter and Dr. Evelyn Baldwin of "repeatedly" seeing other clinics' patients. Rose, it should be noted, was a direct competitor of Potter's and Baldwin's, as he was the outpatient department's designated specialist in diseases of women.[86]

The number of cases referred to Potter and Baldwin by the inpatient staff never was large. From November 1902 to November 1903, for example, Potter was listed as the attending physician in twenty-three cases, averaging fewer than two per month. Baldwin saw only eight. Five other women physicians were listed in the hospital patient register for that same period, although none was a regular member of the staff. Altogether they handled fifteen cases, probably as replacements for Potter and Baldwin, making a total of forty-six patients who requested, or were assigned to, women physicians in the twelve-month period. During the same months, Dr. Edward Mulligan alone saw sixty-nine cases. Hospital records show that Potter cared primarily for female patients. She treated pneumonia (Dr. Baldwin was once her patient), hemorrhoids, and gastric ulcers, but gynecological operations for cystitis, retroflexion of the uterus, endometritis, and lacerations of the cervix and perineum accounted for two-thirds of her work. (She did treat one man for a dislocated shoulder and another for rheumatism.) Mulligan's cases, in comparison, ranged from pregnancies and gynecological surgeries to fractures, hernias, wounds, and, many cases of the surgical specialty du jour, appendectomy. The majority of his patients also were women.[87] By 1908 Dr. Mulligan had a much larger caseload than Dr. Potter. While she attended only twelve cases for the entire twelve months, Dr. Mulligan, a general surgeon, attended more than ninety patients during the same period.[88] At the same time, as was noted earlier, Potter's private practice was booming. It seems likely that many more patients requested Potter's services than the senior staff actually referred to her.

In 1908 and 1909 Potter made one last effort to bring her hospital rank into line with her standing in the community. She wrote to Dr. Dewey requesting that she and Dr. Evelyn Baldwin "be styled gynecologists or gynecological assistants." The request was "tabled."[89] By that same year, all three physicians who had been listed with Potter in 1901 as "Junior Assistants in the House" (and were still on the staff) had been promoted to the regular inpatient staff—but not Potter. Significantly, all had been interns at City Hospital between 1889 and 1891. (None of the women who applied for internships at the hospital before 1920 was accepted.) From 1909 to 1917 Potter was the only physician listed at the rank of "Assistant in the House." Dr. Rose, who, like Potter, lacked a formal internship, was by then listed as a member of the junior surgical staff, and was so listed until his departure from the hospital in 1911.[90]

The outpatient department did not allow women physicians much scope for professional development. The clinic's major department, General Medicine and Surgery, Diseases of Women and Children, saw patients five mornings per week, and only male physicians were assigned to it. A second department, designated Diseases of Women (Woman Physicians) and headed by Drs. Evelyn Baldwin and Cornelia White-Thomas, was open only two afternoons per week. When the outpatient department was reorganized following World War I, the women physicians' clinic was eliminated entirely.

Women Physicians and Hospitals after World War I

The gradual increase of medical, and especially surgical, authority in hospital management after 1910 quickened its pace during the years following World War I. In part, the trend reflected successful efforts by the American College of Surgeons (established in 1913) to impose uniform standards of organization, asepsis, laboratory analysis, and patient record keeping through a national process of hospital inspection and grading. As Rosemary Stevens has argued, the American College of Surgeons could count on support from veterans of overseas military-base hospitals, as their experiences and esprit de corps had forged an enduring commitment to rationalized planning and industrial efficiency.[91]

American volunteer hospital units in France during World War I were modeled on a prototype of self-contained, elite units of medical and surgical personnel. Surgeon General William Gorgas wanted "groups of physicians and surgeons who were accustomed to cooperative endeavor and who might be expected to work well together under the trying circumstances of war."[92] He invited a small number of hospitals, including Rochester General Hospital, to put together such teams well in advance of America's formal entry into the war.[93]

The success of the American base hospitals left an indelible mark on the institutional values and organization of postwar American hospitals. At least until after World War II, surgeons emerged as the authoritative figures in hospitals.[94] In Stevens's words, an ideology of "technological efficiency, impersonality of treatment, regimentation of patients, dependence on diagnostic tests, importance of [standardized] records, and focus on immediate results [were imprinted] as the ideal basis for the practice of hospital medicine." In 1918 the ACS began a decade-long effort to inspect, grade, and standardize hospital-based surgery as a precondition for its ultimate goal, the elimination of most surgery by general practitioners. Within a short time many hospitals began acknowledging the legitimacy and desirability of the ACS standards. More than 90 percent of the larger hospitals (defined as those having more than 100 beds) had been graded by 1929.[95]

Secondarily, but of enormous importance for women physicians, anyone who had not been in on "the action" during the war was far less likely to prevail during postwar planning for hospital reorganization. Since women were excluded from the Army Medical Reserves, and few had been involved with the overseas military hospitals, they largely stood on the sidelines. The ACS's preferred mechanism for hospital reform, internal reorganization of the medical staff, often placed power in the hands of those physicians and surgeons most unsympathetic to the older, patient-centered values associated with general practitioners, including the majority of women physicians. The gender-based professional niche—in which women physicians were explicitly assigned to the treatment of women patients—disappeared from hospital organization charts. Wards were arranged according to patients' medical conditions, payment status, sex, age, and race, but not according to their preference for a male or female physician, which was deemed medically irrelevant. Women physicians, too, now argued that modern, coeducational medical training prepared them to take their place in a gender-blind health care system.

Events at Rochester General illustrate the pattern. In 1919 the medical staff appointed a subcommittee on reorganization, anticipating the need to comply with ACS standards. Early in 1920 the subcommittee's reorganization plan was approved. Six months later, however, an ACS inspection revealed that patient records including histories, diagnoses, and treatments existed for only 40 percent of the hospital's cases. In a follow-up letter to the hospital superintendent, ACS director John G. Bowman, M.D., made it clear that Rochester General would not make its list of "acceptable" hospitals unless it took definite action to establish permanent records for every patient, and to outlaw fee splitting.

As a result, the staff was divided into consulting and visiting staffs, the latter being organized into medical and surgical divisions subdivided into senior and associate levels, with ten special departments, including obstetrics and pediatrics. Gynecology, apparently still a contested domain at the hospital, was not listed at all. The lowest level was designated the junior staff, which included an undifferentiated outpatient service. Dr. Potter and her designation as "Assistant in the House for women patients," as well as the outpatient department for diseases of women, were eliminated. The greatly enlarged medical staff now included only two women, both assigned to the outpatient department, Dr. Cornelia White-Thomas and a younger graduate, Dr. Lucy Baker. Two years later the staff, now numbering more than a hundred, still listed only two women, both assistants in the outpatient department.[96]

Potter's implicit response to her rebuff at Rochester General may be gauged by her increasing efforts to organize women physicians and by her heightened interest in public health. In this, her career mirrored the professional strategies adopted by many women in medicine during the early decades of the twentieth century. She maintained a high profile in the Parent-Teacher Association, the Women's Christian Temperance Union, the Business and Professional Women's Club, in prosuffrage groups, and by making public health radio broadcasts. By deliberately taking charge of the health and hygiene committees of such organizations, she combined a busy professional life with professionally appropriate social activism. Finally, she acted as a mentor to many younger women, albeit from without the realm of hospitals and specialties. As part of her work for women in the profession, she served as editor of the *Woman's Medical Journal*, president of the Women's Medical Society of New York State, and a member of the Committee on Women Physicians of the Council of National Defense during World War I. She began publicly campaigning for a national women's medical society in 1908, and cofounded the Medical Women's National Association in 1915.[97]

For women physicians like Potter, modernization and marginalization went hand in hand in twentieth-century American hospitals. Of course, the local staffs of many hospitals resisted to varying degrees the authority of the American College of Surgeons. Further, many male physicians, especially general practitioners, those educated before the turn of the century, and those of "lower-status ethnic backgrounds" such as Jews, Italians, and African Americans, also were shut out of the wards and operating rooms of local hospitals.[98] But the effects of change seem to have been disproportionately severe for women. By the 1920s, women physicians had lost the traditional niche within medical institutions formerly assigned to their control. Like many general practitioners, they faced marginalization within the newly dominant realms of hospitals and academic medicine. The devaluation of a gendered conception of women's practice and the elimination of the professional outposts from which earlier generations of women physicians had made their mark disrupted patterns of recruitment and advancement for women physicians. Previously women physicians, and not merely those who taught at women's medical colleges, had attracted professional successors through networks of personal and professional reputation. By the 1920s they could claim few heirs, not only because women's admissions to medical school were low but also because, in a world where internships, residencies, and hospital privileges replaced gender-specific medical institutions as avenues to professional prestige, senior women physicians were rarely able to help young women colleagues up the career ladder.

Potter no doubt regretted her exclusion from the new professional power centers and the denial of her bid to be acknowledged a specialist in gynecology. But her commitment to women's health seems to have enabled her to reemphasize her private practice and public health activism. Many women retained this strong commitment to the older ideals of nineteenth-century practitioners. Unlike the exceptional Mary Putnam Jacobi, the anatomist Florence Sabin, and the bacteriologist Anna Wessel Williams, most women doctors either chose not to pursue or were discouraged from pursuing careers as biomedical scientists.[99] To understand the constraints on women physicians' development as specialists during the first third of the twentieth century, one must acknowledge the powerful effects of deliberate exclusion. Potter's and Van Hoosen's careers reflected the professional limits imposed on women of that generation and their creative responses to those limits. They also reflected the fragility of women's place in America's medical institutions and the ease with which women could be marginalized by the professionalizing initiatives of hospital medicine.

5

Getting Organized: The Medical Women's National Association and World War I

> The remedy for segregation is organization . . . Organization outside
> the profession has accomplished much; organization of women
> within the profession will be equally effective and will hasten the day
> when it can be truthfully said, "There is no sex in Medicine."
>
> Bertha Van Hoosen, *Editorial*, Woman's Medical Journal,
> *May 1916*.

BY WORLD WAR I, the decades-long efforts to reform medical education, reinstitute state licensure for physicians, and attract practitioners into a reenergized American Medical Association came to fruition.[1] In the face this "consolidation of professional authority," as sociologist Paul Starr terms it, a few influential women physicians, including Bertha Van Hoosen and Marion Craig Potter, became convinced that the needs of women in medicine were being overlooked.[2] Between 1908 and 1915 they quietly urged their female colleagues to found a national women's medical association to bring about professional equality while preserving women doctors' traditional claim to a uniquely feminine approach to the health of women, children, and the family.

The annual AMA Women's Banquet, organized in 1908 by the Women's Medical Societies of Chicago and the State of Illinois, provided an excellent venue for Van Hoosen and Potter to begin this effort. (Potter's sister, Dr. Sarah Buckley of Chicago, also was among the organizers.) Potter gave the after-dinner speech on the subject of organization. Drawing on her experience organizing the Women's Medical Society of New York State in 1907, she spoke "eloquently . . . in favor of organization." Unfortunately, audience sentiment was "evenly divided for and against national organization." Fearing the stigmatizing effects of separatism, significant numbers of women physicians shunned all-women's associations.[3]

Many wondered, why formalize the segregation of women physicians at the very moment when their professional integration seemed assured? Why create a separate national organization for women? The basic assumptions about the nature and role of women in American society were shifting. Victorian gender discourse, which emphasized the distinctive features of masculine and feminine cultures, no longer commanded unqualified support. The movements for higher education, coeducation, and woman suffrage were expanding—some would say exploding—the bounds of "woman's sphere." Professionalism, formerly linked to gendered character norms such as "strength of character" and "tender care," now was defined by "objective" standards of education and expertise.[4] The very idea of a woman's sphere, which had formed the cultural matrix for nineteenth-century professional women, lacked currency in the second decade of the twentieth century.[5]

Many women physicians, moreover, believed they had finally "arrived." Women had been attending the annual meetings of the AMA since 1876, when the first woman, Dr. Sarah Hackett Stevenson, was allowed to attend as part of her local medical society delegation from Illinois. In 1909 Rosalie Slaughter Morton was named chair of the AMA Committee on Public Health Education among Women. In 1913 Dr. Lillian H. South of Kentucky was named an AMA vice president; the following year Dr. Alice Hamilton was named third vice president. And by 1915 a few women occasionally appeared on the program of the AMA's scientific assembly.[6] Finally, by World War I the settings for medicine's "separate sphere"—women's medical schools, hospitals, and medical societies—were in decline. The majority of women physicians educated before 1890 had attended women's medical colleges. By 1900, however, most women were graduating from coeducational medical schools.[7] If one ignored the persistent exclusion of women from internships and hospital staffs, a women's medical association could seem like a step backward. By 1915, the idea of a women's national medical association could no longer count on the unquestioned support of its "natural" constituency, women physicians.

Women physicians thus faced an unsettling cultural moment. "Assimilation" had replaced "integration" as the goal of many women in medicine. Gender-specific institutions had lost their rationale. Yet women physicians continued to work in medicine's separate sphere long after the turn of the century—despite the prevailing rhetoric of professional assimilation.[8] Further, many were unwilling to abandon their claim to a distinctively feminine gift for healing and a primary devotion to the care of women and children.[9] All in all, the transition to the culture of modern professionalism was far from smooth for women doctors.

Potter and Van Hoosen both grasped this situation. They were confident that organization would increase women's professional visibility. Van Hoosen had been skeptical of the need for a national association until she understood that women acting alone could never equal the force of many unified voices.[10] By 1914 she acknowledged the need to found a national women's medical society. Like Potter, she had seen the necessity for medical women to organize in their own behalf rather than wait passively for their male colleagues to recognize their worth.[11] In her experience, the AMA annual meeting was a "dreary" experience for women doctors. Once united on the outside of organized medicine, women now sat on the inside—alone and isolated. Their influence was "nil."[12]

Potter and Van Hoosen realized, however, that many women physicians did not wish to call attention to the issues of gender difference. A national association of women physicians risked alienating the very constituency they hoped to attract. Therefore, when Van Hoosen invited a group of twelve women to meet at the Chicago Women's Club on November 18, 1915, during the week-long fiftieth-anniversary celebration for Mary Thompson Hospital, she selected only those whom she knew to be "enthusiastic for organization." Marion Craig Potter took the chair. As her first action, she moved that Van Hoosen be named acting president of the fledgling association. By the end of the day concrete plans had been laid for the Medical Women's National Association, the first nationwide organization of women physicians in America and the culmination of seven years of effort. (The organization took its current name, the American Medical Women's Association, in 1937.)

A shrewd plea for unity was offered by Chicago philanthropist Mrs. George Bass. She urged the MWNA "not to be afraid of grouping yourselves together, not to antagonize or to fight the men in your profession, but to obtain fuller opportunity, wider recognition, and greater success."[13] Thus the National (as the MWNA was known to its members) emphasized only positive goals: "to bring Medical Women into communication with each other for their mutual advantage, and to encourage social and harmonious relations within and without the profession." Its theme was harmony not dissonance, or, in the words of the *Woman's Medical Journal*, "Amalgamation, not Separation."[14]

The original officers included Van Hoosen as president, Eliza Mosher, the first dean of women at the University of Michigan, as honorary president, and Marion Craig Potter as first vice president. The organization made the *Woman's Medical Journal* its official publication and its editor, Mrs. Margaret Rockhill, corresponding secretary. Membership was divided into three categories: dues-paying members with voting privileges (dues were two dollars), associate members, and honorary members.[15]

By the time of the National's first annual meeting in June 1916, Van Hoosen knew they faced an uphill climb. After its first six months, the organization had received only $306 in dues from—at most—153 women.[16] With approximately 5,200 women physicians in practice in 1916, this meant a membership of fewer than 3 percent. Not only indifference but also outright opposition had surfaced. One physician, for example, asked that her name be removed from the list of the organization's governing council. At the AMA's annual convention in 1916, several others, notably from California, where women physicians already had achieved a high degree of acceptance, circulated a petition opposing the MWNA as divisive and retrogressive.[17]

Van Hoosen's strategy was straightforward: to convince women physicians, first, of the need for cooperative action by women in medicine; and, second, that the National could be a positive force for women in the profession. As a first step she organized committees on women's hospitals, internships, postgraduate work, and scholarships. Beginning in 1917 the Committee on Internship (later renamed the Committee on Medical Opportunities for Women) surveyed hospital internships open to women. The committee documented for the first time the scarcity and uneven quality of postgraduate training for women. Its sobering findings persuaded some that only by acting collectively would women increase their effectiveness and credibility.[18] Nevertheless, resistance to the National remained strong. America's entry into World War I, which precipitated a professional crisis for women physicians, created an opportunity for the MWNA to prove its worth.

The American Women's Hospitals Service

War service attracted women doctors with a promise of adventure, service, patriotic duty, and professional advancement.[19] At the same time, it underlined the crisis of professional culture faced by women physicians. The war appeared to offer an ideal opportunity for professional assimilation and advancement through work in overseas military hospitals.[20] Many women doctors were eager to cast off the constraints of Victorian gender norms and join their male colleagues in the military medical corps. Yet their bid for war work went unheeded whenever they failed to invoke traditional "feminine" values of self-sacrifice and service.[21] The outbreak of World War I thus accentuated stresses already present for female physicians. As were feminists at the time, historians today, are divided over whether the war furthered or retarded the cause of modern feminism. What seems clear, however, is that the cultural conflation of masculinity, aggression, impersonality, and technological efficiency so characteristic of wartime made this a particularly problematic time for women physicians to claim the mantle of assimilation and abandon the code of woman as healer.[22]

Although America did not enter the war until April, 1917, many American doctors joined the Allies after the outbreak of fighting in August 1914. Besides the wish to join the fight and help the wounded, opportunities to learn new techniques in surgery and bacteriology attracted many.[23] Women were no exception. As Seattle surgeon Mabel Seagrave told a reporter for *The Woman Citizen*, "[War surgery] will give the surgeon a chance to demonstrate things which have heretofore been more or less experimental . . . [m]ilitary surgery in France today is . . . an opportunity all surgeons must covet."[24] Women as well as men volunteered to work as ambulance drivers, orderlies, nurses, and physicians with the British, French, Belgian, and Serbian agencies of the Red Cross.

After the United States officially joined the combatants, however, American women physicians found it much more difficult to work overseas. Physicians were primarily recruited through the Army Medical

Reserve Corps, which, throughout the war, legally excluded women. Directly after President Woodrow Wilson's war declaration, women's associations in California and Colorado began petitioning and lobbying to convince Secretary of War Newton D. Baker to reinterpret the rules of eligibility for the Medical Reserves. California women physicians, who had generally been suspicious of a women's medical association, asked the MWNA to petition the government to commission women doctors for the Medical Reserve Corps. The issue of commissions gave the MWNA a chance to prove its usefulness. Van Hoosen, president of the MWNA, immediately cabled President Wilson to offer the services of women doctors.[25]

Unfortunately, Secretary Baker, a peacetime supporter of women reformers, upheld the army's refusal to admit women physicians to the Medical Reserve Corps.[26] Only two paths were open for women doctors who wished to work directly for the government: they could apply for a position as a bacteriologist or sanitarian for the Public Health Service, or they could sign on with the army as a "contract surgeon" for a specified salary and length of service.[27] The surgeon general envisioned women contract surgeons as "Anaesthetists, Radiographers, Laboratory Workers, and Sanitarians," primarily substituting for male physicians on overseas duty. This was an offer few women could accept. Becoming a contract surgeon entailed "sacrificing their practices, performing the same services as their brothers, but with no rank, no promotions, no standing; when discharged, no bonuses or pensions, and, if injured, no disability provisions for themselves or their dependents."[28]

The War Department's refusal to accept women physicians meant that women would have to find war work through private philanthropy, not the military. Opportunities for overseas war service were available through privately funded women's organizations and the civilian relief agencies of the American Red Cross. With the exception of the few women physicians working in French military hospitals, however, medical women working for the Red Cross were assigned to treat civilians, especially women and children.

Caught between the potentially conflicting goals of professional assimilation and feminine altruism, the MWNA sought to reconcile the two by establishing a voluntary overseas service known as the American Women's Hospitals (AWH).[29] In offering a full range of medical services, the AWH was designed to allow women physicians to serve the war effort, enhance their professional prestige, and preserve their traditional claim to a superior capacity for sympathy.[30]

Van Hoosen opened the discussion on war service for women physicians at the 1917 MWNA annual meeting in New York. Realizing that this meeting would make or break the organization, Van Hoosen sought a powerful speaker for the keynote address. She later recalled, "I had been drawn at some of the AMA's women's banquets to a charming young doctor whose personal appearance, ability as a toastmistress, wide acquaintance with important people, and a certain restless ambition made her outstanding."[31] She chose the Virginia-born New York surgeon Rosalie Slaughter Morton. By the end of the meeting Morton had agreed to chair the National's newly formed War Service Committee. Within weeks, under her leadership, the committee reconstituted itself as the American Women's Hospitals.

The choice of Morton as speaker and committee chair proved momentous for the MWNA. Morton was an aggressive and well-known surgeon who also conveyed in her manner more than a hint of the genteel, romantic, chivalric culture of antebellum Virginia.[32] Years of personal and professional experience lay behind her interest in war work. Born into one of the old families of Lynchburg—a family of doctors—Rosalie Slaughter had defied her father's demand that she carry on the traditions of the southern "lady" and instead chose to attend Woman's Medical College of Pennsylvania. At her graduation in 1897 she won prizes for both the best invention of surgical equipment and the best clinical case report (on pernicious anemia).[33] For the next two years she carried out postgraduate work in Philadelphia, first as an intern at Philadelphia City Hospital and then as a resident physician at the Alumni Hospital and Dispensary. Like her physician grandfather and brothers, Slaughter set aside two more years to attend clinics in surgery, internal medicine, and nervous diseases in Berlin, Vienna, and Paris. After her return to the United States in 1902, she settled in Washington, D.C., and opened a private practice in which she specialized in gynecology.[34]

Slaughter became an immediate success, something she credited to having inherited "a small income" and to her "hereditary urge" to practice medicine. In reality, her practice drew on a wide circle of influential friends and relatives who were settled in Washington. After her marriage in 1906 to George Morton, a lawyer, the couple moved to New

York. Although she lacked the ready-made professional network that had been available to her in Washington, she achieved even greater professional success in New York. Morton soon became active in the state and local medical societies. In 1909 she persuaded the AMA to form a Public Health Education Committee to organize lectures by women physicians around the country. Morton became the committee's first chair. In 1912 she joined the faculty of the New York Polyclinic as a part-time instructor of surgery and gynecology. In 1916 she became attending surgeon in general surgery at the Vanderbilt Clinic of Columbia University's College of Physicians and Surgeons, the first woman on the staff. By 1916, when she left for six months of wartime volunteer work in Serbia, she had built a successful New York practice in surgery and gynecology with a wealthy and devoted female clientele.[35]

Despite her professional successes, Morton's personal life was in disarray. In 1913 she was profoundly shaken by the sudden death of her husband. She was thirty-seven, and they had been married only seven years. As she later acknowledged, "If my parents had been living, if we had had children, if there had been domestic duties, I would have found palliative comfort in them; but with [my husband's] going my domestic life was . . . absolutely demolished." Morton sought refuge in her work and filled her time with socially useful "motherly diversion," such as leading a boys' history club.[36]

In 1915 Morton began to immerse herself in medical philanthropy. She spent the summer on the coast of Labrador operating on poor fishermen in the charity hospitals run by Dr. Wilfred Grenfell. From the outbreak of war in Europe, she was drawn to war work by the example of the Scottish Women's Hospitals and the British Women's Hospital Corps, voluntary units run by suffragist women physicians. After her return from Labrador she learned of the American Red Cross Sanitary Commission's work in Serbia to control a typhus epidemic, and she volunteered to travel to Serbia for the Red Cross. A volatile mix of personal emotion and professional ambition fueled Morton's decision. As she later wrote, "Serbia had been made the scapegoat to receive the calumny of the world. My sympathy for the under-dog flared up . . . I had no parents, husband or children. I had everything to give."[37]

Morton dearly wanted to act heroically.[38] Admiring the leaders of the British and Scottish women's voluntary hospitals, she hoped to equal their achievements. Nearly forty years old, widowed, and childless, Morton saw no reason to sit on the sidelines while others grasped glory. Although her ostensible assignment was to deliver sixty cases of Red Cross supplies to the Serbian army, her intentions were far more grand: to survey and compare the organization of the British, Scottish, Canadian, and French hospital units with an eye to eventually establishing a unit of her own. Toward this end she volunteered her services at the Salonika front for the summer.[39]

At the front, Morton inquired at the Scottish Women's Hospitals unit in Macedonia about equipment, supplies, and organization. Her resulting three notebooks, "filled with condensed details," became the basis for the plans she sketched out before the MWNA in 1917. When she landed back in New York in the fall of 1916, she was determined to head a Serbian expedition: an American women's hospital unit "complete . . . from admission cards to ambulances."[40] Morton immediately gained the interest of other women. Aided by a growing national demand for "war readiness," she lectured widely on the plight of Serbian soldiers and civilians. If America remained neutral, Morton intended to organize her own volunteer unit for the Serbians. She also helped plan a Women's Army General Hospital, approved (but only in principle) by Surgeon General William Gorgas, for wounded American soldiers sent back to America.[41]

Van Hoosen and the 1917 MWNA meeting thus provided Morton with an ideal platform. Wearing the uniform and medals she had received from the French government for her summer's work in Serbia, she was an impressive figure and a smashing success. Morton surveyed several exotic subjects—gas gangrene, facial surgery, the treatment of burns—as well as the work of the Scottish and British women's hospital units. She also answered questions from the floor. The meeting ran overtime and was adjourned until the next day because the members had not decided on a course of action. The next day the MWNA adopted a petition urging commissions for women in the Army Medical Reserves and voted to form the War Service Committee, placing responsibility for the petition in the committee's hands. Van Hoosen offered the chair of the new committee to Morton, who at first declined. She had no stomach for administrative detail and feared that organizational work would force her to "set aside any hope of returning to the front." But when Van Hoosen publicly offered her "carte blanche" to lead the committee, she reluctantly agreed to take it on.[42]

In spite of the tension between her personal goals and the needs of the MWNA, Morton accomplished a great deal. Yet she also had many limitations. These are better assessed in comparison to the career choices and values of the AWH's other important early leader, Dr. Esther Pohl Lovejoy. Lovejoy, who became chair of the AWH in 1919 and held the post for more than forty years, was implacably hostile to Morton's personal and professional values. Her background, organizational experience, and professional style contrasted sharply with Morton's. Against Morton's traditional, pure-but-heroic "new woman" of the 1890s, Lovejoy's brand of new woman stepped smartly to the rhythms of twentieth-century America.[43] Having been born in a logging camp in 1869 in Washington Territory, she neither knew, nor

Dr. Esther Pohl Lovejoy (1869–1967), chair of the American Women's Hospitals service. Undated photograph, ca. 1920s. (Courtesy of the Archives and Special Collections on Women in Medicine, MCP Hahnemann University, Philadelphia.)

seems to have missed, the genteel comforts of the Victorian home. She was brought up like a boy. Adventure, not respectability, colored all of Lovejoy's career. After a haphazard early education and a year's work in a Portland department store, she entered medical school at the University of Oregon in 1890. She won an award for academic excellence and graduated in 1894. After some postgraduate obstetrical training in Chicago she married a former classmate, Dr. Emil Pohl. They set up practice in Alaska, where they often traveled by dogsled to visit patients.

Always independent, Esther Pohl soon returned to Portland and would visit her husband during the summer. In 1901 they had a son, who was cared for by Esther Pohl's mother. Within a few years she joined the Portland Board of Health. When she was appointed its director in 1907, Pohl was the first woman to run a major municipal health department. She also became active in the suffrage movement in Oregon. Like Morton, Pohl was briefly traumatized by personal loss: the deaths of her son in 1908 and her husband in 1911. She threw herself completely into her work, particularly her work for women's suffrage, giving the Oregon campaign an "extra push which helped put over woman suffrage" in Oregon in 1912.[44] She also got married again, this time to a businessman, George Lovejoy, but she divorced him several years later. Lovejoy was forging a public self. She spent little energy on the personal and private.

With America's entry into World War I, Lovejoy's interests again began to shift. She left Portland and traveled east to work with the MWNA on medical war work. Bertha Van Hoosen made her the National's liaison to the Women's Committee of the Council of National Defense, a coordinating body of voluntary organizations committed to furthering the war effort. As an unpaid delegate of the MWNA, she traveled to France in the fall of 1917, where she simultaneously worked for the Red Cross and scouted the medical scene for the National. She returned early in 1918. For the next year and a half she lectured, wrote a book about her experiences in France (*The House of the Good Neighbor,* 1919), and performed occasional services for the MWNA. In 1919, Lovejoy was made interim chair of the AWH. After a brief return to Portland in 1920 to run—unsuccessfully—for Congress, she was asked by the National to become permanent chair of the AWH, and she returned to New York. With that position, she found her true vocation: combining medicine and politics.

Lovejoy and Morton both wrote a memoir of their wartime and post-war work. The contrasts between Lovejoy's *Certain Samaritans* (1927) and Morton's *A Woman Surgeon* (1937) reflect the transformation of women's professional culture between the 1890s, when both women entered the profession, and the new century in which they came to professional maturity.[45] Morton was caught between the old and the new, between the older idealization of womanhood and the new culture of expertise and efficiency. In a late Victorian mode, Morton began her autobiography with classical allusion, filial piety, and social-Darwinist determinism: "Since the day in 1620 when my father's ancestors came from England to settle in Virginia, seventeen of their direct, and fifty-two of their collateral descendants had followed in the footsteps of Aesculapius."[46] Her memoir consistently invokes the influences of heredity and social class on her accomplishments. Without such cultural supports, Morton could not make sense of her own achievement. As she wrote an old friend in 1922: "I must confess I enjoy the social world. It seems natural to live the sort of life I did as a girl and I can now see why my parents did not want me to study medicine, for they realized that having been born in a circle which gave me the opportunity [for] comradeship . . . with the best products of evolution, . . . they could not understand their daughter wishing to push it all aside to become a self-supporting woman and to [choose] the hardest profession."[47] Rather than acknowledge the cold realities of modern professionalism, Morton interpreted her life through a veil of romanticized, feminine gentility.

Lovejoy's memoir shunned romantic self-reflection. She braced her descriptions of AWH relief work in Serbia, Greece, and Armenia with brisk self-confidence and a not-so-gentle irony. In contrast to Morton's sentimentality toward Serbia, Lovejoy's first sentence read, "The Balkan Peninsula lies between the Devil and the deep blue sea." Continuing, she wrote, "This service has not been a bed of roses. Sometimes it has been a bed of straw in a box car . . . or a cot in a typhus camp." Lovejoy wholly approved of the new woman, twentieth-century-style: "The [AWH] chauffeuses were the youngest group in the unit, and manifestly ladies of the new school. They were not sitting in balconies gazing at the sympathetic stars and longing for the hero to return. No, indeed, they were following him in a motor car." Lovejoy's wisecracking never faltered. She scorned the elevated rhetoric of the Victorian woman, employing the underinflated, no-nonsense prose of a woman used to getting results. She was angered not by injustice—she had come to expect it—but by inefficiency.[48]

Rosalie Slaughter Morton and the AWH

When Morton took on leadership of the War Service Committee in June 1917, the MWNA was determined to pursue two distinct goals: to change the law governing the Army Medical Reserve Corps and to send women physicians overseas as volunteers. It directed Morton's committee to do both.[49]

Morton was unsuccessful in the campaign to change the law. From the beginning, MWNA's petition faced key opposition from the War Department and from the Red Cross.[50] When Morton presented it to Dr. William Lucas, head of the Red Cross Bureau of Women and Children in France, he disapprovingly called it a piece of "local [i.e., feminist] propaganda." Hastily Morton backed off. She attempted nothing more in the campaign for women's commissions.[51] Now only a women's volunteer hospital could enable the MWNA to send women overseas and reap some credit for women physicians. This was probably Morton's intention all along. On June 28, at Morton's urging, the committee created a new entity, the "American Women's Hospitals, organized by the War Service Committee of the Medical Women's National Association." The committee's officers were then reconvened as the Executive Committee of the AWH.[52]

Perhaps Morton's gravest obstacle was the reluctance of the American Red Cross to affiliate with the AWH. Sending voluntary hospitals overseas required Red Cross consent. Furthermore, the agency's support also was crucial for fund-raising. Initially, however, the future looked bright. Probably at the suggestion of Surgeon General William Gorgas, her ally in organizing the AMA public health lectures in 1909, Morton was invited to Washington on June 24 to describe the War Service Committee's plans before the Council of National Defense's General Medical Board, chaired by Dr. Franklin Martin of Chicago.[53] Gorgas himself was present. A copy of her remarks was left with Colonel Jefferson Randolph Kean, a fellow Virginian on leave from the Medical Corps to direct the Red Cross Department of Military Relief. In July, Franklin Martin invited Morton to join the General Medical Board. She was to chair the Medical Board's new Committee on

Women Physicians and, more particularly, to register all women physicians and catalogue those willing to volunteer their services to the government.[54]

Morton now found herself caught in a whirlwind of committee work both in Washington and New York. For the General Medical Board she oversaw the compilation of a complete census of women physicians, actually compiled by Marion Craig Potter and completed by the end of November 1917. (Of the 5,322 women physicians recorded in the census, about two-fifths agreed to register for possible government service.)[55]

The key to the success of the AWH, however, was not the General Medical Board but the American Red Cross. Morton hoped to send four hospital units to France, one to Serbia, and possibly one to Russia. The French units were to comprise one large, central hospital surrounded by mobile dispensaries. Moreover, ever mindful of its professional prerogatives, the AWH insisted on control of its own personnel. To accomplish all this, Morton had to find a way to raise sufficient money. Yet unless the Red Cross gave public backing to the AWH overseas units—staffed by the AWH and bearing the insignia of the AWH—Morton knew her fund-raising efforts would attract scant public interest.[56] For months, lack of Red Cross support constrained Morton from making a public appeal. Instead she quietly approached wealthy laywomen such as Mrs. Andrew Carnegie, whom she recruited for an auxiliary board of the AWH. She also organized women physicians into state committees for fund-raising among medical men and women.[57]

Unfortunately, by the end of 1917 these methods had produced only about $11,000, far too little to support Morton's plans. After the first six months, the organization's only tangible contribution to women's medical war work consisted of the outfitting of two women physicians for work with an American Red Cross unit in Vodena, Greece. As matters stood at the end of 1917, the AWH could not even contribute to the women's salaries. If this impasse could not be broken, the AWH and the MWNA, its parent body, would soon have to admit failure.[58]

One commentator has suggested that Morton herself was the primary obstacle to AWH success. For one thing, her loyalty, energies, and time were divided between the AWH in New York and the Committee on Women Physicians of the General Medical Board in Washington.[59]

In addition, Morton's involvement with the AWH was a deeply personal matter. Once she had renounced her plan to create her own Serbian hospital, her need to give tangible aid to war victims (and to appear heroic herself) began competing with the MWNA's need to take credit for the advancement of women physicians.

For example, many of Morton's early decisions as AWH chair tended to dissociate the AWH from the Medical Women's National Association, both in the public mind and in her own. Consciously or not, she began disengaging the AWH from its mission of professional advancement; instead, she aligned it with the "overwhelming impulse for service and sacrifice" with which she personally identified so strongly. It was Morton, after all, who disavowed the MWNA's petition to Dr. Lucas of the Red Cross with the reassurance that the AWH was "not interested in any [feminist] propaganda; that we stood purely for war-service for the relief of suffering."[60]

The "platform" adopted by the AWH Executive Committee in July 1917 clearly articulated the tension between Morton's underlying premises and those of the MWNA. It virtually abandoned petitioning for Medical Reserve Corps commissions. Instead it "requested" that women physicians be accepted into the Medical Reserves but did not "make it a condition of service, for we realize that by so doing we handicap our opportunity for immediate service, at home and abroad, which is our main desire in volunteering."[61] Finally, Morton's highly personalized leadership may have exacerbated the organization's external difficulties. Overidentifying with "the life and growth of my child," as she later described the AWH, she was swept up in a frantic effort to nurture and protect it. As a sympathetic friend acknowledged, "No one loves [the AWH] as much as Dr. Morton—it is her war baby in truth."[62] Reluctant to delegate authority, unable to yield the spotlight, she alienated valuable supporters.[63] Frustrated and impatient with Morton's apparent ineffectiveness, some members of the Executive Committee grew restive under her leadership.[64]

Ultimately, however, the fault for AWH's lack of progress lay with the Red Cross, which was beyond Morton's control. The AWH's insistence on control over its own personnel, in its view a matter of professional prerogative, proved the sticking point for any agreement with the Red Cross. World War I was the first major conflict involving the United States in which the Red Cross exercised the exclusive right to coordinate nonmilitary medical assistance. It was in no hurry to give untested organizations the right to share this responsibility. As an official of the Red Cross Medical Advisory Board explained to Dr. Purnell of the AWH early in August 1917, "It would be impossible for the Red Cross to send out units of other organizations or to recognize officially the existence of other organizations." Beneath the surface lay another source of Red Cross reluctance to affiliate with the AWH. As Dr. Gertrude Walker, chair of the AWH Finance Committee later wrote, the Red Cross balked at supporting AWH units because, she had been privately informed, "the ideals of the American medical women's organization were parallel to those of the Scottish Women's Hospitals." The latter group was firmly identified with the cause of woman suffrage. This could have only one meaning: suspicions of women's rights "propaganda"—first evident in the Red Cross's reaction to the MWNA's petition—made the agency hesitant to affiliate with the AWH.[65]

The Red Cross did, however, use the AWH as a clearinghouse for women physicians willing to work directly for the Red Cross. As early as July 1917 the Red Cross had written to Morton requesting such assistance. According to the estimate of Dr. William Lucas, director of the Red Cross Bureau for Women and Children in France, by May 1918 nearly 50 percent of all Red Cross physicians in France were women, many of them affiliated with the AWH.[66]

Yet for the AWH such cooperation hardly seemed worth the effort. The visibility and authority of women employed in Red Cross units were bound to be diminished if they were scattered all along the western front with no source of recognition other than their Red Cross insignias. The whole point of American Women's Hospitals was to demonstrate the medical and administrative abilities of women physicians, and particularly those of the MWNA. Hence the frustration felt by Morton's colleagues when, by the end of 1917, they had little to show for their labor.

That frustration was directed at Morton. She was by this time overextended and overtired from fourteen-hour days of private and clinical practice, teaching, and war work. Also, her efforts for the General Medical Board in Washington caused her to miss AWH meetings in New York. Word began to spread from members of the AWH Executive committee to Van Hoosen that Morton was suffering from overwork. Perhaps, it was implied, she was not well enough to be an effective leader. In March, Morton herself wrote privately to Marion Craig Potter that "the long strain is being very seriously felt by me."[67]

To make matters worse, communication between Morton and leaders of the MWNA had all but evaporated. In a letter inquiring about Morton's health, Van Hoosen gently took her to task for not keeping her informed of AWH business. The AWH legal counsel had urged strongly that the AWH be legally incorporated under its own name as a distinct entity, to protect against misrepresentations by others. The following February, on the eve of its first large fund-raising campaign, the AWH even drew up a separate constitution. Van Hoosen squelched both these plans, reminding Morton that American Women's Hospitals was not an independent agency but a part of the Medical Women's National Association. Morton replied, humbly enough, "We have no idea of separation, but emphasize daily M.W.N.A."[68] Nevertheless, opportunities for loss of trust in Morton's leadership were multiplying. Whatever the state of Dr. Morton's health and the character of her motives in the early months of 1918, her fortunes and those of the AWH began to diverge.[69]

From the perspective of twenty years later, Esther Lovejoy bitingly noted that "it was clear to . . . straight-thinking women, that nothing worth-while could be done without adequate funds." To that end, without waiting for Red Cross approval, early in 1918 the Executive Committee voted to hire a professional fund-raiser, Mrs. Elizabeth Currier. The committee members hoped to raise about $300,000. Despite Morton's fatigue, the Executive Committee insisted she devote two full weeks of her time to the campaign. She wearily agreed.[70]

The campaign was conducted from March 26 until April 6. In preparation, photographs of Morton, decked out in her uniform with hat, medals, and AWH insignia, appeared in major newspapers alongside public relations puff pieces. The drive opened, according to one account, "with a dinner at the Biltmore Hotel." Fifty "teams of doctors and laywomen" canvassed for funds. Pennants were awarded to the leading teams each day. Morton spoke at Town Hall and India House and solicited contributions from as many of her private patients as she could.[71]

In keeping with the organization's diverse objectives and ideological roots—one part each social feminism, suffragism, and professionalism—the campaign organizers played on several distinct themes. The teams were exhorted to do their utmost for the "little boys you are going to save over there." At the same time, Harriet Stanton Blatch (daughter of suffrage pioneer Elizabeth Cady Stanton), assured them that their efforts would help to win "complete enfranchisement" for the women of America. Dr. William Polk, dean of the Cornell University Medical School, encouraged them to demonstrate the equal abilities of male and female physicians. Finally, they were reminded of the desperate need of the women and children of France. On those notes they went forth to collect what they could.[72] By April 6, the last day of the campaign, $95,937 in cash and $140,795 in pledges had been collected. (Unfortunately, only a small portion of the pledges were realized.) Nevertheless, including money donated apart from the campaign, by June 1918 the AWH had taken in a total of $192,800, which was enough to send a unit overseas.[73]

Ironically, money had become less of an issue for the AWH by the time the campaign ended. Behind-the-scenes negotiations with the Red Cross had resumed early in 1918, and by mid-February they were beginning to pay off. In the first place, official Washington was beginning to change its mind about suffrage.[74] Even more crucial, the Red Cross had had sufficient time to evaluate both the enormous need for medical personnel at the front and the high caliber of the women physicians already overseas.[75] Finally, by February of 1918 the Allied forces appeared to be in peril in the war. Revolutionary Russia had dropped out of the fighting. The Allies expected a major German offensive as part of a final campaign to win the war, and in March it began, at the Somme.[76] American physicians, like American soldiers, were badly needed at the front. All these factors made the Red Cross much more receptive to affiliating with the AWH.[77] On February 17 the Executive Committee secretary recorded that it was "definite" but "confidential" that the Red Cross was considering an affiliation. Almost as important, the Red Cross would allow the AWH to keep its name and thereby "retain its identity." A well-timed announcement of the imminent agreement came during the fund-raising kickoff at the Biltmore.[78]

According to the agreement signed by the AWH and the American Red Cross in April, the AWH was to finance its own administrative expenses in America, while the Red Cross would equip, maintain, and pay the staffs of all AWH hospital units requested for overseas duty.

They would be known as the American Women's Hospitals Unit No. 1 of the American Red Cross. AWH personnel would be allowed to wear the AWH uniform. Although the Red Cross retained the right to transfer personnel from one unit to another as needed, only physicians acceptable to the AWH would be employed in its units or wear its uniform. Thus the agreement acceded to virtually all of the AWH's important demands.[79] For its part, the AWH donated $30,000 to the Red Cross for a children's hospital and dispensary in Blois, France.[80]

At the same time the AWH forged another alliance, with the American Committee for Devastated France (ACDF), founded by Anne M. Dike and Anne Morgan, youngest daughter of financier J. P. Morgan. Since 1917 the ACDF had raised money to restore the devastated Aisne region northeast of Paris. The ACDF would continue with its planned work of reconstruction and replanting, while the AWH would provide medical and surgical services.

In June the first of two AWH hospital units was ready. Its female staff consisted of ten physicians, one dentist, six nurses, five "robust" chauffeuses (ambulance drivers), three volunteer nurse's aides, and a general factotum, Mrs. Emilie K. Lehman, who served as administrative assistant, purchasing agent, and liaison to the ACDF and the French. (Two more dentists were added later to accommodate the vast need for dentistry.) At least two of the staff, one general practitioner and one dentist, had chaired AWH subcommittees prior to volunteering for overseas work.[81] In July 1918 they arrived at the chateau of Neuf-Montiers, moving later during the Allied counteroffensive to a chateau at Luzancy in the Aisne district near the Marne. They were in business, at last.

AWH Unit No. 1

Unit No. 1 evolved through three distinct incarnations, first at Neuf-Montiers, then at Luzancy, and finally at Blérancourt. Neuf-Montiers, a hospital for "summer illnesses" was open from July 28 to August 17, 1918. Under the unit's first director, Dr. Barbara Hunt, a general surgeon from Bangor, Maine, the unit got off to a slow start. Although AWH publicity described Hunt's work in glowing terms, a letter from an insider told a different story. "It seems Dr. Hunt was simply not competent to take charge of the *work*. There was great criticism and lack of harmony among the women doctors." As this correspondent explained, Hunt was unable to mobilize her resources for the trying conditions surrounding her, and the figures bear out this charge. At the end of her three months as director, the staff had completed only fourteen surgical operations; of those, nine were tonsillectomies performed by a temporary member of the staff. Hunt's correspondence to AWH headquarters in New York also reinforces this impression. In a tone totally uncharacteristic of most AWH workers, Hunt complained about the quality of her unit's equipment (four cars, two ambulances, and a Ford touring car: "All in bad shape") and requested that the Executive Committee send her the latest copies of the *Journal of the American Medical Association*, the *Journal of Obstetrics and Gynecology*, and other "first class surgical journals." After three months, Hunt resigned as director to work in the French military hospitals. She was replaced by Dr. M. Louise Hurrell, a general practitioner from Rochester, New York, who had taken over as assistant director before Hunt's actual resignation.[82]

Under Dr. Hurrell's leadership (from October on), the staff finally hit its stride. In August, during heavy fighting, the unit had moved into a chateau in Luzancy on the Marne river. The lovely chateau had been used as a school before the war. After the outbreak of war, the French had made it into a military hospital; the AWH took it over from them. The AWH unit remained there from August until the end of March 1919. By September, however, the only soldiers remaining were the twelve Americans buried in the cemetery nearby. As described by a visiting Rochester physician on her way to the Red Cross children's hospital in Dinard, the chateau had "stone floors, cold and damp and down by the water where fogs collect . . . The house is pretty well established, the big room of the chateau—high as a church—with two long tables—where they eat . . . was full of shadows from the kerosene lamp. One fine oil painting was left on the walls . . . The big open windows were barred . . . They've been bombed all the time." Comparing the women's unit to the Rochester-General and Bellevue-sponsored base hospitals in Vichy, she continued, "All I could say was praise for [the AWH's] fine establishment . . . I think it compares very favorably, in a small way." She noted, however, the absence of the laboratory equipment that was available to the military physicians.[83]

Supplies and equipment were constant concerns. Despite the Red Cross's pledge to supply all AWH units, the strain on the AWH's resources must have been extreme. The AWH generously allotted $50,000 for the operation of Unit No. 1, but this could not keep the hospitals going without reinfusions of cash and supplies. (In fact, the total expenditures for AWH Unit No. 1's three hospitals from July 1918 to January 1920—including salaries, equipment, food, barracks, and publicity—eventually amounted to more than a million dollars.)[84] Competition for continued Red Cross support was keen. Dr. Hurrell was unstinting in her praise of her assistant, Dr. Inez Bentley, for her successful extraction of supplies from the Red Cross in Paris. After the unit had moved on to its third and last location, Hurrell wrote the Executive Committee about three badly needed automobiles allotted to

Dr. M. Louise Hurrell (1871–1958) in her office at American Women's Hospitals Unit No. 1 at the Luzancy château in France, 1919. (Courtesy of the Archives and Special Collections on Women in Medicine, MCP Hahnemann University, Philadelphia.)

the unit; after waiting for them for months, she had finally dispatched Mrs. Lehman to find them. Find them she did, in two feet of water in the Bordeaux harbor, still in crates.[85]

As Dr. Hurrell quickly learned, Luzancy was where "the epidemics of dysentery, typhoid, grippe, and pneumonia were fought and where the neglected surgery of the Aisne received attention." The people of the surrounding region had been literally devastated by the war. Cottages and fields (as she wrote, in the words of a village *"maire,"* "only fourteen kilometers from the enemy, constantly under the menace of bombardment") were destroyed in the fighting. Husbands, sons, and fathers were all gone, either dead, wounded, or, at best, stationed far from home. The French physicians were gone, too, many of them casualties of the war. The earlier military hospital in Luzancy had been the only medical installation in the region prior to the arrival of Hurrell and her staff, and it had been off-limits to civilians. Thus the AWH unit quickly had its hands full running medical and surgical wards as well as covering thirteen different dispensary routes.[86]

Patients ranged from newborns and young children to aged men and women close to death from starvation, illness, and despair. Occasionally, the staff treated soldiers who had been wounded nearby and brought to the hospital as emergency cases, as when a soldier who had been in a railroad accident was brought in for extensive repairs to his hand. During seven months at Luzancy, 420 hospital, 254 surgical, and 472 dental cases were cared for. At the same time, the staff treated 3,344 dispensary cases from a total of 1,552 different patients and made 3,626 house calls to 1,218 different patients. Surgery among civilians was a very different matter from the cases handled by the staff of a large military hospital. More than one-fourth of the AWH unit's surgical cases (67), following the accepted medical wisdom of the time, involved the removal of tonsils and adenoids. Herniotomies accounted for the second largest number of procedures (11). The remaining miscellaneous procedures included trephining, appendectomy, Caesarean section, hysterectomy, and removal of a toenail.[87]

The house calls and dispensary routes were the real heart of the work. Twice a week dispensary ambulances were sent out with a doctor, a nurse, a chauffeuse, a sterilizer, a stove, and all necessary medicines. Patients either left word for the doctors at the local mayor's office or traveled to one of the satellite examining stations set up in the region for dispensary care. In this way, a relationship of trust between the villagers and *les dames Americaines* was established—and none too soon. In September, just after Dr. Hurrell became de facto director, twin epidemics of typhoid and influenza broke out in the Aisne district. Although Hurrell and her staff could do little to counter the influenza, they responded with vigor to the threat of typhoid, immunizing nearly everyone in the region. After sending out an announcement through local officials, the dispensary physicians arrived in the villages. All villagers consenting to the procedure would line up in the square for the great event. According to one account, a record was set at one village, where seventy-five immunizations were given in twenty-five minutes.[88]

Hurrell's prompt response to the crisis proved effective. On March 30, 1919, the unit's last day in Luzancy before moving on to its final hospital site in the village of Blérancourt, the mayor and villagers gathered to honor Dr. Hurrell and her staff. All eighteen of the staff members still stationed in Luzancy received a gold medal from the French government—the Medaille de Reconaissance Française, with Palms—for stamping out the typhoid epidemic. Along with the directors of the American Committee for Devastated France, they were also made honorary citizens of Luzancy.[89]

Beneath this apparent bonhomie, however, lay the roots of a thorny ethical dilemma. In truth, the AWH physicians had worn out their welcome. The armistice of November 11, 1918, had occurred about midway through their residence at Luzancy. By the beginning of 1919, demobilized French physicians were slowly making their way back home. The sight of American physicians (and women, at that) poaching on their practices did nothing to sweeten their homecoming. As Dr. Hurrell wrote to the AWH Executive Committee in New York, explaining her decision to leave Luzancy and move on to Blérancourt, "We can say that if the French in some matters are slow, in the matter of their clientele they are particularly swift." At the same time, she understood their position. As she wrote the committee a few months later, it was a matter of professional ethics. "You will comprehend . . . that your AWH in France did not come to take the work of French physicians, nor to be their competitors."[90]

The AWH's cosponsor, the ACDF, did not appreciate the issue of professional ethics or etiquette. The well-to-do, philanthropic laywomen of the ACDF, were most interested in seeing results for their money. They insisted that the AWH follow through on its agreement to run two French hospitals at least until December 31, 1919, despite Dr. Hurrell's ethical discomfort. In fact, ethics and economics were commingled in Dr. Hurrell's analysis of the situation at Blérancourt. To her experienced eye, there was simply not enough work to justify the expense of maintaining the unit's surgical ward. Hurrell's solution was to reorganize the tent-and-barracks hospital at Blérancourt for preventive, rather than acute, care. In Hurrell's words, "Preventive medical care has been the primary object—vaccination for typhoid and smallpox, isolation for scarlet fever and measles, examination of schoolchildren, dental work, and removal of adenoids and tonsils."[91] The AWH continued to run the Blérancourt unit from April 1919 to January 1920, although Dr. Hurrell herself left for home in September 1919.

It is quite clear that little good could have been accomplished by the American Women's Hospitals without the presence of a few extraordinary leaders. Decades later, Dr. Hurrell's vigor, self-possession, competence, and humor still speak out through the lines of her reports home. For one thing, she loved the work. As she wrote to Marion Craig Potter in Rochester at the end of the typhoid epidemic, "This unit is made up of a collection of the most wonderful women ever assembled, and we're a happy harmonious crowd." Her staff repaid her affection in full. A visiting AWH commissioner reported back to the Executive Committee in New York, "The spirit of the group here has been remarkable. Dr. Hurrell is an efficient and beloved leader, and she and her whole staff have conducted the work of the hospital with the greatest efficiency and foresight."[92]

Hurrell was replaced by the equally competent Dr. Hazel Bonness, who later served the AWH in a postwar unit in Serbia. Even after curtailing services at the hospital, the Blérancourt unit saw a total of 1,145 patients and performed 161 major operations and 614 tonsillectomies. In one quarter alone (from July 1 to September 1, 1919) its physicians averaged about 150 house calls and dispensary visits a week. They also opened a maternity ward where "prophylactic work and the teaching of hygiene became a part of the routine care of each mother."[93] When in January 1920 the AWH completed its work in France, the hospital and most of its supplies were turned over to the ACDF to be run by French physicians.[94] A few of the AWH physicians, such as Dr. Bonness, moved on to a new unit in Serbia. The rest returned home to civilian life.

～ In June 1918, before AWH Unit No. 1 sailed for Europe, Dr. Rosalie Slaughter Morton had just completed her first year as chair of the organization. Despite her success, the incoming president of the MWNA, Dr. Angenette Parry, did not reappoint her to chair the AWH. At the MWNA's annual meeting, Morton's supporters tried to force her reappointment from the floor—to no avail. The humiliation and her exhaustion took their toll, and the next day Morton suffered a serious nervous collapse. Confined to a sanatorium for the summer, forbidden to see professional friends, Morton attempted to save face. She resigned from the chairmanship for reasons of ill health. Dr. Mary Walker became interim chair for about a year, after which she was succeeded by Esther Lovejoy.[95] Lovejoy would direct the AWH until her death in 1967.[96]

Over the next twenty years, little love was lost between Lovejoy and Morton. After she returned from France in 1918, Lovejoy formed a low opinion of Morton's leadership. Accustomed to taking personal and political risks, comfortable subordinating the private to the public self in her political and professional life, insistent on efficiency, she had little patience either for Morton's ambivalence or for the polite diplomacy of home-front organizing. The AWH clearly needed a full-time manager who—unlike Morton—would be neither too busy nor too unbusinesslike to run a full-scale charitable enterprise. Fund-raising and efficient management were too important to be taken lightly.[97] Morton's troubles can be traced to several sources. Initially she was forced to pursue inharmonious, if not incompatible, goals: gender-blind professional equality and gender-linked sociomedical service. Yet her leadership suffered from self-aggrandizement, ambivalence, and divided attention. Her temperament, unlike Lovejoy's, proved unsuited to the needs of a modern professional organization.

Under Lovejoy, the AWH outgrew its mandate and threatened to overshadow its parent body. In the years following the war it became an oversize jewel in a very fragile crown. Throughout the 1920s and 1930s the MWNA struggled to maintain its constituency, its revenues, and its sense of purpose, while the AWH possessed all three in abundance. In

1924, for instance, the AWH raised more than two million dollars for nine AWH hospitals in Greece, the Balkans, and the Near East. That same year, an editorial in the *New York Times* praised AWH relief work alongside that of the Red Cross and the Rockefeller Foundation.[98] But Lovejoy, who was elected president of the MWNA in 1932–33, was careful to give the MWNA its full measure of publicity and credit in her annual reports. And, more important to the evolving professional image of women physicians, her clear-eyed, underinflated style could never be mistaken for the unprofessional, emotional heroics of an earlier era.

Morton's leadership, in contrast, was a disquieting harbinger of the conditions facing women physicians after the war, in particular, their still-unfulfilled ambition of professional integration. The AWH was founded by the Medical Women's National Association both to bolster the professional esteem of its members and to serve the women and children of Europe. Yet by the end of the war, the organization's directors knew that its unique contribution to medicine would be medical care for the needy, particularly women and children, women physicians' traditional, culturally sanctioned mission. For AWH units in France, and later in Serbia, Russia, Turkey, and, more recently Latin America, public health and preventive medicine were central to their mission, essential to their success. An organization that began as an exercise in compensatory professionalism—almost a demonstration project for professional equality—emerged from the war with a distinct health care mission. Perhaps it is ironic that the very successes of the AWH were dependent on the continued viability of a "woman's sphere" in medicine, precisely what its mission was intended to abolish.

6

New Directions: The Eclipse of Maternalist Medicine

> State medicine is to my mind an ideal, and the sooner it changes from an ideal to a practical reality, the better off the human race will be.
>
> *Dr. S. Josephine Baker*

> I wish to say that the Sheppard-Towner Bill, fortunately or unfortunately is not working in the State of Illinois . . . Personally it is nothing to me, whether it functions or not.
>
> *Dr. Lena K. Sadler*

FOR WOMEN IN the professions, the decades between the two world wars were years of faltering momentum and "elusive promise."[1] Between 1910 and 1930 the proportion of women among professionals rose from about 41 percent to nearly 45 percent, but during the Depression it declined again to pre-1910 levels. The percentage of women in medicine declined during the 1920s from 5 percent to 4.4 percent and remained at 4.6 percent through the Depression.[2] Women physicians had expected their high-profile war work to position them for postwar prominence, especially in maternal and child health. Passage of the 1921 Act for the Promotion of the Welfare and Hygiene of Maternity and Infancy, better known as the Sheppard-Towner Act, one year after ratification of the woman suffrage amendment, seemed to presage a breakthrough for maternalist medicine and for women physicians. But the moment was short-lived. A postwar political backlash (abetted by anti-Bolshevik anxieties) unleashed a period of retrenchment, a "politics of normalcy" that, for women, signaled the end of an era in feminist politics and progressive reform.

By the midtwenties, opposition from congressional conservatives and the AMA to federally funded social initiatives and the U.S. Children's

Bureau revised the political landscape. Moreover, changing professional priorities among private practitioners now delivering preventive care also dampened support for the Children's Bureau.[3] Women physicians responded to declining support for maternalist medicine in diverse ways. Many were quick to distance themselves from politics entirely. While some continued in public health departments or the Children's Bureau, many pursued maternalist goals through private practice and local volunteerism. A few forged a new kind of career for women physicians, combining an interest in maternal-child health with advancement in academic medicine. Their careers and professional philosophies reflected the same diversity as those of their male colleagues. Professionalism, not feminism, was ascendant.

The Sheppard-Towner Act and Women Physicians

For women, the postwar political picture looked bright initially. In 1920, a powerful remnant of the suffrage movement created the Women's Joint Congressional Committee (WJCC) to lobby for social and feminist reform. The WJCC consisted of organizations such as the League of Women Voters, the National Congress of Mothers/PTA, the General Federation of Women's Clubs, the National Consumer's League, the Women's Trade Union League, and, in 1924, the Medical Women's National Association. Virtually the first bill for which the committee lobbied was the pioneering Sheppard-Towner Act to establish prenatal and child health centers for maternal education and well-baby exams for all mothers and mothers-to-be. Congress was briefly attentive to its new female constituency and passed the bill in November 1921. Sheppard-Towner was the culmination of years of public health organizing, and it captured the support of the entire spectrum of Progressive health reformers, from sanitarians and settlement workers to public health organizers, academic pediatricians, women's groups, and, especially, the U.S. Children's Bureau, with which it was identified.[4]

This coalition worked well so long as Children's Bureau activities did not jeopardize the separation between large-scale preventive measures, the domain of public health, and the treatment of individual patients, the province of private physicians.[5] For the first few years after its founding in 1912, the Children's Bureau explicitly tried to avoid blurring these unwritten boundaries, taking special care not to impinge on the interests of private practitioners. Most of its efforts went into disseminating reams of advice literature to thousands of mothers. Julia Lathrop, the bureau's first director, and her successor in 1921, Grace Abbott, also gathered data on the mortality and morbidity of childbirth. For this they relied both on Bureau physicians and on a group of widely known obstetricans and pediatricians, many with academic affiliations.[6]

However, the Bureau's investigations of infant mortality, conducted between 1913 and 1923, did bring it into conflict with the territorial interests of private practitioners, the Public Health Service (PHS) and, eventually, organized pediatrics. Before the bureau began its research Lathrop asked for—and received—assurance that the PHS was not engaged in any similar work. With this assurance in hand, she directed the Bureau toward what became its most visible role prior to Sheppard-Towner, gathering data on maternal and infant mortality and morbidity across the country. *Maternal Mortality from All Conditions Connected with Childbirth*, written in 1917 by Dr. Grace L. Meigs, director of the Child Hygiene Division, was the first federal study to compare America's maternal mortality rates (dismal) with those of other industrial countries. As study after study associated a dearth of pre- and postnatal care with increased risk of infant and maternal death, the Bureau considered establishing screening clinics to target the mothers and infants most at risk. Meigs was aware that in investigating medically preventable maternal deaths, the Children's Bureau was on precarious grounds, because its mandate directed it toward population-based public health measures, not medical care for individuals. But, she noted matter-of-factly, the health of newborns was tied directly to the health of their mothers.[7]

The Bureau began emphasizing the sort of hands-on public health work that would enlist the work of nurses and physicians, building on the experience of physicians like S. Josephine Baker and public health nurse Lillian Wald.[8] In 1917 Drs. Frances S. Bradley and Florence Brown Sherbon published *How to Conduct a Children's Health Conference*, one of the earliest and most widely read guides to conducting a well-child exam. The next year, in concert with the General Federation of Women's Clubs, Meigs organized a series of child health promotions for a "Year of the Child."[9] Thousands of pamphlets were distributed to child welfare committees across the country, urging their participation in child health fairs, healthy baby contests, and the like. The Bureau also distributed 5 million record cards on which well-child examiners could record height and weight.[10] In 1918, capitalizing on momentum from the Year of the Child, Lathrop invited Representative Jeanette Rankin of Montana to introduce a bill for maternal and child health work under Bureau auspices. Representative Horace Towner of Iowa and Senator Morris Sheppard of Texas reintroduced it in 1920 and 1921.[11]

The Children's Bureau's new level of medical activism, particularly the proposed Sheppard-Towner Act, led it and the women physicians who carried out its medical mission into direct conflict with private practitioners and the Public Health Service. To physicians, especially those in private practice, the Bureau's child health services seemed to directly challenge the unwritten rule dividing public health from private practice medicine. To those identified with the PHS, on the other hand, these activities cried out for a national department of health to bring federal health care activities under the unified, medical control of the PHS. Introduction of the Rankin and Sheppard-Towner bills coincided with postwar budget tightening for the PHS and challenged its ambition to bolster its shrinking role on army bases with expanded duties in preventive medicine for civilians.[12]

Public health at the state and national levels was in a surprisingly undeveloped condition prior to this period. Historian John Duffy has noted that state boards of health prior to the 1920s were little more than "weak, ineffective agencies, staffed largely by volunteers."[13] In an alliance with the Conference of State and Provincial Health Authorities of North America, a professional association of state boards of health, the PHS began developing a long-range plan for consolidation of all federal health care activity into a single, physician-directed U.S. Department of Health. In 1919 representatives of the state health officers, the PHS, and the AMA held a joint meeting to discuss ways to bring about such consolidation. The resulting federal agency, it was hoped, would both expand and rationalize federal public health initiatives under a civilian, nonpartisan administration that would encompass the functions of state boards of health, the Public Health Service, and any other federally funded medical services, such as that "portion of the Children's Bureau dealing with matters of health." Thus, at congressional hearings for Sheppard-Towner in 1920, PHS surgeon-general Cumming supported the bill but urged that it be placed under the administration of the PHS.[14]

But other interests overshadowed those of organized medicine and the Public Health Service. For the first time, the "woman's bloc" was in command of both necessary votes and the public's imagination. Sheppard-Towner had strong enough backing to overcome the opposition of conservatives and organized medicine. Senator Reed Smoot, a Republican from Utah, went on record as preferring "a childless woman at the head of the bureau to a childless man, who would direct [Sheppard-Towner] if it were under the Public Health Service."[15] On November 23, 1921, the measure finally was passed into law for a term of five years. It would come up for congressional renewal in 1927.

At its heart, Sheppard-Towner was meant to provide education—primarily from public health nurses—in personal and child hygiene, which was broadly interpreted to include nutrition, exercise, proper clothing, disease prevention, and personal cleanliness. Again, no direct medical care of individual patients was funded by the bill. Typically, a physician and several nurses would set up a prenatal and well-baby screening clinic, sometimes on a temporary basis, sometimes on a permanent one. Great effort was expended to send these teams out into rural, underserved sections of the country. Matching funds were given to any state that would put up $5,000 of its own money, and the program usually was administered through state child health bureaus. Massachusetts, Connecticut, and Illinois refused to participate, but the latter two states established their own maternal-child health programs. Most states used their Sheppard-Towner money to organize health conferences, promote birth registration, distribute literature, and, in thirty-one states, to train midwives. Ultimate authority for the act was placed in the hands of the Federal Board of Infant and Maternal Hygiene, a newly established body consisting of Grace Abbott, the PHS's Hugh S. Cumming, and the U.S. commissioner of education. Actually its administration lay in the hands of the Bureau's Division of Maternity and Infancy, established in 1922 for this purpose, under division chief Dr. Blanche M. Haines, and after 1924, Yale pediatrician Martha M. Eliot.[16]

Although the final draft of the Sheppard-Towner bill did not require that women be appointed to key positions, this had been one of the draft provisions and it was, in fact, one of its results. A disproportionate number of physicians involved—employed either directly by the Children's Bureau or as directors of the state child health divisions with which the Bureau interacted—were women. By 1927, forty-three states had established child health bureaus or divisions using Sheppard-Towner funds. More than a third of these bureaus were headed by women. Forty of the eighty-nine full-time physicians employed over the life of the program were women. Some of the best known, including Drs. Dorothy Reed Mendenhall, Frances Sage Bradley, Florence Brown Sherbon, Ella Oppenheimer, Anna E. Rudé, Grace Meigs, Ethel Dunham, and Martha Eliot, worked directly for the Children's Bureau. Public health nurses, funded either directly through Sheppard-Towner or by the individual states, carried out most of the day-to-day work.[17] Perhaps to combat opposition from private physicians, the Children's Bureau at first relied heavily on local help from the General Federation of Women's Clubs, which initially supported the act. Throughout the life of the bill, volunteers continued to play a central role in the program, but the influence of its trained professional staff gradually became dominant.

Between 1924 and 1929, when the law lapsed, the Children's Bureau conservatively estimated that approximately three thousand prenatal and child health centers were established using at least some Sheppard-Towner money. From 1925 to 1929, 4 million infants and expectant mothers used these services. Almost 20,000 classes were held, often taught by public health nurses, to instruct midwives, mothers, and girls in child health and hygiene.[18] Sheppard-Towner's effectiveness was not easy to assess, however. During its legislative life, infant mortality (after the first day of life) declined but maternal mortality, particularly deaths from eclampsia and puerperal sepsis, did not. While the latter conditions, for example, were sensitive indicators of obstetrical skill, as many leading obstetricians realized, trends in infant mortality and morbidity reflected a constellation of socioeconomic factors, of which prenatal education was but one important factor.[19] Finally, because funding was distributed through state health departments subject to regional variation in racial attitudes, those counties known for concentrations of African American or Native American populations received a disproportionately smaller allocation of funds and trained nurses, despite the Bureau's having targeted for special attention the rural counties where most African Americans lived.[20]

Throughout the life of the legislation, the Children's Bureau relied on the expertise of well-known, university-affiliated obstetricians and pediatricians, notably J. Whitridge Williams of Johns Hopkins, Robert L. De Normandie of Harvard, and Edwards Park of Yale.[21] This reinforced the perception of many private practitioners that the Bureau's work was a threat to their own. In 1925 the Children's Bureau published *Standards of Prenatal Care: An Outline for the Use of Physicians*, based on a consensus conference of its Obstetric Advisory Committee of government, private, and university-affiliated physicians.[22] Around 1926 Grace Abbott called on the committee to conduct a comprehensive survey of maternal mortality, in part to demonstrate the need to renew Sheppard-Towner. From January 1927 until 1933 the committee organized and advised the Children's Bureau in running a study of maternal mortality in fifteen states. A total of 7,537 deaths were investigated through death certificates and interviews with the attending physician or midwife. Puerperal sepsis was one major cause of death, often the result of ill-judged or botched surgical interventions by physicians. Such findings, inadvertently or not, magnified the distance between elite and nonelite practitioners, and reinforced the latter's suspicions of the Sheppard-Towner Act.[23]

A good example of the potential for conflict between the Children's Bureau and private practitioners can be found in the New Haven Rickets Demonstration Project, conducted from 1923 to 1926 for the Children's Bureau by Drs. Edwards Park and Martha M. Eliot, both of the Pediatrics Department at Yale. This important study established the requirements, appropriate dosages, and optimal means of administration of vitamin D for children. Eliot established an elaborate protocol to ensure cordial relations with the local physicians of the children in the study: If "the birth certificate of [a baby] born in the [study] district . . . is signed by a local physician, a call is made by one of the Children's Bureau physicians, to explain to him the purpose of the demonstration and to ask his cooperation. If the physician has already been seen, a letter is written telling him that the birth certificate has been sent to the Children's Bureau, and that the mother will be urged to bring the baby for examination. No new baby is brought to be examined without the knowledge and consent of the family physician."[24] If the mother agreed to participate, she would bring in her baby to be weighed, measured, examined, and have a roentgenogram (x-ray) taken once a month for up to a year. It is easy to see how this might have been viewed as unfair competition by the local physician. Thus, despite the accomplishments of Sheppard-Towner, opposition to the act—and to the Children's Bureau—grew more pronounced as the decade advanced. Congress permitted the act's renewal after a bitter debate in 1927, but only for two years. In 1929 Sheppard-Towner was allowed to lapse.[25]

Maternalist Medicine and Its Critics

> The loss of Sheppard-Towner . . . interrupted the drive by women to be recognized as having a particular expertise and a broader outlook about preventive health care for themselves. In addition, [it] meant that new preventive services would be available mostly for women able to consult private physicians.[26]

Who were Sheppard-Towner's opponents, and what was the impact of its demise on women in medicine? The strong presence of MWNA leaders in the effort to pass Sheppard-Towner attests to women physicians' support for the act's maternalist philosophy.[27] No fewer than four future MWNA presidents, including Drs. Kate Campbell Mead, Ellen C. Potter, Esther Pohl Lovejoy, and S. Josephine Baker, testified on behalf of the bill. At the urging of Mead, the National passed a resolution at its annual meeting in June 1921 to "urgently recommend" the law's passage.[28] Having always moved freely between the realms of prevention and therapeutics in the case of women and children, women physicians played a disproportionately large role in the administration of Sheppard-Towner. As a result, the Children's Bureau and the women physicians who directed its medical programs were caught in the middle of the competing interests focused on the new legislation. The Sheppard-Towner Act, the Children's Bureau, and, by association, women physicians were accused by organized medicine of harboring "communistic" sympathies. The AMA also charged that the clinics were second-class institutions run by second-class practitioners for second-class citizens.[29] Between World War I and 1930, the nation rejected both government-funded maternity care and national health insurance. The legacy of these decisions was the marginalization of public health care—both those who received it and those who provided it.[30]

Some physicians who had initially supported free public health clinics as a way to build a client base for private practitioners soon changed their minds. The AMA had only recently consolidated its position as the defender of the beleaguered practitioner. Now it challenged the need to rely on Sheppard-Towner clinics. Not wanting to be seen as opposing preventive medical care, the association set forth its own positive approach to preventive medicine. For one, it launched a new magazine, *Hygiae*, aimed at wooing the laity. The AMA also began promoting "periodic health exams" for children and adults (known today as the yearly physical), which had long been championed by physicians in the child hygiene movement but were now to be carried out by private practitioners, whether GPs or pediatric specialists.[31]

In 1922, the AMA's Council on Health and Public Instruction, chaired by New York public health physician Haven Emerson, began preparing to educate the membership in how to conduct a periodic health examination. Emerson, who was New York City's commissioner of health from 1915 to 1917, the chief epidemiologist of the American Expeditionary Force during World War I, and head of Columbia University's School of Public Health from 1921 to 1940, had begun his career combining private practice and research on communicable diseases. He stressed the need for physicians to set aside ample time—at least forty-five minutes per patient—for well-patient exams, and he urged doctors to be forthright when advising patients on hygiene, nutrition, and exercise. Thus, with the backing of organized medicine, general practitioners adopted techniques previously available to patients mainly at public health clinics.[32]

Not only general practitioners but also many office-based pediatricians felt threatened by the renewal of Sheppard-Towner. The specialty of pediatrics had been slow to develop, especially within the medical curriculum. Even after the development of pediatric subspecialties, it was not among the high-status fields in the profession. Much like other specialties devoted to the "whole patient," such as general practice, pediatrics valued competence in psychosocial skills, catered to children and, usually, their mothers, and commanded only modest fees. Moreover it could not yet boast technical exclusivity, and it attracted a disproportionately high percentage of women practitioners.[33]

Ironically, one consequence of the success of Sheppard-Towner was an increased demand for pediatricians. The number of child care specialists increased fivefold between 1914 and 1923 and by more than 150 percent between 1923 and 1934. Rising demand, in turn, legitimized office-based pediatrics as a distinct specialty within medicine. In 1933 residency certification was established by the American Board of Pediatrics.[34] As pediatricians carved out a new niche as "primary-care specialists," they increasingly opposed extending the life of the Sheppard-Towner Act. Office-based pediatricians, following the AMA, opposed renewing Sheppard-Towner because they no longer needed any help in developing their private practices.[35]

Academic pediatricians, on the other hand, strongly supported Sheppard-Towner and the Children's Bureau's central role in its administration. In their view, government-sponsored child health programs would both foster pediatric research and enhance the prestige of the specialty in the public's eye. In 1915, for example, the American Pediatric Society formed a committee to "cooperate" with the Children's Bureau. By the time of Sheppard-Towner's passage, the APS had endorsed increasing involvement in child health activism. At the request of Julia Lathrop, in 1919 it appointed one of its members as liaison to the Bureau's Hygiene Division. In 1923 APS president L. Emmett Holt lamented that the APS had "neglected" child health activism in the past, ruefully observing that "we have left the subject of popular health education too much to the nurse, the social worker, and the nutrition worker, and some of these groups, largely owing to our neglect, have gotten somewhat out of hand."[36] By the 1920s, Holt's comment suggests, pediatric specialists understood the importance of their presence in child health policy forums. Thus, in contrast to office-based pediatricians, the academic and research-minded pediatricians who made up the membership of the APS were among Sheppard-Towner's strongest advocates. Their support of the Children's Bureau endured into the 1940s.[37]

The Public Health Service, on the other hand, supported continuation of the Children's Bureau's medical divisions but insisted their administration be transferred to the PHS. Surgeon General Rupert Blue of the Public Health Service and his successor, Hugh S. Cumming, who became surgeon general in 1920, perceived the Bureau as competing for congressional health care money. Blue hoped to have the agency transferred to the Public Health Service, where it might become the nucleus of a unified national Department of Health, an objective dating back at least to 1919. Cumming's objective, to transfer the entire Children's Bureau from the Department of Labor to the Public Health

Service, was supported even more strongly in the Hoover administration by Dr. Ray Lyman Wilbur, secretary of the interior and a physician.[38] Wilbur won Herbert Hoover's support to create a broad, centralized federal health care agency under the auspices of the PHS. Those aspects of the Children's Bureau concerned with medical care, Wilbur believed, ought properly to be transferred to the PHS.[39]

Another obstacle to Sheppard-Towner's renewal was the waning support of middle-class Americans for government-funded public health. At the height of roaring Twenties' prosperity, many more middle-class families than before could afford the services of a private doctor. The loss of "cross-class support" by middle-class women can be seen, for example, in the General Federation of Women's Clubs' decision in 1924 to back away from most of its political agenda of the previous three years, including active support for the Sheppard-Towner Act. By 1928 the federation was no longer affiliated with the Women's Joint Congressional Committee. The fact that the act was not revived after Hoover's election in 1928 was in part due to the absence of support from such grassroots women's organizations, which had been crucial sources of popular support in 1921. The fate of Sheppard-Towner was thus not an isolated phenomenon. From the moment postwar prosperity overtook postwar inflation, middle-class Americans lost interest in supporting organized labor, immigrants, indeed any of society's "have-nots." In this context, it is not surprising that the majority of American physicians, including many women, now turned their attention to the cultivation of private practices, not public health.[40]

The 1930 White House Conference on Child Health and Protection provided a forum in which the professional tensions of the past decade surfaced sharply, publicly, and dramatically. Through the intercession of Ray Lyman Wilbur, President Hoover decided to use the conference to lobby for a new Department of Health that would be under Public Health Service control. The divisions of the Children's Bureau with primarily medical functions would be transferred to the PHS. Hoover appointed Wilbur to head up the Conference planning committee in consultation with the PHS and with representatives of the medical profession. The president was inclined to favor not only the organizational goals of the PHS but also the desire of organized medicine to consolidate the standing of private practitioners. One of the conference objectives, therefore, was to put the White House imprimatur on the role of the private practitioner in preventive health care.

As a result the 1930 White House Conference, unlike its predecessors, was dominated by physicians (all male) in public health, pediatrics, and obstetrics. Children's Bureau leaders such as Grace Abbott, who had been working toward a child health conference of her own, were shut out of the conference planning phase. In fact Abbott was barely consulted, although many conference subjects were those to which the bureau had been devoted for years: "problems of dependent children; regular medical examination; school or public clinics for children; hospitalization; adequate milk supplies; community nurses; maternity instruction and nurses; teaching of health in the schools; facilities for playgrounds and recreation; voluntary organization of children; child labor and . . . allied subjects."[41]

The conference's initial official report was printed and distributed to delegates at the outset of the meeting. One of its most controversial recommendations, authored by PHS surgeon general Hugh Cumming and Haven Emerson, chair of the subcommittee on Public Health Service and Administration, called for transferring the maternal and infant health programs of the Children's Bureau, all of which were directed by women physicians, to the PHS. Grace Abbott's minority report, which argued strenuously for maintaining the organizational integrity of the Bureau, was not included in the copies distributed to delegates, a fact that infuriated long-time Children's Bureau supporters. Programs for infant and maternal health, the report argued, should be run by municipal health departments, but "some activities now included in the program of most health departments can be transferred gradually to the general practitioner of medicine. In the interests of child health, the family physician should become a practitioner of preventive as well as curative medicine."[42]

Haven Emerson's subcommittee, which included Abbott, met on the first day of the conference to endorse this majority report. To Emerson's consternation, two hundred observers appeared for the meeting, many of them prestigious figures with years of experience in child health—all of them vocal, angry, and determined to have their say. Abbott's supporters were not about to stand by for the dismemberment of the Children's Bureau. Her advocates included Lillian Wald, Edith

Abbott, Sophonisba Breckenridge, Dr. Martha Eliot, and Dr. Alice Hamilton. Emerson unsuccessfully tried to shoo away the overflow crowd. Finally he was prevailed upon to read aloud Abbott's minority report, which received the support of the audience. Yet at the plenary discussion the next day, Emerson presented only the majority report. Surgeon General Cumming announced that no votes would be taken. Speaker after speaker, however, rose to endorse Abbott's report and the Children's Bureau's continued administration of the child health divisions. Supporters included Josephine Goldmark on behalf of the dean of Yale's School of Medicine; representatives of national women's organizations; Dr. Dorothy Mendenhall, who spoke "for women physicians"; Florence Kelley; Dr. De Normandie; and many others. Telegrams of support from eminent pediatricians such as Edwards Park, now chair of pediatrics at Johns Hopkins, were read before the large assembly. In view of the fractiousness of the meeting, President Hoover sent word through Secretary Wilbur that the recommendations for the Children's Bureau should be set aside. No further action to dismember the bureau was taken for eight or nine years. But the scars from this bruising battle ensured that the issue would not be forgotten.[43]

Indeed, Haven Emerson wrote an immediate—intensely angry—letter to Dr. Park, inquiring if he "was aware of the public use made of your name . . . by Miss Lillian Wald and Miss Josephine Goldmark and others? You were quoted as representing informed and expert medical opinion in a matter in which so far as I am aware you have had little if any professional experience" (November 21, 1930). Park immediately wrote for advice from his former protegée, Dr. Martha Eliot, director of the Maternal and Child Health Division of the Children's Bureau since 1924. To Park, as a pediatrician, the crux of the matter was that Emerson's committee was "composed almost entirely of Public Health men" who had no understanding of maternal-child health, and thus the committee itself was "entirely responsible" for the response it received. A portion of Emerson's reply to Edwards Park, the last letter in the series, deserves quotation. Noting that he was not likely to be "impressed by arguments of clinical pediatrists or their personal friendship for Misses Abbott, Goldmark, or Wald," he charged that "you, from the cloistered walls of Academe, and seeing in a rosy glow the halo of a besieged princess Goldilocks, have rushed, in an endocrinological manner to the aid of embattled feminism, with the idea that public health administration is an emotional or intuitive art upon which the dispenser of baby bottles and the expert in variegated and diapered stools is competent to give constructive and discriminating advice. Go to!"[44] Emerson's letter expressed many of the submerged professional anxieties on which maternalist medicine ran aground, including disdain for the preventive and educational holism advocated by maternalist health reformers—lay and professional. The entire episode vividly demonstrated, too, the dangers attendant to any woman physician who stood too close to the "rosy glow" of maternalist medicine.[45]

Women Physicians and the Decline of Maternalist Medicine

It is tempting to interpret the struggle over the fate of the Children's Bureau as a contest between female reformers and male physicians. Yet it oversimplifies the conflict to conceptualize it, dichotomously, as a confrontation between "women" and "doctors." Not all male physicians were unsympathetic to the bureau's leadership, as we have seen. Nor did all women physicians support it.

Most women doctors, like men, were private practitioners with predominantly middle-class patient populations. Their support for government-funded public health was highly ambivalent. Such support, particularly among members of the Medical Women's National Association, actually declined after 1924. Initially this seems puzzling. Theda Skocpol, Richard Meckel, and Stanley Lemons claim that within the medical profession, women were the most persistent supporters of Sheppard-Towner. Certainly those who were employed by the Children's Bureau or by state or local health bureaus supported it with enthusiasm. Dr. Martha Eliot, who became assistant director of the Children's Bureau in 1935, was one of the three authors of Title V of the Social Security Act, which helped perpetuate many of Sheppard-Towner's key provisions in the form of Aid to Families with Dependent Children.[46] But among women physicians in private practice, public health was losing its cachet. By the 1930s, few women headed public health departments, in part for the lack of interest, something Dr. Josephine Baker ruefully reported to the MWNA in 1932. Her report on women's work in public health attempted to reinvigorate a sense of responsibility among women physicians—especially those in private practice—for work that, traditionally, had been theirs for the taking.[47]

Baker understood that some women practitioners, no less than their male colleagues, viewed government-sponsored clinics as unfair competition.

In short, by the mid-1920s even the MWNA reflected the new political climate. Previously the organization had identified strongly with the maternalist, social-housekeeping goals of Progressive-era reformers. In this context, the MWNA initially chose not to support the National Woman's Party's proposed Equal Rights Amendment (ERA). Initially it took the advice of Dr. Alice Hamilton, Hull House resident and pioneer industrial toxicologist, who opposed the ERA as a threat to the protective legislation for working women long supported by Progressive reformers. In 1924 the MWNA's leaders still linked the identity of the woman physician with protection of the sick and powerless. They voted to oppose the ERA. At the same time, however, the organization's values were beginning to shift. The MWNA's growing detachment from the Sheppard-Towner Act during the 1920s, while never as sharp as the AMA's, reflected an ideological drift toward the individualism of the private practitioner.[48] Thus, although historians have contrasted the AMA's hostility toward the Sheppard-Towner Act with the MWNA's uncompromising support, the National's position was both more complex and less unchanging than often has been noticed. For Josephine Baker, it is true, the prospect of what she called "state medicine" held no terrors. Rather, as she later wrote, "State medicine is to my mind an ideal, and the sooner it changes from an ideal to a practical reality, the better off the human race will be." Dr. Louise Tayler-Jones, chair of the MWNA's Committee on Legislation, also supported Sheppard-Towner to the end. But Baker and Tayler-Jones occupied only one end of the spectrum of opinions held by MWNA leaders.[49]

In reality, as opposition to government intervention in health care grew more pronounced, the MWNA's position became more ambiguous. Faithful as ever to the principles of health education and preventive medicine, it nevertheless aligned itself with the claims of private physicians. It actively promoted preventive medicine, but through the combined auspices of women's clubs, medical societies, private doctor's offices, and public health departments. And the MWNA was not alone. Pediatricians such as Josephine Kenyon, for example, writing advice columns for mothers in magazines such as *Good Housekeeping*, assured their audience that reading a monthly medical column was no substitute for regular checkups with one's private doctor.[50]

Reflecting the shifting stance of the profession as a whole during the twenties, MWNA support for state-funded programs now took a backseat to private medical care for the middle class. The National looked to alliances with state and local women's clubs to introduce programs of preventive medicine. Many MWNA members used the activities generated by the Sheppard-Towner Act to cement ties to the middle-class public. Lena Sadler of Chicago articulated this emerging position at the May 1925 annual meeting: "I wish to say that the Sheppard-Towner Bill, fortunately or unfortunately is not functioning in the State of Illinois . . . Personally it is nothing to me, whether it functions or not, because I believe my record is behind me . . . I believe in every educational feature of that bill, but when organized medicine [the AMA] is against it in my state, I must do something with the [women's] clubs to take its place." Sadler, the Illinois Department of Public Health, the Federation of Women's Clubs and the state medical and dental societies began a jointly sponsored program of examinations for preschool children. "It is a five-year program," Sadler commented, "and by that time we hope to make it the custom for parents to have their children examined before entering school." Sadler likely envisioned these future examinations taking place in the offices of private physicians.[51]

Thus when the fight to renew Sheppard-Towner was under way, MWNA support already had been redirected into other channels. At the annual meeting in 1926, the National's representative on the WJCC, Louise Tayler-Jones, proposed a new resolution of support for Sheppard-Towner, but no action was taken. In October 1926 the MWNA *Bulletin* published an anonymous attack on the law. Arguing that "maternity education should be directed only by physicians," it insisted that the government has no more right to subsidize health care with tax money than "it has the right to make Rockefeller 'divide up' with [Socialist] Eugene V. Debs." (A signed rebuttal by Tayler-Jones appeared in the next issue.) When a resolution supporting Sheppard-Towner was presented to the 1927 annual meeting, the matter was tabled because it was "not at issue at present, having been already settled by Congress."[52] This tepid support reflected the National's strategy of allying with moderate, middle-class women's groups to cultivate support for women in private practice.[53]

The MWNA's retreat from social welfare activism intensified during the Depression, when private practitioners everywhere faced the prospect of declining income. But even before 1929 many professional women, including women physicians, had become disenchanted with governmental initiatives such as protective labor legislation after several states began including white-collar, professional jobs in the categories covered by proposed wage-and-hours laws designed to protect women workers.[54] In 1937 the MWNA voted to support the ERA, and it was not the only women's professional association to do so. As Lemons has shown, the National Federation of Business and Professional Women's Clubs, the Osteopathic Women's National Association, the National Association of Women Lawyers, and the Association of Women Dentists were just some of those who were won over to free-market individualism during the 1930s.[55]

On the issue of birth control, too, individualistic professionalism proved a stronger force than the tradition of social reform for women doctors. Here also, political values reflected women physicians' middle-class patient base. Carole R. McCann argues that the majority of feminists in the 1920s, excluding the small number who were socialists, never gave much attention to the issue of birth control.[56] Since most women physicians were well aware that women doctors had often been suspected of performing illegal abortions, few were prepared to be publicly identified with the cause of family planning. Although the MWNA came out in favor of birth control in 1928, nearly a decade before the AMA, it addressed the issue in the context of maternal mortality and medical competence, not feminist politics. A resolution passed at the 1928 annual meeting mainly concerned the need for better prenatal care: "In consideration of the high death rate of woman due to childbirth, and the high infant mortality from poorly managed deliveries and from criminal abortions; and since these deaths are due in large part to the ignorance of women of the need for proper care during maternity and to the prevailing practice of criminal abortion," the group recommended that all Federated Women's Clubs "give some time on their programs each year to papers which shall emphasize the need of better care during pregnancy and greater regard for the lives of women and unborn children." It also resolved to prepare a bibliography and a list of speakers—drawn from MWNA members—for use by the federation. The resolution concluded by asserting that birth control clinics be promoted only under the proper medical auspices.[57] In 1930, incoming MWNA president Olga Stastny of Omaha urged more knowledge of "normal" sexuality and birth control.[58] This was not a radical stance. By

1935 hundreds of organizations had come out for relaxation of the Comstock laws which, since 1873, had banned all traffic in birth control information and devices. In 1937 even the AMA passed a "qualified endorsement of birth control," but only insofar as it was addressed within the doctor-patient relationship.[59]

Despite the overwhelming climate of professional disapproval, some women physicians did take a strong public stand in support of birth control, especially in making it available to the poor. Those physicians most identified with birth control clinics—both in the public eye and in the eyes of the medical profession—were women. Dr. Dorothy Bocker was medical supervisor for Margaret Sanger's Birth Control Clinical Research Bureau (CRB) from 1923 to 1925. Her successor, Dr. Hannah Stone, continued at the CRB until 1941. They kept their activities within the law by incorporating the CRB into their private practices as a research project. Patients accepted for CRB examination had to be unable to pay the fees of private physicians and to have some health-related condition for which pregnancy prevention would be an appropriate treatment. (Those who could afford to pay a physician's standard fee, or for whom no medical condition could be found, were referred to the private practices of Bocker and Stone.) Indeed in 1924 Stone was forced to give up a hospital affiliation because of her birth control activities. African American women physicians also played a crucial role in the CRB. After 1930 Sanger received partial support from the Julius Rosenwald Fund for a coalition between the CRB and members of the Harlem community to found the CRB Harlem Branch; it operated until 1935. Dr. May Chinn, an African American physician practicing in Harlem and the first black woman graduate of Bellevue Hospital Medical College, and Mabel Staupers, executive director of the National Association of Colored Graduate Nurses, were members of its advisory council. Other African American health professionals, such as Dr. Harold Ellis and nurses Lou Thompson and Dorothy Ruddick, worked at the Harlem Branch, at times with white women doctors such as Stone and Dr. Marie Levinson. Another African American physician, Dr. Mae McCarroll, worked for Planned Parenthood Federation of America around 1942 as a liaison to African American professional organizations.[60]

African American women physicians, nurses, and midwives generally saw government-sponsored public health work as a scarce, desirable resource—not as undesirable competition. The black women's community, especially in the rural South, maintained whatever contacts it could with the Children's Bureau and, overall, invested considerable time and money in efforts to develop health care partnerships between voluntary and government organizations.[61] In the late nineteenth century, when Reconstruction gave way to Jim Crow laws throughout the South, black women's clubs and churches had constituted the driving force behind black social welfare activism. But by World War I, black leaders such as Booker T. Washington of the Tuskegee Institute sought partnerships between the voluntary sector and the government, especially for public health reform. In 1915 Washington launched National Negro Health Week (NNHW) to focus and promote these efforts. In 1932 the Public Health Service agreed to create an Office of Negro Health Work, in part to coordinate NNHW activities. The PHS's new Office of Negro Health Work was headed by Dr. Roscoe Brown, a dentist and public health educator. Although the campaign was endorsed by the National Medical Association, many black physicians, like most white doctors, were struggling to build their private practice and had little time for voluntary public health work. (According to one study, by 1930 about 3,700 physicians in the United States, or 7 percent, were African American.[62])

Increasingly, Depression-era public health work among African Americans, no less than among whites, was structured and stratified by gender. The executive committee of the NNHW campaign, for example, included Mabel K. Staupers and Modjeska Simpkins, leaders in nursing and health education, respectively, but it was dominated by male physicians. None of the sixty or so African American women physicians currently in practice was appointed to the committee. Even the Children's Bureau, which was funded by the Julius Rosenwald Fund to hire a black physician to oversee obstetric and pediatric continuing education for southern black physicians, hired a man for the job. There were, however, notable exceptions, such as Dr. Zenobia Gilpin, president of the Richmond (Virginia) Medical Society, who during the Depression worked with the local Urban League and many Richmond churches to develop a highly successful program of health talks and free clinics.[63]

At the local level, day-to-day leadership, organizing, and public health field work usually were carried out by black laywomen, women doctors, nurses, and midwives. One of the most remarkable of these programs was created by Alpha Kappa Alpha (AKA), a national sorority founded at Howard University, for black college women and graduates. Searching for ways to assist poor rural blacks during the worst years of the Depression, AKA's leadership decided to sponsor a multifaceted public health program in Mississippi from 1935 to 1942. Organized by AKA president Ida Louise Jackson, a teacher, the Mississippi Health Project, as it was named, was headed by Dr. Dorothy Boulding Ferebee. Ferebee was a private practitioner in Washington, D.C., an instructor in obstetrics at Howard University Medical School, a wife, a mother, and an AKA member. In 1929, when she was just starting to build up a practice in Washington, Ferebee had organized a social settlement, Southeast House, for the black community in the southeast sector of the city by successfully appealing for funds to the new District of Columbia Community Chest. From 1949 to 1968 she was director of the Howard University Health Service.[64] Jackson and Ferebee decided that the Mississippi Health Project should focus on child health for the sharecroppers living in the Mississippi Delta. For the first summer's work, Ferebee chose AKA members with professional experience in medicine, dentistry, education, social work, and, especially important, nursing. Perhaps the most experienced was Mary E. Williams, a public health nurse who had long worked in the South; had directed the Tuskegee Institute Health Center, a settlement house in New Orleans; and had been awarded scholarships to study public health at Simmons College in Boston and at the Harvard School of Public Health. With Williams's public health experience and broad knowledge of local politics in the deep South, the AKA clinics garnered support from some county health officers while sidestepping the opposition of others. The clinics operated for seven consecutive summers in Holmes and Bolivar Counties. Despite the obstacles imposed by suspicious plantation owners, white ministers, and even some of the local black community, the AKA volunteers gained sufficient trust to take health histories, provide immunization against diphtheria and smallpox, treat malaria and venereal diseases, do routine dental work, and conduct public health classes in nutrition and personal, social, and dental hygiene. They also learned firsthand that, as Ferebee commented in 1941, "the standard of health is indissolubly linked to all the socio-economic factors of living."[65]

For Ferebee and other black public health activists, another basic lesson was the difficulty of achieving lasting change on their own. To uproot deep-seated patterns of social inequity required sustained legislative, financial, and institutional support from the federal government, but they were unable to persuade either federal or state agencies to take over or even indirectly assist the project. At Ferebee's prompting, the AKA joined with the National Council of Negro Women (NCNW), led by Mary McLeod Bethune, in trying to persuade the government to commit more resources to African American citizens' health. In 1938, at the NCNW conference titled "The Participation of Negro Women and Children in Federal Programs," Ferebee, Williams, and more than sixty other women delegates addressed Eleanor Roosevelt, the directors of the U.S. Children's Bureau, the Women's Bureau, and other federal administrators. Their message was direct: that African Americans, as citizens, were underrepresented in proportion to their need, and as government employees were underrepresented in proportion to their experience and skill. In 1943 Martha Eliot's office at the Children's Bureau alerted the Mississippi Health Project to the existence of Title V funds, but as the project lacked the requisite matching funds from Mississippi, the Bureau's ability to help was severely limited.[66]

⌒ In contrast to the AKA and the Children's Bureau, however, as most women physicians moved more fully into the medical mainstream, they shunned institutions at the periphery of professional respectability. The changing political tenor of the MWNA typified the decreasing political activism of women physicians overall. Even during the 1930s, when MWNA membership was declining in response to the hard times, delegates rejected measures for wider provision of health care as inroads to "socialism." In 1933, in the depth of the Depression, the National's annual meeting undertook an open discussion of the minority and majority reports of the reformist Committee on the Costs of Medical Care. Opening the discussion, Bertha Van Hoosen began by reminding those present that "the only way women physicians got a foothold [early in their careers] was through charity work until they had gained the confidence of the public." She suggested that young physicians be given the cases of "those who are not able to pay a great deal for medical care," since these fledgling physicians "have the time to care for them." But even she stopped short of suggesting that medical care for the poor become a permanent responsibility of the government.

Most other speakers supported measures such as those taken in Milwaukee, Dayton, San Diego, and Portland to assure access to care by the poor without sacrificing the autonomy of the practitioner.

The American Women's Hospitals Service, still led by Esther Pohl Lovejoy, provided the only exception to the National's remove from publicly subsidized health care. In 1931 the AWH responded to a plea by MWNA president Rosa Gantt to establish the Mountain Medical Service to benefit mothers and children in Gantt's native Appalachia. The first center was directed by Dr. Hilla Sheriff in Spartanburg, South Carolina. Over the next two decades, attending physicians included Sheriff, Hallie Rigby, Lonita Boggs, and Gertrude R. Holmes. The program's centerpiece institution was the Greenville Maternity Center, originally founded by AWH registered nurse Emily P. Nesbitt. The center offered prenatal, well-baby, and maternal health clinics, as well as nutritional counseling to combat pellagra. In 1935 it also began offering contraceptive counseling. Gantt herself, along with Nesbitt, expanded the work into North Carolina. Another unit was established in Kentucky. In addition to AWH support, the Mountain Medical Service soon attracted funding from community chests and other local sources. Its physicians combined their subsidized medical care for the poor with private practice.[67]

For the next thirty years, however, the MWNA's leadership generally retreated from its vanguard position on Sheppard-Towner. In 1944 it went on record as "opposing that part of the Wagner-Murray-Dingle Bill which specifically refers to compulsory health insurance." It did endorse medical-society supported "medical plans" (such as Blue Cross–Blue Shield), but only to prevent further "socialization" of medicine. Shamefully, on the issue of racial equality in the profession, the MWNA did no more than keep pace with American health care institutions in general, most of which remained segregated until the 1960s. In 1924 an application for membership from "colored women . . . who belong to the American Medical Association" was referred to the Executive Committee and tabled. As late as 1939 the organization (since 1937 renamed the American Medical Women's Association, or AMWA) continued to reject applications for membership from otherwise eligible African American women physicians. Two years later Bertha Van Hoosen threatened to resign from AMWA if it did not change this policy, but no action appears to have been taken until at least 1942.[68]

Women and Academic Medicine before 1950:
The Case of Pediatrics

Between the 1930s and the 1950s, when medical specialization was imprinted on the profession, pediatrics offered women the best opportunity to reconcile professional advancement with a commitment to maternal-child health. The unusual confluence of public and professional acceptance and women physicians' own interests made pediatrics an ideal specialty for women to pursue. Following World War I, as shown earlier, women were about a generation behind their male colleagues in achieving the outward signs of specialization. They were absent from the American Pediatric Society, for example, until 1928, when Dr. Ethel Dunham of Yale was elected the first woman member. But between 1928 and 1966, women made up 6.5 percent of APS members, even exceeding their representation in the profession as a whole. During the next twenty years women accounted for 9 to 10 percent of the society's membership. Indeed, the percentage of female pediatricians has always been disproportionately high. In 1966, for example, almost twice as many women as men received board certification in pediatrics. In 1987, 45 percent of those seeking board certification in pediatrics were women, although the percentage of women seeking board certification in all fields constituted less than 12 percent. According to a survey completed in 1980, a total of nine women held positions as chair of pediatrics between 1952 and 1980, exclusive of those at the Woman's Medical College of Pennsylvania. The first woman to chair a pediatrics department at a coeducational medical school was Dr. Katherine Dodd, in 1952, at the University of Arkansas Medical Center.[69]

Several factors may explain women's relative success in the specialty. Since the late nineteenth century, women physicians had enjoyed public acceptance as physicians for children. Prior to the end of the century, most children's hospitals were women's and infant's hospitals rather than centers dedicated exclusively to the care of children. Several such institutions were founded by women, including the Blackwells' New York Infirmary for Women and Children (established in 1857), Dr. Marie Zakrzewska's New England Hospital for Women and Children (1862), Dr. Mary Thompson's Chicago Hospital for Women and Children (1865), the Children's Hospital of San Francisco (1875), and the Babies Hospital of the City of New York (1887).[70] In the early twentieth century, widespread recognition of women physicians' child health work in voluntary societies and municipal agencies, the still shaky status of pediatrics as an academic specialty, and the unstinting support of several well-placed male academic pediatricians helped ease the way for women to make inroads in the new field. Emmett Holt's 1923 presidential address to the APS, for example, envisioned pediatrics as a tripartite field consisting of academic researchers, practicing pediatricians, and pediatricians in public health—all three, avenues that were, or soon became, well traveled by women.[71] Women in pediatrics and closely related subspecialties, such as pediatric cardiology and anesthesiology, were able to move ahead sooner than women in other academic specialties. Physicians such as Ethel Dunham, Martha Eliot, Emmett Holt's colleague Martha Wolstein, Dorothy Anderson, cardiologist Helen Taussig, anesthesiologist Virginia Apgar, and the founder of Yale's rooming-in project, Edith Banfield Jackson, exemplified the opportunities pediatrics could provide for the post–World War I generation of women doctors.[72]

Women who made a name for themselves in pediatrics and closely related areas had several characteristics in common. Generally speaking, they all received substantial boosts early enough in their career to propel them beyond the lower academic ranks. But in addition, they consistently performed at high or even "super achiever" levels.[73] All were educated at excellent coeducational medical schools. They were admitted to high-quality internships and residencies and continued in academic medicine directly after finishing their training. Much of their forward momentum was the result of their own gifts and preparation, of course, but it also resulted from astute and committed mentoring by male faculty who recognized their skills and went out of their way to alert them to career opportunities. Frequently they created and became identified with innovative clinical approaches, sometimes working at the border between medical specialties. In developing these innovative subspecialties, they were praised for their ability to create effective and respectful teams of nurses and physicians, usually including both male and female physicians. They also were given a fair degree of administrative latitude early on in their careers. Several attained the rank of full professor, although none became a dean, department chair, or hospital administrator. Prominent women specialists also were notable (and publicly praised) for maintaining an interest in the "whole patient," despite their commitment to a particular specialty. Ironically, however, career success for women pediatricians in the 1930s often precluded their marrying or having children of their own: Ethel Dunham, Martha Eliot, Helen Taussig, Edith B. Jackson, and Virginia Apgar never married. Dunham and Eliot, however, shared a household most of their adult lives. The other three women lived alone.

The careers of Eliot and Dunham illustrate the general conditions required for a woman physician to succeed in an academic specialty during the 1920s and 1930s.[74] Both women came from prominent New England families willing and able to provide them with an undergraduate education at an elite school. Dunham's father was president of the Hartford Electric Light Company; Eliot's grandfather had been the first chancellor of Washington University. Dunham attended Bryn Mawr College after several years of European travel. She began her lifelong friendship with Eliot when Eliot took a year off from Radcliffe to study at Bryn Mawr. Both women graduated in 1914 and enrolled at the Johns Hopkins Medical School. They graduated in 1918. During their four years at Hopkins (where they also attended woman suffrage rallies) they made lifelong friends and supporters of Edwards Park and Grover Powers, young members of the pediatrics faculty and future leaders in the field. The chair of pediatrics, John Howland, accepted Dunham as the first woman pediatrics intern at Hopkins. Since Howland refused to have more than one woman on his service, Eliot served first as a junior intern at Bellevue in New York and then, because of "manpower" shortages during World War I, completed her internship at Peter Bent Brigham Hospital in Boston. In 1919, at Park's suggestion, Eliot began a pediatrics residency at St. Louis Children's Hospital. Dunham, after completing her internship at Hopkins, became a house officer at Yale in 1920 at the invitation of Park, the new chair of pediatrics there. She was named an instructor in 1921, in charge of the newborn nursery and the outpatient clinic. That same year, Eliot was invited by Park to become the chief resident and an instructor in pediatrics. These were remarkable opportunities for both women. Park also hired Grover Powers at the rank of associate professor. Both women benefited from the mentoring of these well-respected academics. Park, besides being a lively and remarkably democratic colleague (Eliot wrote home in amazement that he had taken her to a Yale football game one Saturday), was an unusually active mentor. One colleague reminisced that "Park was famous for choosing and guiding young members of his staff into new subspecialty careers where they gained great eminence, for which he denied all responsibility." Powers explicitly paid tribute to Park as a "vigorous proponent" of the careers of women.[75]

Dunham and Eliot benefited from the move to Yale, especially as it facilitated their moving out of Howland's orbit and into that of Edwards Park. Howland's insistence on bench research resulted in a deliberate deemphasis of the social welfare aspects of the field. Park's "activist vision" for the specialty, however, closely matched Eliot's and Dunham's own aspirations. At Yale Park created a department that combined clinical research with a broad vision of pediatrics' place in maternal-child health education, a commitment continued by Powers, his successor after Park's 1927 departure to chair the pediatrics department at Johns Hopkins. Between 1920 and 1935, when Dunham and Eliot held faculty positions at Yale, Park and Powers enthusiastically supported their holding simultaneous appointments with the Children's Bureau, beginning in 1923.[76]

Ethel Dunham's research focused on the care of the newborn, especially the physiological challenges of prematurity. Her systematic study of morbidity and mortality among 1,000 newborns at the New Haven Hospital was presented to the American Pediatric Society in 1933. She was soon named the chair of a new APS committee on neonatal studies. By 1943 she had spearheaded the establishment of uniform standards of hospital care for newborns, both full term and premature, published under the auspices of the Children's Bureau. Her own compendium, *Premature Infants, a Manual for Physicians,* was published in 1948. In 1957 she became the first woman to receive the American Pediatric Society's highest honor, the John Howland Medal.[77]

Of the two, Eliot was much better known. In 1923, two years after Edwards Park invited her to become chief resident at Yale, she and Park began a three-year, community-based, prospective study of rickets prevention among a representative New Haven population that included a large African American cohort. They presented their results before the AMA Section on Diseases of Children in 1925, but Eliot alone published the initial report of the study in *JAMA* several months later.[78] Eliot became director of the Children's Bureau Division of Maternal and Child Health in 1924. Ten years later, she, Grace Abbott, and Acting

Chief Katherine Lenroot drafted the text of Title V of the Social Security Act, including provisions for aid for families with dependent children, crippled children, and tubercular children, thus reviving many provisions of the Sheppard-Towner Act. When Lenroot, a social worker with strong bipartisan political ties, was appointed Abbott's successor, she asked Eliot to become assistant director. As a result, both Eliot and Dunham moved to Washington in 1935 and began their full-time affiliation with the bureau. Eliot retained the title of clinical associate professor at Yale until 1950. Dunham became the bureau's head of research on child development, also holding the title of lecturer in clinical pediatrics at Yale. From 1951 to 1956 Eliot was the director of the Children's Bureau. In 1957 at age sixty-six, rather than retiring she became professor of maternal and child health at the Harvard School of Public Health. Both women then moved to Cambridge. In 1948 Eliot received the prestigious Lasker Prize of the American Public Health Association, and in 1967, like Dunham a decade before, she was awarded the John Howland Medal by the APS.[79]

The career of psychiatrist and pediatrician Edith Banfield Jackson, like those of Eliot and Dunham, also profited from the cooperation of male faculty mentors at Yale.[80] Jackson was born in Colorado Springs, Colorado. She graduated from Vassar in 1916 and from Johns Hopkins Medical School in 1921, three years after Eliot and Dunham. She completed a general internship at State University of Iowa Hospital and a residency in pediatrics at Bellevue in New York City. In 1923 she joined the Children's Bureau, working on the New Haven rickets project with Drs. Eliot and Park. Jackson, however, was more interested in the principles of childhood behavior than in community-based preventive medicine, and she moved to Washington, D.C., to train in child psychiatry at St. Elizabeth's Hospital. From 1930 to 1936 she lived in Vienna, where she studied psychoanalysis and entered into a training analysis with Sigmund Freud. At the same time she, Anna Freud, and Dorothy Burlingham established an experimental all-day nursery to facilitate the psychoanalytic study of childhood. While in Vienna Jackson also assisted many Austrian refugees in their efforts to escape the Nazis and to obtain American visas.

After her return to New Haven in 1936, Jackson held a joint clinical appointment at Yale in pediatrics and psychiatry and also became director of pediatric psychiatry in the department of pediatrics at the Grace–

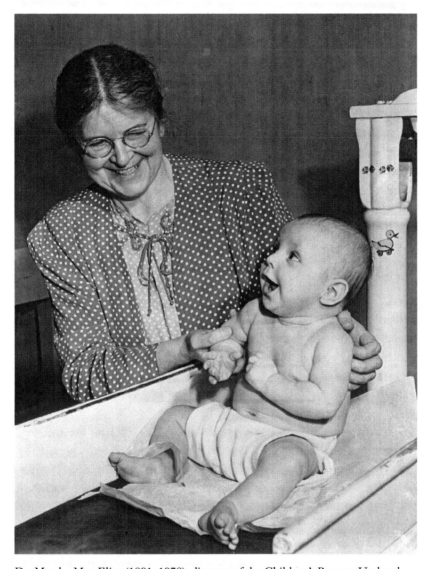

Dr. Martha May Eliot (1891–1978), director of the Children's Bureau. Undated photograph, ca. 1940s. (Courtesy of the Schlesinger Library, Radcliffe College, Cambridge, Massachusetts.)

New Haven Community Hospital (now Yale–New Haven Hospital). Over the next ten years she gradually directed her attention more to the psychological needs of newborns and their mothers, from 1946 to 1953 creating and directing the country's first hospital rooming-in unit. Rooming-in was intended to provide a calm setting in which mothers, infants, and, to some extent, fathers could learn each others' personalities and needs and, thereby, successfully initiate breastfeeding. The revival of these traditional ideas was not original to Jackson. Many pediatricians attributed the origin of infant eating disorders to hospitals' rigid schedules for the feeding of newborns. Child psychiatrists such as Therese Benedek partly attributed emotional disorders of children to disruptions of postpartum maternal-infant bonding. Yale pediatrician Arnold Gesell and his colleague Frances Ilg suggested the concept of rooming-in (and coined the term) in their book, *Infant and Child in the Culture of Today* (1943). But Jackson set out to put the idea into practice. She saw it as "a measure for wholesome family integration and avoidance of anxieties" that would combine the security of the hospital setting with the intimacy of the home. The project was staffed on a regular basis by Jackson, her medical fellows, a nurse-supervisor, and rotating groups of nursing students. Many of the staff jokingly referred to the project as the "RIP," and for many mothers, to "rest in peace" was exactly what they wanted.[81]

The unit's success was hardly guaranteed. Dr. Jackson, the nursing students, the medical fellows, and, most of all, the patients were subjected to not-so-gentle resistance from the other staff and met with deep reluctance on the part of the hospital administration. Nevertheless, by developing an early form of "primary care" nursing (assigning one nurse to each mother-child unit), Jackson and Kate Hyder, a newly hired assistant professor of obstetrical and gynecological nursing, kept the unit running smoothly. Jackson also received timely support from Dr. Herbert Thoms, the obstetrician-gynecologist-in-chief, from the chair of obstetrics and gynecology, and from the dean of the nursing school, Elizabeth Bixler. The new hospital director, Dr. Albert W. Snoke, first thought the scheme "as crazy an idea as I've ever heard," but he soon became enthusiastic. Perhaps most influential, in supporting rooming-in, however, was Dr. Grover Powers, now chair of pediatrics.[82]

Jackson, like Eliot and Dunham, owed part of her success to her ability to create respectful and effective teams that crossed traditional professional lines of authority, especially in working with nurses as colleagues. But Jackson was also dependent on support from influential colleagues to back her innovations. Jackson was promoted to clinical professor of pediatrics and psychiatry in 1949, yet four years later, in 1953, she retired from Yale at the age of fifty-eight. The previous year the Rooming-In Project had been closed down without provision for further financing either from the hospital or from outside foundations. Kate Hyder no longer worked there, and Jackson's old ally, Dr. Powers, had retired. When Jackson could not bring about continuation of the project by herself, she too retired—but only temporarily. She returned to Colorado and became visiting professor of pediatrics at the University of Colorado School of Medicine, directing a rooming-in unit at Colorado General Hospital in Denver from 1962 to 1970. Jackson died in 1977, at age eighty-two.[83]

Rooming-in was ahead of its time in its patient-centered approach to maternal-neonatal care. Few nurses and administrators supported the organizational and professional realignment required to implement rooming-in units on hospital obstetrical wards. Many physicians feared that an inherently nonsterile environment would subject vulnerable newborns to high infection rates; some resisted the parent-centered approach to childbirth. Academic physicians were more committed to studying pathological conditions than the apparent physiological and emotional "normality" associated with natural childbirth and rooming-in. Finally, obstetricians and pediatricians may have distrusted the tone of intimate feminine warmth that informed the rooming-in concept. Silberman, for example, cites the taunt of some Detroit physicians who labeled their pro–rooming-in colleagues the "big bosom boys."[84]

Helen Taussig, best known as the co-developer of the Blalock-Taussig shunt (commonly called the "blue-baby" operation), also began her career during the 1920s. The daughter of a Harvard professor of economics, she attended Radcliffe for two years but then completed her college education at Berkeley "to get out of the shadow" of her father's reputation. When she told him she wanted to study medicine he suggested public health as an alternative. The School of Public Health at Harvard, however, like the Harvard Medical School, refused to award women a degree. Instead Taussig studied anatomy with Alexander Begg at Boston University, focusing her work on a study of the heart muscle.

Begg suggested she apply to medical school at Johns Hopkins, where she was accepted as one of ten women in a class of seventy, and she graduated in 1923. As a first-year medical student, Taussig was assigned to assist with electrocardiograms at the hospital's heart station, something she continued for her entire four years. The medical service at Hopkins in those days never accepted more than one woman intern. Upon her graduation, Taussig was the second choice for the Hopkin's position, so instead she was offered a fellowship in cardiology based on her four years' work at the heart station. (She titled her memoir of these early years, "Little Choice and a Stimulating Environment.")[85]

Luckily, Edwards Park became chair of pediatrics at Hopkins during her fellowship and developed a series of new subspecialty divisions, including one in pediatric cardiology. Park invited Taussig to take an internship in pediatrics for a year and a half and then appointed her an assistant in pediatrics. He also named her physician in charge of the new pediatric cardiology division. Taussig's decision to study one subgroup of her cardiology patients, those bearing the symptoms of tetralogy of Fallot, was partly the result of Park's persistence. These infants and children were known as "blue babies" because of their cyanotic tinge, caused by a congenital heart malformation that resulted in an inadequate supply of blood moving to the lungs for reoxygenation. Such patients at the time were considered untreatable. Moreover, the symptomatology of the condition was extremely murky. Park was ambitious for the department to become a center for clinical pediatrics research. Except for work being done by Maude Abbott at McGill University in Montreal, nothing was currently known about congenital heart disease in children. He urged Taussig to specialize in an area that was entirely new, one in which contributions to both pathophysiology and therapeutics might result. Taussig, uncertain if such intense specialization would suit her, asked her father for advice during a visit home to Cambridge. Reportedly he told her, "'Helen, for a woman, recognition will come only through specialization.'"[86]

Soon Taussig's clinic became a mecca for chronically oxygen-deprived children and their families. Taussig, who was hearing impaired, developed powerful observational skills and a keen sense of touch.[87] By correlating her clinical observations with fluoroscopy studies of her young patients' hearts, she slowly discerned meaningful patterns in their symptomatology. She was also able to dissect many of those hearts after the patients died, thus confirming her observations. In 1938 she sought out Maude Abbott's help in classifying the many anomalous heart structures she had examined. By the early 1940s, she realized the anomaly that caused blue-baby syndrome might be corrected with an artificial channel, or shunt, designed to bring more blood into the lungs—an entirely untried surgical procedure. For this she enlisted the help of Hopkins cardiovascular researcher and surgeon, and the chair of the department of surgery, Alfred Blalock. He and his associate, the African American experimentalist and laboratory technician Vivian Thomas, took up Taussig's experimental challenge in the fall of 1943. Blalock and two surgical residents, with Taussig observing and Thomas helping guide the procedure, operated—successfully—on the first blue baby about a year later. Taussig and her cardiology fellow Dr. Ruth Whittemore monitored the baby girl (not quite a year old) continuously for days until it was clear she would survive.[88]

Patients flocked to Taussig's clinic. Within a year the operation had been performed fifty-five times. Taussig and Blalock became medical celebrities. Her major concern, apart from ensuring that her patients received good care, was to encourage the founding of other centers around the country for the training of pediatric cardiologists. Of her own fellows, at least two out of seven were women. In 1947 Taussig was elected to the APS, received the Mead Johnson Award, and published a textbook, *Congenital Malformations of the Heart*. She became a professor of pediatrics at Johns Hopkins in 1959, four years before her retirement.

Taussig enthusiastically fought to educate the public about issues affecting scientific progress and public safety. She once demonstrated her support for responsible animal-based biomedical research, such as Vivian Thomas's initial use of dog models to design the heart shunt, by appearing at a Baltimore City Council meeting with many former patients, some accompanied by their own pet dogs. Along with Dr. Frances Kelsey of the U.S. Food and Drug Administration, she successfully campaigned to keep thalidomide off the market in the United States in 1962. She also testified widely on behalf of Sherry Finkbine, a pregnant woman who had taken thalidomide early in her pregnancy and was seeking a legal abortion. According to Whittemore and others, Taussig was unusual in her refusal to distance herself from her patients and, reportedly, was like an aunt to many. She herself never married.[89]

Virginia Apgar, an anesthesiologist best known for devising the Apgar score as a means to assess neonatal viability, is another example of a woman physician who rose to prominence by combining academic specialization and dedication to maternal and child health. Her career, too, demonstrated that women who occupied a stereotypically "feminine" niche in medicine did not necessarily conform to stereotypical feminine behavior. (As an undergraduate at Mount Holyoke, Apgar supported herself in part by "catching stray cats" for the comparative anatomy labs.) Of her decision not to marry, she explained, "It's just that I haven't found a man who can cook."[90] In 1929 she became one of the few women admitted to the College of Physicians and Surgeons at Columbia University. After graduating in 1933 she gained a prestigious and— for a woman—rare opportunity to intern at Columbia-Presbyterian Medical Center in surgery. But after participating in hundreds of operations, she decided she could not successfully support herself as a surgeon. Despite her disappointment, she turned to anesthesiology, a field in which she felt she might do pioneering work. She began her training under the nurse-anesthetists at Columbia and then studied with two pioneering anesthesiologists, Dr. Ralph M. Waters at the University of Wisconsin and Dr. Emery A. Rovenstine, Waters's protégé, at Bellevue. Five years after she received her M.D., Columbia-Presbyterian invited her to become the director of the division of anesthesiology. She was the first woman to head a division at the hospital. In 1949, after a decade during which she transformed the division into an academic medical department for the training of residents and medical students, she was promoted to full professor of anesthesiology, a first for a woman at Columbia and an indication of the new importance of the field. (Anesthesiology's certification board had been incorporated just the previous year.) In 1952 she capped years of clinical observation of newborns by creating the Apgar score. Her rating for neonatal vitality gave two points each for "respiratory effort, reflex irritability, muscle tone, heart rate, and color." It is still considered the "best clinical tool" to evaluate neonates.[91]

In her late forties Apgar decided to follow her deepening interest in maternal and child health. After receiving a master's degree in public health from Johns Hopkins in 1959 (she was nearly fifty), she changed course and spent the rest of her life at the National Foundation–March of Dimes, at Cornell University Medical College in New York, and at Johns Hopkins, specializing in teratology.

⌒ The failure to renew the Sheppard-Towner Act in 1927 seemed to complete the dismantling of an explicitly gendered niche for American women in medicine. To be sure, the decline of gender specificity in medical institutions was not in itself a bad thing; generations of women physicians explicitly looked forward to the time when, in Bertha Van Hoosen's words, there would be "no sex in medicine." Yet the majority of women in practice were nearly a generation behind their male colleagues in preparing for careers in academic medicine and in the emergent office-based specialties.[92] All practitioners, male and female, felt the effects of the new professional hierarchies. But women faced other, less visible barriers to advancement.

Nevertheless, during the interwar decades a vanguard of women physicians began to infiltrate the elite bastions of the profession—academic careers focused on research and specialized clinical care. The careers of these few, favored women physicians demonstrate both the opportunities for and the limits placed on the best trained medical women of the period. This sturdy vanguard managed to reconcile an interest in maternal-child health or public health, the traditional interests of women doctors, with recognition as researchers and academic physicians. But their careers predicted the future rather than typifying their own day.

7

Resisting the "Feminine Mystique," 1938–1968

In the past sixty years we have come full circle and the American housewife is once again trapped in a squirrel cage. If the cage is now a modern plate-glass-and-broadloom ranch house or a convenient modern apartment, the situation is no less painful than when her grandmother sat over an embroidery hoop in her gilt-and-plush parlor and muttered angrily about women's rights.

Redbook, *1960*

THE OUTBREAK of World War II and the continued exclusion of women doctors from the military reserves revitalized the activism and organizational élan of American women physicians. The postwar era, however, confronted them with a more insidious challenge, an enervating entreaty to return to home and family. Postwar insinuations that women physicians were *under*productive and unreliable, moreover, added a jarring note to such siren calls. Women physicians were no more immune to the determined domesticity of the fifties than other women professionals, but in the American Medical Women's Association they at least had the advantage of an organization with long experience in the struggle for equal opportunity.[1] Even during the baby boom, with its collateral encouragement of "momism," AMWA attempted to reinvigorate the movement of women into medicine. As women physicians confronted the major dilemma of all women professionals in twentieth-century America—the demand to be productive professionals *and* "well-adjusted" wives and mothers—they hesitantly began the journey back to political awareness and activism.

Women Physicians and World War II

Two decades after they had failed in a similar attempt, the Medical Women's National Association, which reincorporated as the American Medical Women's Association in 1937, renewed its efforts to win commissions for women in the Army Medical Reserves. At the same time, AMWA tried to address another pressing obligation, to assist women colleagues trying to escape from Europe. Of the approximately six thousand physicians who fled Germany for the United States, an estimated 12 percent were women.[2] But AMWA's leadership was far more interested—and effective—in working for military commissions. Along with Dr. Minnie Maffett, President of the National Federation of Business and Professional Women's Clubs in 1941 and a tenacious fighter for women's rights, AMWA conducted a well-organized, skillfully publicized campaign.[3] In 1939 two AMWA representatives interviewed the surgeon general of the Army and learned that a change in the laws pertaining to military service probably would be necessary before women could be commissioned. But Secretary of War Harry Woodring refused to meet with them. Emily Dunning Barringer, a veteran of the American Women's Hospitals Executive Board in 1917, an AMA delegate from the Medical Society of New York State, and president elect of AMWA for 1939–40, made the question of commissions a top priority. In 1940 AMWA petitioned the AMA for support in changing the law regarding the Medical Reserves. During a chance encounter with *JAMA* editor Morris Fishbein, Barringer asked him why the AMA held a different position toward nurses—given military rank since World War I—and women physicians. Fishbein replied that, "'Nurses are well supervised.'" Barringer blandly asked him to put his comment in writing, but he prudently declined.[4]

Undeterred, representatives of AMWA, as well as Dean Martha Tracy of the Woman's Medical College of Pennsylvania, cabled President Roosevelt to volunteer the services of women physicians and faculty, sending a copy to Surgeon General James C. Magee. He replied with thanks and the tepid assurance that the army likely would need only "a small number of women physicians as civilians on a contract basis."[5] Nevertheless, AMWA kept up its effort to gain AMA and War Department backing.

On the morning of December 7, 1941, AMWA's board decided to

appeal directly for public support. Barringer fretted that it would be hard to gain support in a period of uncertainty about American participation in the war, but the afternoon's news of the attack on Pearl Harbor erased all doubt. AMWA telegraphed an offer of its support to the White House that day. The board also agreed to hire noted New York judge and lobbyist Dorothy Kenyon, one of the few women members of the Association of the Bar of the City of New York and a self-described defender of the underdog, to assist the association in Washington.[6] Throughout the winter and spring of 1942, Kenyon and Barringer were in close communication. Kenyon lobbied in Washington while Barringer gathered support from a wide variety of women's business and professional groups. Barringer also wrote an open letter to the *New York State Medical Journal*, asking physicians for their support. (One reply, from a retired military physician, revived an old slur from World War I. He asked whether women physicians were aware that "a considerable part of the duties of a medical officer in the services consists in the inspection of the men for venereal disease . . . Are they ambitious to assume this duty?") Still, public opinion was moving in the women's direction. When, in June, the AMA once again rejected AMWA's petition for support, the *New York Herald Tribune* issued an editorial supporting AMWA. As Kenyon observed, "We have at last got our case out into the Court of Public Opinion." But the verdict was not easy to predict. In one hotly contested incident, for example, Dr. Alice McNeal, a Chicago anesthesiologist considered an integral part of an operating room team that was called up for U.S. Base Hospital No. 13, was requested by her medical chief to mobilize along with the rest of the group. But the army refused to commission her.[7]

In addressing the AMWA Inaugural Banquet in June 1942, Kenyon unveiled a two-pronged argument that would be the basis of her approach. First she drew on the basic, democratic justice of the women's claim: physicians should be used based on their qualifications, not their sex. "A male obstetrician for instance had better stay at home and bring babies into the world rather than take care of wounded soldiers at the front while a woman surgeon had better do just the opposite." Second she argued on legal grounds: "Army regulations state that persons, competently trained and qualified are eligible to the Medical Reserve Corps." "Simple and unambiguous" words such as "person" are legally understood according to their "plain and natural meaning," she argued, and must therefore be assumed to refer to both men and women.[8]

Within six months AMWA began to see results. More than by rational argument, the military was persuaded by the pressing shortage of doctors. In the words of James Burrow, "Military medical care often stood perilously close to the crisis stage." Between 1940 and 1945 new draftees swelled the army's ranks from 267,000 to 8,266,000, a more than thirtyfold increase. Physician supply could not keep up with this gigantic demand. Even the American Legion, alarmed at the possibility that American soldiers might not receive adequate care, became receptive to admitting women into the Medical Reserves.[9] At the end of 1942, the surgeons general of the army and navy withdrew their objections to female physicians serving in the Medical Reserves, although the new secretary of war, Henry Stimson, continued to oppose the idea.

In December, Congressman Emmanuel Celler of New York introduced legislation to permit commissioning women in the military. His arguments closely paralleled Kenyon's. Hearings were held before the House Committee on Military Affairs in March 1943. Barringer, Kenyon, representatives of the government's Procurement and Assignment Service for Physicians, and members of the military all testified in favor of commissions for medical women. Congressman Celler, borrowing from Kenyon's testimony, explained, "I think women doctors have reached a situation where they should not be judged by sex; they should be judged by accomplishments and skill." Besides, he remarked (undercutting the principle in his own argument), the military was beginning to "scrape the bottom of the barrel" to find sufficient physicians to meet its needs.[10]

At the request of the American Legion, John Sparkman, congressman from Alabama, introduced a narrower bill specifically aimed at commissioning women physicians for the reserves. The bill, H.R. 1857, cosponsored by Edwin C. Johnson in the Senate, was approved unanimously by the House. After its passage in the Senate, President Roosevelt signed it into law on April 16, 1943. Four days later, Dr. Margaret D. Craighill, formerly dean of WMCP, was given the rank of major, the first woman physician to be commissioned into the Army Medical Reserve Corps. She was stationed in the surgeon general's office. Dr. Achsa Bean, commissioned as a lieutenant in the Naval Reserves, was stationed at the Naval Training School located at Hunter College in New York. Ultimately, 119 women received commissions in the army, navy, and Public Health Service.[11]

"Manpower" and the Woman Physician

During the war, many medical schools faced a shortage of male students and increased their admission of females accordingly. Women's graduation rates for the classes of 1948 to 1951 were the highest they had ever been or would be again until the 1970s, peaking at 12.1 percent in 1949. By 1950 women comprised 6.1 percent of American physicians. AMWA's campaign for military commissions substantially heightened its professional visibility, and membership doubled between 1933 and 1943. In 1945 AMWA tried to capitalize on its increasing recognition by establishing its own monthly journal, the *Journal of the American Medical Women's Association (JAMWA)*.[12]

Despite these gains, the culture of postwar America—America of the 1950s—was deeply ambivalent toward women in medicine and other professions. Murmurs that women professionals had been used during the war only as a last resort demonstrated the stubborn persistence of old prejudices. At a Washington, D.C., conference in 1942 titled War Demands for Trained Personnel, Dr. R. H. Spencer, medical director of the National Institute of Health, had spoken on the "Future of Women in Medicine." He had predicted an "enlarged" need for women physicians in the postwar period to handle an increased demand and a new emphasis on preventive medicine. But in his view they would be employed as

> general practitioners, as specialists in the diseases of women and children, as hygienists in schools, as x-ray technicians in treatment and diagnosis, as medical technologists in . . . general hospitals . . ., as surgeons, especially, though not exclusively in hospitals for women and children, as consultants in industrial hygiene, as teachers in medical schools, and as independent research workers in nutrition, the control of infectious diseases, sanitation, venereal disease control, psychiatric services, and numerous other public health activities.[13]

Government planners, in short, envisioned medical women occupying their traditional "feminine" niche. Indeed, women physicians were given a highly contradictory message. On the one hand, they were urged to contribute to the fullest in the face of a feared shortage of physicians. During the Korean War from 1950 to 1953 and after the launch of the first Soviet Sputnik spacecraft in 1957, women were encouraged to train for jobs in science, engineering, and medicine. The Office of Defense Mobilization's Defense Manpower Policy No. 8, published in 1952, stressed the need for more scientists and engineers and, democratically, urged that no distinctions be made on the basis of sex or race. In 1954 the Rockefeller Foundation produced a report, *America's Resources of Specialized Talent: A Current Reappraisal and a Look Ahead*, directed by Dael Wolfle. In one of the first national inventories of the available professional workforce, the Wolfle report stressed the need to make better use of women professionals. On the other hand, some commentators insisted that women doctors were not as productive as their male colleagues nor, given the demands of family life, would they become more productive any time soon. In this vein, a report titled *Womanpower* (1957), cosponsored by the Ford Foundation and Columbia University, emphasized women's future role as teachers and nurses rather than as scientists and doctors.[14]

These contradictory expectations intensified two years later, when the surgeon general published a report on "health manpower." The 1959 report predicted a severe physician shortfall and warned that only by increasing the supply of new medical graduates by an average of 3,600 per year over the next fifteen years could a health care crisis be avoided.[15] The surgeon-general's report on the medical workforce sent shock waves throughout the health care system. Calamitous pressures on the health care system could be avoided only if, in the words of the report, "present medical school facilities [were] increased substantially and new schools [were] established . . . at once."[16] The discussion of physician productivity and the threat of a scarcity of doctors underscored a new conceptualization of medical work as a commodity and physicians as its producers. Not only must more physicians be trained but also those already in practice must work to their fullest.[17] The emphasis on physician productivity—which emerged within this context of projected scarcity—had a paradoxical effect on the careers of women doctors.[18]

The idea of measuring "physician productivity" is a twentieth-century application of the Victorian obsession with efficiency and the consequent determination to measure the productivity of human labor.[19] By the turn of the century, the concept of labor productivity was being applied to many sectors of the workforce, including such white-collar work as university science and engineering.[20] Hospitals, too, were subjected to productivity assessments. Through the efforts of efficiency engineer Frederick Taylor, such assessments were even adapted to hospital nursing.[21] Hospital administrators spoke of their institutions in the new industrial idiom. As one editorial in *American Medicine* put it, "What the 'foot pound' is to dynamics, what the 'volt' is to electricity, such is the 'bed-day' to the hospital. It is the unit of work."[22]

Nevertheless, the work of physicians continued untouched by such measures as input, output, and time-and-motion studies until the 1940s. Occasional efforts by prominent physicians, such as surgeons E. A. Codman of Boston and gynecologist Robert L. Dickinson of New York, to measure "end results" and standardize surgical procedures made little impression on their colleagues.[23] Medicine's claims to professional expertise, ethics, and altruism, coupled with reminders of the uniqueness of each doctor-patient encounter, effectively delayed attempts to increase medical efficiency by standardizing the clinical day. As one surgeon claimed in 1917, "We all know you cannot standardize an art."[24] Well into the twentieth century, physicians were still assessed according to their character, the general quality of their relationships with patients, and, to some extent, the efficacy of their interventions. With the exception of wartime hospital duty, their work was not viewed as "output" to be judged according to quantitative measures of productivity.[25]

The profession was not entirely unconcerned about productivity, particularly in light of persistently low physician earnings. But both phenomena were understood as functions of oversupply and underutilization rather than the productivity of American doctors themselves. When William Pepper, M.D., provost of the University of Pennsylvania, addressed the issue of reform of medical education in 1877, for example, he cited figures to show that the ratio of physicians to the population as a whole in America was 1 to 618, compared with 1 to 1,193 in Canada and 1 to 1,672 in Great Britain.[26] As Gerald Markowitz and David Rosner have observed, creating fewer but better (and more affluent) physicians was a major goal of the profession at the turn of the twentieth century. The reduction in the number of medical schools from a high of 166 in the 1890s to a low of 85 in the 1940s, as well as the reduction in class size at the remaining schools, was designed to foster this goal.[27] In 1932, the final report of the Association of American Medical College's Commission on Medical Education, which began work in 1925, concluded that the number of physicians in America was still increasing at a rate faster than that of the general population. Thus, prior to World War II, the medical profession was concerned not with underproductivity but with oversupply.[28]

Several factors, however, transformed the perception of surplus into predictions of disastrous physician shortages. First an overall increase in the population was projected for post–World War II America. Second, the gradual aging of the population, accompanied by the replacement of acute infections with subacute, chronic illnesses, contributed to a heavier demand for physician services. Finally, a boom in hospital construction, fueled by passage of the Hill-Burton Hospital Survey and Construction Act (1946), quickly expanded the need for medical graduates to fill internships and residencies. By 1957, medical schools were graduating just under 7,000 physicians a year, yet hospitals needed to fill 12,000 internships annually.[29] Both the NIH and the Bureau of Health Professions sought to raise the number of academic medical centers, the one to expand medical research, the other to increase the number of practitioners.[30]

The postwar emphasis on health care "manpower" focused on increasing both the number of new physicians and the productivity of those already in practice—especially women. Calculations of physician "productivity" thus became necessary for accurate forecasts of the need for health care personnel, including doctors. Studies were conducted to assess and compare the productivity of physicians according to specialty, sex, age, and other variables.[31] Health care analysts turned their attention to physician practice patterns rather than the public's use and payment patterns.

As questions of physician supply and demand were linked to measures of productivity, the idea of *under*productivity also began to crystallize. And from the beginning, underproductivity was associated with women physicians. Given the cultural anxieties during the 1950s over women's dual role as homemaker and professional, women physicians simply were not expected to assimilate into the new workforce culture of managed efficiency and commodified professionalism. For example, an early AAMC study, "Survey of Women Physicians Graduating from Medical School, 1925–1940," initiated in 1953 and published in 1957 by Roscoe A. Dykman and John M. Stalnaker, was specifically intended to size up the potentially negative impact on productivity of the slight rise in the number of women medical graduates after World War II.[32]

At the time of the study, 87.5 percent of the female respondents were active in full- or part-time practice. Only 0.4 percent had never practiced since graduation. The study found that 90 percent of men and 68 percent of women physicians worked at least three-quarters of the time, where full-time was defined as a forty-hour week. Dykman and Stalnaker found that 33 percent of the female respondents had been out of practice sometime in their career, on average for 4.5 years. Of this group, 71.5 percent were married and most had "terminated" their practice for a few years after having children, until their children began school. Ten percent of male physicians also had stopped practicing, for an average of 2.1 years, primarily because of physical disability. In the authors' opinion, "The most conspicuous difference between women and men [was] the number of women not in practice." One byproduct of this observation was the claim that medical women were "poor investments" for the profession—what Mary Walsh referred to as "the stereotype of the dilettante woman physician." As early as the 1920s, medical educators had argued against admission of more than a few women to the profession, on the grounds that marriage and childbearing would force them to drop out. Although Dykman and Stalnaker actually concluded that "women physicians make a substantial contribution to the nation's health," their findings were widely used to justify low expectations for women in medicine. In the context of the perceived future shortfall of physicians in practice, the finding that disproportionate numbers of women worked part-time had important implications for assessing their future place in the profession.[33]

Profession vs. Family

The impact of the medical "manpower" scare in the early 1960s was of great significance for the future of women in the medical profession. It provoked a debate over women's minority status and contributed to a much-needed repoliticization of AMWA's leadership. But it also cast the question of women physicians' productivity as a challenge to their claim that marriage and career could be successfully combined. AMWA was first alerted to the problem when Medical Education Committee chair Mary K. Helz attended the National Health Council Meeting on Manpower Shortages in the Field of Health in October 1959. AMWA delegates also attended the AAMC annual meeting in November on the theme of "Physicians for a Growing America." The delegates were given information packets that included "startling predictions of the future," especially indications of future health care shortages.[34] Between 1950 and 1960, women as a percentage of practicing physicians had remained almost unchanged at about 6 percent. AMWA could not ignore the need to attract greater numbers of young women into medical school and, once they were there, to address the question of how to balance career and family.[35]

In the conservative political climate of the 1950s, however, women physicians had few conceptual tools with which to address fundamental concerns of prospective students. As Rossiter has elaborated for women scientists, the 1950s were years of unrelenting enthusiasm for the "traditional" household in which Father worked for money and Mother worked (at home) for the emotional well-being of her family. Despite the fact that wives and mothers were working (outside the home) in greater numbers than ever before, an array of social arbiters—from the *Ladies' Home Journal* to the AMA's *Hygiae* to the ubiquitous pediatrician Dr. Benjamin Spock—all solemnly urged mothers to stay home and raise their children.[36] AMWA's agenda during the 1950s generally had reflected the private, apolitical values of the period. The association's 1957 annual meeting in Dallas, for example, featured a keynote address titled "Woman's Greatest Enemy—Fatigue" and workshops on "The Emotional Health of the Family." Subtopics included education for marriage, for homemaking, and for parenthood. An evening session was devoted to "Today's Teenagers—Tomorrow's Homemakers."[37] Not for another decade would AMWA's leadership understand the real obstacles to equal opportunity for women: the linked assumptions that women ought to be the primary caretakers for home and family and that they would be less productive in their profession.

Clinging to its conservative individualism from the late 1920s and 1930s, AMWA continued to oppose government-subsidized medical insurance for Social Security recipients, a measure it identified with "ultra-liberals" and "healthcare handouts."[38] Locked into a conception of professionalism that equated disinterested "objectivity" and scientific legitimacy, AMWA's leaders were constrained from pleading a "special case" for women based on unequal opportunity. Having never learned, or else forgotten, the tools to analyze gender discourse in Western society, they could only wonder why so few women pursued their theoretically "equal" opportunity for a career in medicine. Some idea of prevailing attitudes can be gleaned from the generally progressive 1954 Wolfle report, *America's Resources*, which stated that "the professionally ambitious woman is doubly handicapped in the attainment of her goals, handicapped by the prejudice and competition of men *and* by the lesser professional ambitions of most women and the employment policies which take into account that lesser ambition" (emphasis added). In the face of a widespread assumption—internalized by not a few women physicians themselves—that most professional women would sacrifice their career for the sake of their family, women physicians frequently insisted that for those who did wish to advance in their profession, discrimination was not much of a problem.[39] Consider the recollections of Dr. Bertha Offenbach, president of AMWA's New England Branch 39 in 1962. Offenbach recalled her early efforts in the 1950s to drum up interest in AMWA. She was told consistently by women colleagues on the staff of Massachusetts General Hospital, "'I have *no* interest in joining . . . a women's medical association—there is no sex in medicine.'"[40]

Widespread predictions of a physician shortage and the concomitant debate over physician productivity provoked AMWA to rethink its assumptions about gender discrimination in the medical profession. The issue influenced its change of heart for three main reasons. First, it alerted AMWA leaders to the problems of married women medical students and physicians, problems that frequently interfered with their ability to complete their training, to practice their profession to the fullest, or even to practice at all. Second, the identification of social rather than strictly individual deterrents to women's success as physicians slowly moved AMWA's policymakers from a position of conservative individualism to an acceptance of the need for underlying structural changes in the organization of medical education and health care. Third, the fear of a physician shortage provided the impetus to increase enrollments at existing medical colleges and to construct forty-one new medical schools between 1960 and 1980. As the surgeon general's report *Physicians for a Growing America* had made clear in 1959, increasing the supply of physicians required both increasing and *"equalizing opportunity"* through expanded facilities and more abundant scholarship money.[41]

For the first time, national policymakers were countering organized medicine's longstanding goal of keeping the number of physicians as low as possible. This somewhat more welcoming climate proved to be a powerful magnet for prospective applicants and a significant factor in the increasing numbers of women who entered medicine during the 1960s. Twenty-two new schools enrolled their first classes between 1961 and 1971, the decade during which the number of women students finally began to climb. Between 1960 and 1970, the combined "push" of the women's movement and the "pull" of increased enrollment opportunities resulted in a tripling of the number of women applicants at a time when the number of male applicants increased only twofold. The number of women medical students rose from an enrollment of 1,745 in 1961–62 to 4,733 in 1971–72. As a percentage of all medical students, women increased from 5.8 to 10.86 percent, nearly doubling their percentage in ten years. Even more startling, AAMC figures show that in schools newly opened between 1970 and 1980, women filled 65 percent of the first-year slots.[42]

Expectations for women physicians changed more slowly than their medical school admission rates, however. Alongside calls for a greater role for women, criticism of their lower practice rates, based on the Dykman and Stalnaker study, persisted. A 1961 editorial in *Medical Economics* titled "The Case against the Female M.D." began, "I get sick and tired of ridiculous statements about helping solve the alleged physician shortage by having more women physicians . . . Why not solve the problem at hand more efficiently by having more *men* physicians?" Referring to the Dykman and Stalnaker study, the author asserted that "women gave much less time to practice than the men."[43] As late as 1968, at a Josiah Macy, Jr. Foundation conference titled "The Future of Women in Medicine," many speakers reiterated the claim that women "do not devote as much time to medical activities after graduation as men." One, a vice president of New York Hospital, cited a study claiming that "two female graduates between them will devote as much time to medicine as a single male graduate," an unsupported assertion. As Carol Lopate's research on women physicians during the 1950s makes clear, residual prejudice against women's capabilities clung to all their efforts to achieve professional equality. One of her respondents described her tactics for coping with closed-minded male colleagues: "'If you're feminine and let them think *they're* making the decisions, you'll get along very well. Men don't like aggressive, crusading women.'"[44]

This double bind, the conflicting mandate to be successful but also "feminine," was most extreme for the few women entering the field of surgery. A martial, "iron surgeon" ethos pervaded this specialty for much of its history; even today its professional culture seems at odds with traditional feminine norms.[45] An interview with Dr. Sally Abston of the surgery department at the University of Texas Medical Branch (UTMB) Galveston, made this abundantly clear. Abston was born in Refugio, Texas, in 1934 and graduated from Rice University with a degree in English. Her years of medical training extended from 1956, when she entered medical school at UTMB, to 1967, when she completed her residency. (In those days, majoring in English put her at a disadvantage in her preclinical courses; she dropped out of medical school at the end of her first year to go back and take additional college science courses.) Today the Annie Laurie Howard Distinguished Professor in Burn Surgery, she has been director of the emergency room, director of the burn unit, and medical director of John Sealy Hospital. She is known for her gruff intensity, her unyielding commitment to her patients, her loyalty to residents and junior colleagues, her demanding perfectionism with students and residents, and her ability to curse like a marine drill sergeant. According to Abston, to succeed in surgery during the fifties and sixties, she had to be "one of the boys." Surgery's macho culture decorated the metaphors that studded her recollections. The residency was "sort of like being in the Marine Corps, and we thought of ourselves as the best doctors in the house, the hardest working . . . we were the 'meanest man on the mountain.'" Adopting a masculine affect was not optional: the chairman of surgery, with whom she feuded intermittently, was known for his "dislike and distrust of women professionally." Although Abston tried for two years to get into the orthopedic surgery program, a notoriously difficult subspecialty for women to enter, she finally contented herself with training as a general surgeon. Dr. Abston recalled that it was little or no strain for her to fit in with "the boys"—her lifelong love of sports made for easy camaraderie. "Being one of the gang meant eating at their table, doing their operations, talking their professional talk . . . Cursing may have been part of that, I don't know. I think I probably cursed more than any of my male colleagues." But in Abston's view, she was also held to a double standard. She believes that nurses, medical students, and residents resented her obscenities far more than they would have in a man. On the other hand, when she, an unmarried woman whose lifelong partner was another female physician, adopted two children, her relations with the nurses quickly became "far more supportive." After a while, Abston even brought her children in to the hospital when she was on call at night. "A sleeping kid wasn't any problem."[46]

But not many women had the fortitude to bring a child onto the hospital floor—and certainly not as a member of the junior faculty in surgery in 1969. Those who attempted to pursue an academic career while maintaining a marriage and raising children faced formidable obstacles and not a little prejudice. Dr. Ruth Lawrence, professor of pediatrics at the University of Rochester School of Medicine and Dentistry, a national authority on lactation and breastfeeding, and a participant in some of the early Macy Foundation surveys of women physicians, described in an interview how she coped with the multiple demands of an academic career, marriage to another physician, and motherhood.[47] Dr. Lawrence entered medical school at the University of Rochester in 1945 as one of thirteen women in a class of seventy. Of the men in the class, only twelve were civilians; the rest, including her future husband, had been sent to medical school by the military.

As an undergraduate at Antioch College, Lawrence had worked as a research intern for the president of Ciba Pharmaceuticals, and she benefited from his strong support of her desire to become a doctor. Using animal models for research, she learned routine surgical procedures that she would need during her years as a medical student. The idea of prejudice against women in medicine "did not occur" to her. Nor does it seem to have been much of a problem—until the question of internships and residencies arose. By 1948, when she began to think about it seriously, it was clear she might have a problem when she finished medical school: not one of the six women graduates from the class preceding hers had gotten an internship. Indeed, she knew of some— especially in surgical residencies—who failed to be renewed in their programs in part because large numbers of male physicians who had entered medical school during the war were given priority, even over women already in the program. Complicating matters for her, she and her fiancé wanted to intern in the same city. In the end, he was accepted in surgery at Rochester (he later became an anesthesiologist), and she was accepted by Dr. Grover Powers for a position in pediatrics at Yale—one of the premier pediatrics positions in the country. As Lawrence commented, Yale in the late 1940s was the site of Dr. Edith Jackson's Rooming-In Project and they were doing "key clinical work in the field of lactation, which of course ended up being an important part of my own work." Indeed, Yale was "breast-feeding Mecca." Lawrence and her fiancé married a month after graduating. Two months into her internship, she discovered she was pregnant. Her husband had been recalled into the army and was in Korea when their son was born. She and the baby lived across the street from the hospital, and for a few months she breastfed him. During the day, the baby stayed with another new mother, a nurse, and her baby. During her on-call nights, he "got plunked on my hip and I went back to the hospital." Dr. Lawrence weaned her first child after a few months ("I was really blazing trails at that point"), but she breastfed all the rest of her nine children for two years each. Following her residency, she and her husband began their long careers in Rochester.

How did the responsibility of childbearing and child rearing affect Dr. Lawrence's career? For one, it made her become extremely well organized and intolerant of wasted time. Here is her description of the years when she worked three part-time jobs at the medical center (without revealing her part-time status to most of her colleagues), while also handling the majority of the child rearing duties at home: "I just got organized at home; had everything set out. When I came [to work] I had my lectures lined up, rounds lined up, and I just kept right at it. Then I just packed up my memos, my telephone calls, my mail, my stuff-for-tomorrow kind of thing, and I worked [at home] . . . And you lined everything up at night so morning goes as smooth as humanly possible. And, you get places on time and you do your job."

As Lawrence's recollections suggest, having the primary responsibility for child rearing added a significant hurdle to the course of women's academic advancement. Not until 1960, after eight years at the medical school, did Dr. Lawrence become board-certified in pediatrics. She was then given a full-time position at the medical school—chief of the pediatric service at Highland Hospital, one of the hospitals affiliated with the main teaching hospital, Strong Memorial. Simultaneously, she developed the neonatal intensive care unit at Strong, and was director of both the Poison Control Center and the blood bank. Her rank? Senior instructor. She attained tenure only after another decade. It took even longer for her to eke out the time and lab space for research. Today few junior faculty would consider this "mosaic" a reasonable career load. Dr. Lawrence herself urges junior faculty to get research training during a fellowship year before seeking a full-time academic post. "If you haven't got your foot on a [lab] bench, you'll never get it." What took her so long to begin her research? "In fact there was just no discussion about it. No one ever told you how to do it, or what to do, or when to do it." Just as important, however, was her perceived status as a "captive wife": "They knew I wasn't going anywhere. They knew I'd be here tomorrow. They didn't have to worry. They didn't have to make it appealing to me at all."

Perhaps the most intractable obstacle for women physicians in these decades was their own deeply rooted ambivalence about their "right" to a career at all. Dr. Lawrence commented on what prevented her and others from proceeding directly through the normal stages of career development—indeed, from even taking her boards for eight years after her residency: "Well . . . at that time men were not sharing as they do today—things were not put down the middle of the road . . . I did have the responsibility of keeping things in order and making sure the children were cared for and that meals were on the table . . . Those of us who continued our careers were very pleased to be allowed to do that."[48] Internalizing this attitude, so much a part of the culture of the era, might have affected "productivity"; it certainly hindered career advancement.

If Dr. Lawrence's career path is on one end of the spectrum with respect to balancing family and career, consider the career trajectory of Dr. Mary Ellen Avery, physician-in-chief emeritus at the Boston Children's Hospital and Thomas Morgan Rotch Distinguished Professor of Pediatrics, emeritus, at Harvard Medical School.[49] Avery graduated from Johns Hopkins in 1952, three years after Lawrence graduated from Rochester. In 1974 she became the first woman to become chief of medicine at a Harvard teaching hospital, a position she retained until

1985. In 1994 she was elected to the National Academy of Sciences. During an interview in Boston in 1997, Dr. Avery commented that, like her, most of her fellow women medical students had decided not to marry. Avery seems to have focused on medicine as her career goal from the years when her neighbor and "role model," Dr. Emily Bacon, chair of pediatrics at WMCP, brought her—a high school student—to see the "preemies" at her hospital in Philadelphia. Another exemplar for Avery was Helen Taussig at Hopkins, who was "at the height of her career" while Avery was a student there and who later became a close friend.

Avery is internationally known for her groundbreaking research on the role of pulmonary surfactant in hyaline membrane disease, a major component of infant respiratory distress syndrome. Avery realized that the lungs of infant respiratory distress victims, who die from lack of air in their lungs, did not have the characteristic compliance of healthy

Dr. Mary Ellen Avery (b. 1927), physician-in-chief, emeritus, Boston Children's Hospital, and Thomas Morgan Rotch Distinguished Professor of Pediatrics, Harvard Medical School, Boston, 1997. (Photograph courtesy of Dr. Avery.)

lungs, nor did they exhibit the characteristic "foam" found in the airways of infants who died of other causes. Avery began to wonder about the function and production of the foam, especially its effect on the surface area and surface tension in the lungs. This foam would be named "pulmonary surfactant" by physician-scientist John Clements. Its absence from immature lungs, Avery hypothesized, was a crucial element in retarding lung expansion and in preventing the capacity to retain air at the end of each breath, something Avery and others would shortly establish. Avery's interest in this highly specialized research began in earnest when she was a pediatrics resident at Johns Hopkins, but it was first piqued when she was a medical student. "The pathologists were telling us all about [airless lungs at autopsy of many premature babies], babies were dying, lots of them, and [hyaline membrane disease] was the leading cause of death, and the clinicians didn't seem to know what was going on . . . and this was the hot topic because . . . it was a vast area of ignorance. And that's what excited me." In fact, one of her pathology professors had already questioned the prevailing hypothesis that the hyaline membrane found postmortem in respiratory distress syndrome victims was itself the source of the problem. During her residency Avery was advised to take a fellowship at the Harvard School of Public Health with a well-known researcher on the mechanics of breathing, Jere Mead. When she discovered that research on infants' lungs was also being done at Harvard, in Dr. Clement Smith's lab, Avery applied for—and got—an NIH fellowship to work on the problem in both men's laboratories. By the end of 1959 she had published the critical article that established her name in pulmonary surfactant research.[50]

How did she make so much headway so quickly? And how did a young woman, a clinician and a relative novice in basic research, manage to keep control of such a "hot topic," even retaining the major share of the credit for her insight into the importance of pulmonary surfactant? Once in Boston, Avery eagerly learned the basic physiological, biochemical, and mechanical concepts and techniques needed to study surface tension in the lung. She also had the confidence to approach key figures whose work impinged on hers for their advice and support. She was lucky in that her primary mentor, Jere Mead, was "one of the most rigorous, honest, and decent people . . . He would never claim something that he didn't do." Then, too, "his interests had taken other directions." In his view, her research was "not physiology." Besides, as Janice Kahn's study points out, Avery's ideas ran counter to prevailing ideas about neonatal respiratory distress: for better or worse, Avery was left alone to defend her surfactant hypothesis—if she could. She recalled spending little energy defending her "turf" from more senior researchers. Fortunately, her results were corroborated by others. After her return to Hopkins and appointment as an assistant professor there, in 1960 she was successfully nominated to be the John and Mary Markle Scholar at Johns Hopkins and was the first woman to receive this prestigious award. In 1965 she was given an endowed chair, and in 1969 she left to become the chair of pediatrics at McGill University and physician-in-chief at Montreal Children's Hospital. Five years later, in 1974, she returned to Harvard. In 1991 she was awarded the National Medal of Science and the Virginia Apgar Award of the American Academy of Pediatrics.

For the majority of women physicians during the 1950s and 1960s, academic medicine, on the one hand, was not a straightforward path to success. Those who had not trained at an elite institution, who lacked professional mobility due to marital and child rearing obligations, or who did not immediately identify a specialized research interest found it difficult to generate support for their career development. Women in private practice, on the other hand, may have made a conscious decision to reduce their practice hours to allow time early in their career to care for their children. As Judith Lorber shows, they often missed out on the professional meetings and informal events where colleagues become acquainted. Such women often did not develop the referral networks that make for a successful private practice. Despite documentation showing that most returned to full-time practice after a few years, the general perception held that, on the contrary, women doctors just didn't amount to much.[51]

In 1962 AMWA's leadership responded to the claim that women physicians were underproductive by sponsoring a panel discussion titled "Medical Womanpower." AMWA wanted to know, "Are female M.D.'s as useful in society as male M.D.'s, or are they poor social investments?" Courageously, the association asked whether women were dropping out of medicine with disturbing frequency, as charged, and whether it was possible to successfully combine medicine with what "we, in our culture, consider a successful marriage." One young participant complained, "I think you must make a lot of personal sacrifices. I think you miss a lot of personal time with your family: When the kids come home from school, the little times when you should be together, seeing after your husband, being rested when he comes home from work; getting the house cleaned up."[52] From the floor came an immediate "rebuttal" from Dr. Lillian P. Seitsirse, a graduate of WMCP and the partner and recent widow of a general practitioner. Her comments reflect the other side of the coin, the view held by many married women who entered practice, as she did, during the 1930s: "I was married to a physician for twenty-six years. I practiced medicine all [that time] except two weeks apiece for each of two children when I was confined to the hospital . . . because they didn't let you out before that. You can be a wonderful wife and a wonderful mother and a very good doctor, but you must have qualities for all three. [Applause.]"[53]

Nevertheless, again and again the issue of "mothers' guilt" was raised by the younger women physicians at the conference. As Dr. Rosa Lee Nemir (professor of pediatrics at New York University School of Medicine and a wife, mother, and clinician-researcher), observed, "Women everywhere, who are college educated have listened to the psychologists who have told us for a long time that we must be with our children . . . I think that too often the standards that women hold to themselves are too high. They are much too hard on themselves . . . they are apt to feel guilt when anything goes wrong."[54] Many solutions were offered, such as federal subsidies for child care, day care centers at hospitals, and the creation of residencies with flexible hours for physician-mothers. Not even Dr. Nemir's reassurance and successful example dispelled the atmosphere of frustration. The younger women, those educated in the 1950s, felt keenly conflicted about their roles.[55]

Fortunately, a countertrend was beginning to coalesce. President John F. Kennedy's Commission on the Status of Women, for example, began reporting its findings in 1963, including the recommendation that child care services be expanded and improved for working mothers.[56] In January 1964, to offset the charge that women physicians functioned below optimal levels of productivity, AMWA's then-President Nemir persuaded the AAMC, the AMA, and the U.S. Department of Health, Education and Welfare (HEW) to cooperate with AMWA in a study of women physicians' practice patterns.[57] The resulting study, *Survey of Women Physicians Graduating from Medical School 1935–1960,* was designed to investigate: "1. Factors influencing the entry of women into medicine; 2. The attitudes of medical faculties towards the admission of women to medical schools; 3. The career choices and influence of medical schools on a woman's attitude towards medicine." The Josiah Macy, Jr. Foundation agreed to sponsor it.[58] The Macy Foundation also agreed to host the Conference on Women for Medicine, the first of several foundation projects in support of women and minorities in medicine.

Preliminary results of the combined study were released in time for the Macy conference in 1966. Its findings clarified, rather than contradicted, the conclusions of the Dykman and Stalnaker research. The study found, for example, that twice as many men *and* women graduating in 1956 married before completion of their medical training than had been the case in 1931. More than a third of the women had had children before the completion of their training. When counting 2,000 hours per year as a full-time practice for 1964, 55 percent of female physicians worked full-time, 31 percent part-time; the figures for male physicians were 90 percent and 7 percent, respectively. The influence of family responsibilities on the number of hours in practice was clearly highlighted by the study. In 1964, 39 percent of women with three or more children worked full-time, compared with 86 percent of single women (presumably without children). Indeed "family responsibilities" were the major reason given for curtailment of professional activities by married women physicians. Likewise women physicians saw fewer patients and earned, on average, substantially less than their male colleagues. Finally, regarding specialization and type of practice, women specialized most frequently in pediatrics (33.6 percent), psychiatry (13.7 percent), internal medicine (10 percent), anaesthesiology (9.1 percent), preventive medicine (8.6 percent), obstetrics-gynecology (8 percent).[59]

Participants at the Macy conference, however, were not content simply to look at the statistical profile of women's participation rates. Dr. John Z. Bowers, president of the Macy Foundation and previously the dean of medicine at the University of Wisconsin, "cared desperately" about the need to actualize the potential of women and minorities in the medical profession, an outgrowth of his long-standing interest in international health.[60] Typically, the foundation hosted small, focused research conferences, and then published the results.[61] The 1966 conference, for example, resulted in Carol Lopate's important 1968 study,

Women in Medicine. Indeed, between 1966 and 1984 the Macy Foundation sponsored a total of four different conferences on women in medicine and one on women in science. Between 1965 and 1980, Macy gave more than one million dollars for programs to enhance pre-medical, medical, and postdoctoral opportunities for women.[62]

For the 1966 conference, Bowers wanted participants to look beneath the statistical surface to understand the place of women in the medical profession. His interest, like AMWA's, focused more on the cultural factors that discouraged women from entering the profession than on those factors limiting their participation once they completed their training. Nevertheless, Bowers's published views on the role of women physicians suggests how deeply embedded was the belief that, even with higher utilization rates, women would mainly take up those aspects of medicine traditionally seen as appropriate to and, presumably, desired by women. Bowers, for example, thought women could make a "unique contribution to medicine through their interest in specialties that do not attract adequate numbers of men," such as pediatrics, child psychiatry, and "especially maternal and child health." As he wrote, "The sympathy, affection, and concern that a woman can bring to patient care is a valuable resource." Central to his thinking was the additional assumption that most medical women would want to marry and have children, often during the period of their internship or residency. Bowers believed it to be a responsibility of medical schools and teaching hospitals to make this possible. He was equally concerned to understand the mechanism through which many bright women received the message that "women are not wanted in medical schools."[63]

Conference participant and Radcliffe College president Mary Bunting focused on those factors hindering full participation by women once they had completed their medical training. She saw the problem both as a conflict between profession and family and as a structural phenomenon through which women with children were marginalized in the profession. "Increasing the interest of women in medicine will not of itself bring more physicians into the profession . . . The average woman physician practices somewhat fewer hours than the average man . . . This calls for more medical schools and better organization of health care delivery systems." One physician-mother summed up the conference's conclusions: "We are not the same as men, and never will be. We do not want to compete, but only to do our share. Women have

something of their own to give the profession." It was in society's interest, she believed, to organize its health care in ways that would make use
of those special contributions. To do so, however, would require addressing the issue of the physician-mother.[64]

At the AMWA annual meeting that year, the question was taken up
again through the meeting's theme: "Medicine, Marriage, and Motherhood." Echoing the concerns of President Kennedy's National Advisory Commission on Health Manpower, AMWA sought to determine
"the factors which interfered with women entering the medical profession." For more than thirty years the admissions ratios of males and
females had remained equal, at about 50 percent. Many believed, with
Dr. Glen Leymaster, president of the Woman's Medical College of
Pennsylvania, that "few will claim that significant limitations on opportunities for women physicians are still in force or that individual
women fail to carry their share as well as their male counterparts. The
struggle for equal opportunity for women . . . seems to have been won
during the last century." Why then were women still so vastly in the
minority in medicine?[65] Referring to the Macy conference, Dr. Nemir
pointed to the "discouragement" doled out by guidance counselors,
"primarily based on the difficulty of combining marriage and a career
and on the problem of finances."[66] For its part, in 1966 the Macy Foundation agreed to fund a Radcliffe Institute program to help support
women in part-time residencies, many of them at Boston Children's
Hospital. The Radcliffe program was started in 1961 to fund "baby-sit-
ters, household costs, transportation," and other expenses typical of
residents who were mothers.[67]

Thus, by the late 1960s, feminist physicians were beginning to analyze the place of women in the medical workforce and to link social
policy to the specific circumstances of women in medicine. In contrast
to its social conservatism of the 1950s, by 1966 AMWA supported
income tax deductions for child care; legalized therapeutic abortion "at
licensed hospitals if at least two physicians agree there is substantial
danger to the mother's physical or mental health"; hospital-centered
child care for health care personnel; and development of part-time
programs for women residents and physicians.[68] While AMWA's policy
shift may have reflected its need to attract younger members (in 1965,
more than 50 percent of its members were fifty years old or older), it

also reflected an attempt to counter criticism from women physicians themselves. Only after special prompting from Dr. Mary Steichen Calderone, medical director of Planned Parenthood Federation of America (PPFA), for example, did AMWA energetically discuss family planning. In 1961 it cautiously invited Calderone to address the Pan American Medical Women's Congress in Colombia. But in 1963, Calderone still felt it necessary to write a strong letter to AMWA urging it to address the "terrible inequality that exists in availability of family planning services between income groups" in the United States. "I think," she continued, "it is high time that the AMWA emerge from its timid state."[69] Only after the Macy conference did AMWA broaden its political vision.

Sexuality, Family Planning, and Health Care: Dr. Mary Steichen Calderone

Through her work as a public health educator in the areas of family planning and human sexuality, Mary Steichen Calderone was one of the few physicians in the 1950s to address the relationship between personal and professional responsibilities. As the medical director of Planned Parenthood Federation of America between 1953 and 1964, as a principal founder of the Sex Information and Education Council of the United States (SIECUS), and as an eloquent yet plainspoken expert on human sexuality and public health, Calderone was instrumental in provoking a reexamination of the assumption that a woman's life must be held captive to her reproductive capacity.[70] In the tradition of Margaret Sanger, Virginia Woolf, and others, Calderone understood that a woman's professional and intellectual freedom would be secure only with the freedom to tell "'the truth about her own experiences as a body.'"[71]

Mary Steichen was born in 1904 in New York, the elder of two daughters of the Luxembourg-born photographer Edward Steichen and Clara Smith Steichen, an aspiring singer from Springfield, Illinois. She and her younger sister, Kate, spent most of their early years in France. At the outbreak of the First World War they and their mother returned to New York. Not long after, her parents divorced. Mary, who was devoted to her father, became estranged from her mother. She spent many of the years prior to her departure for Vassar College living

Dr. Mary Steichen Calderone (1904–1998), medical director of Planned
Parenthood Federation of America and cofounder of the Sex Information and
Education Council of the United States. Undated photograph, ca. 1930s.
(Courtesy of the Schlesinger Library, Radcliffe College, Cambridge,
Massachusetts.)

in New York City with Dr. and Mrs. Leopold Stieglitz, brother of her
father's mentor, the noted photographer Alfred Stieglitz. After graduating from the Brearley School in New York and, in 1925, from Vassar
College (with a major in chemistry), Steichen married a young actor,
W. Lon Martin, and had two daughters, Nell and Linda. She worked
briefly with the American Laboratory Theatre in New York, an "off-
shoot of the Moscow Art Theatre," until her first pregnancy. Then, she
recalled, she was told to choose between "baby and work . . . So of
course I chose the baby." After her marriage ended in divorce in 1933,
she decided to return to school to be better able to support herself and
her children. Lacking a clear sense of direction, Mary Steichen Martin
took a series of career aptitude tests developed by the Johnson O'Con-
nor Human Engineering Laboratories at the Stevens Institute of Technology. The results convinced her to accommodate both her love of

science and her "need to deal with people"—by practicing medicine. In 1935, after being accepted into a doctoral program in the physiology of nutrition at the University of Rochester, she persuaded George Whipple, dean of Rochester's medical school, to transfer her to the incoming class of medical students. He agreed—in spite of the fact that she was thirty-one and a single mother. Facing considerable hardship, including the death of eight-year-old Nell from pneumonia and the necessity of living apart from Linda during her last two years of medical school, Steichen nevertheless graduated on time, in 1939, thirteenth in a class of forty-five. But taking an ordinary residency, with its low pay and demanding hours, seemed out of the question, particularly since twelve-year-old Linda was once again living with her. Instead, after an internship in pediatrics at Bellevue Hospital in New York, she accepted a two-year fellowship with the New York City Health Department in 1940 that allowed her to pursue, jointly, a master's degree in public health from Columbia University.[72]

During her fellowship, Steichen met her second husband, Dr. Frank A. Calderone, a district public health officer who was soon to become deputy commissioner of health for New York City. Following World War II, he became chief administrative officer for the World Health Organization, and he was medical director of the U.N. Secretariat until 1954.[73] They married in 1941. Soon after, they moved to Long Island, and Mary Steichen Calderone gave birth to two more daughters. She was unwilling to work full-time until both girls were in school but "kept her hand in" by working for the Great Neck, New York, school system as a school physician. In the late 1940s Calderone also addressed Parent-Teacher Association groups on sexual hygiene and "the goodness of sex" under the auspices of the Nassau County Mental Health Association. Through her husband's public health work, she met many people active in public health circles around the country. When she was asked to accept the "part-time" position of medical director of Planned Parenthood of America in 1953, her youngest child was in first grade and Calderone was ready to take on something more substantial: "The timing was perfect." Besides, she acknowledged, "I had no established practice, and nothing to lose, and no *qualified* male physician would have taken it."[74]

The American medical profession rarely has been at the forefront in matters of family planning and contraception. Not until the 1930s did the AMA form the Committee to Study Contraception and Related Practices. Its first report was made in 1936, the same year the Supreme Court's *One Package* decision exempted doctors from the Comstock antiobscenity laws by allowing them to receive contraceptive materials through the mail. Although the committee recommended that "some responsible group" establish standards for contraceptive materials, the House of Delegates would not endorse its recommendation. Instead it censured "lay bodies organized solely for the purpose of disseminating contraceptive information," such as Planned Parenthood Federation of America. In 1938, a year after giving qualified support to birth control as being part of the doctor-patient relationship, the AMA again limited its support, refusing to give a blanket endorsement to birth control per se, apparently out of determination "to control the delivery of contraceptive care."[75]

At issue was the supervision of freestanding birth control clinics, such as those sponsored by Margaret Sanger's Birth Control Clinical Research Bureau (CRB). The first of Sanger's clinics opened in New York City in 1923. According to the New York State law governing medical dispensaries, no public clinic could operate without a license from the State Board of Charities, which, in accordance with prevailing concerns of leading physicians, favored hospital outpatient departments over freestanding specialty clinics. Sanger's clinics could operate within the law only as the research arm of a physician's private practice. Sanger's freestanding clinics were at a triple disadvantage: they were part of a nonprofessional organization; that organization was dominated by women; and it provided free or at most low-cost medical care. As a case in point, the PPFA's contraceptive clinics of the 1940s became the object of much negative opinion on the part of the AMA and the New York Academy of Medicine, ostensibly for being under lay control and unaffiliated with either a hospital or a department of public health. In a New York Academy of Medicine report in 1945, however, the main attack was directed at the clinics' employment of mostly women physicians, some of whom were foreigners "difficult to understand in ordinary conversation." As a group, the women doctors were said to be "not interested in the underlying medical and social problems. Often, they are not the best physicians available, and their work is frequently superficial." Since the same report noted that these clinics were both overcrowded and understaffed, and that many of the staff physicians were working part-time on a voluntary basis, one can conclude that conditions at the time—principally, a scarcity of alternative contraceptive clinics and the doctors to staff them—were the real culprits. The predominance of women physicians, however, became a target in itself.[76]

When Calderone was offered the job of PPFA medical director in 1953, in her words, "the whole field of family planning was, quite literally, a bootleg operation in the private physician's office, as only well-to-do women could get any kind of birth control." This may exaggerate the situation for the New York metropolitan area—after all, Calderone reports that she was herself a patient of Dr. Hannah Stone, medical director for the CRB, during the 1930s. But it was no exaggeration for most of the United States.[77] Calderone's work with the American Public Health Association (APHA) and the AMA during her tenure at Planned Parenthood played a crucial part in changing the medical profession's attitude toward birth control during the 1950s and '60s. By the time Calderone resigned from PPFA, the APHA in 1959 and the AMA in 1964 both had issued policy statements supporting the inclusion of family planning in general medical practice—in large measure because of her behind-the-scenes lobbying.

Her principal objectives at Planned Parenthood were "to get this whole concept of family planning accepted in the medical field" and to make the organization less of a "propaganda" group and more of a "sound, medically based organization."[78] In her own estimate, she "gave it a real public health orientation which it had never had before . . . [M]ost doctors are interested in whether Mrs. Smith is using a diaphragm. I was interested in broader numbers. I said, if in a given year eighty women become pregnant, if we can lower that to forty, it's a wonderful thing—no matter how they do it. I used to tell the doctors with Catholic patients not to tell them their [rhythm] method is bad, but to teach them how to use it to the best of their abilities."[79]

Calderone learned early on the advantages of gaining the widest support possible for any potentially controversial initiative, an important lesson during the years of Senator Joseph McCarthy's greatest influence. The year following her arrival at Planned Parenthood she was asked to develop a project that would increase the organization's visibility. Calderone, who personally disliked abortion and saw it as a symptom of society's failure to provide rational alternatives for family planning, chose to sponsor an international conference in 1955 on abortion as a public health problem. Her emphasis was not on expanding the availability of abortion (although, like most of those present, she deplored the effects of illegal abortion on women's health) but to reduce the need for it through increased access to contraception and sex education. From her medical and public health perspective, however, she knew that such a project had to be carefully designed to bolster the federation's perceived professional legitimacy. Otherwise she risked increasing its marginality among policymaking bodies such as the AMA Committee on Human Reproduction and the APHA Section on Maternal-Child Health. Accordingly, most of the conference was held at Arden House of the New York Academy of Medicine. Participants included physicians representing the academy, the Rockefeller Foundation–supported Population Council, prestigious medical school departments of obstetrics-gynecology and psychiatry, and municipal public health departments, along with Planned Parenthood officials. The proceedings were published in 1958 in a volume titled *Abortion in the United States*. Calderone also began urging the PPFA Medical Advisory Committee to consider ways for Planned Parenthood to become a recognized source of scientific expertise on the various forms of contraception. By the beginning of 1958 she had persuaded the organization to create the Clinical Investigation Program. Its controlled clinical trials would be designed to assess the "use-effectiveness" of contraceptive methods, comparing the diaphragm (with and without spermicidal jelly) with spermicides alone. She and Dr. Alan Guttmacher, who was then on the Advisory Committee and would become its chair in 1960, began developing plans for PPFA to cosponsor such trials with the Population Council, whose statistician, Christopher Tietze, would design and run them.[80]

At about this time, Calderone and the PPFA Medical Advisory Committee began lobbying the APHA to take a stand on the need for family planning as part of national public health policy. In March 1958, working with physicians active in the APHA, including Martha Eliot, Helen Wallace, William C. Spring, Jr., and Leona Baumgartner (commissioner of health for New York City and president of APHA in 1958–59), Calderone proposed that the APHA Maternal-Child Health Section sponsor a resolution supporting birth control as part of the service offered at all local public health departments and tax-supported hospitals. Her proposed resolution called for family planning to be "widely offered on the same democratic basis as any other public health service, available to all, through such medical or physiological methods as may be in accord with the patient's religious convictions." A year later she learned from Dr. William M. Schmidt, chair of the APHA's Family Planning Committee, that another, similar proposal was in the works. In fact, a resolution on the "population problem" passed by the APHA at its eighty-seventh annual meeting in 1959 closely paralleled Calderone's. Her letter to Schmidt illustrates the thin line she was forced to walk between propagandizing on behalf of Planned Parenthood and collegial work as a public health professional: "Because the idea of such a resolution was my own baby a couple of years ago, I think you should know how particularly pleased I am that the whole thing has been lifted out of PPFA's hands. This is exactly as it should be and what I hoped would happen. When I first came to PPFA my greatest concern was to attempt to remove the stigma of its being a 'propaganda' organization and every action I have taken since has been to that end."[81]

Calderone also began writing regularly to colleagues at the AMA, suggesting that they, too, support family planning as a standard part of good medical practice. In 1963 the AMA board appointed a Committee on Human Reproduction to revise its policy on contraception in relation to the larger issue of "population control." The committee, chaired by Dr. Raymond T. Holden of Georgetown University School of Medicine, included Calderone and Dr. Janet T. Dingle, head of the PPFA affiliate in Cleveland. Its proposals, subsequently adopted by the House of Delegates, stated that "the medical profession should accept a major responsibility in the matter of human reproduction as they affect the total population and the individual family"; that "the AMA will take the responsibility for disseminating information to physicians on all phases of human reproduction, including sexual behavior, by whatever means are appropriate"; and that "the prescription of child-spacing measures should be available to all patients who require them, consistent with their creed and mores, whether they obtain their medical care through private physicians or tax or community-supported health services."[82]

When Calderone began her work for Planned Parenthood in the early 1950s, feminists like Betty Friedan had not yet begun to question the postwar tendency to reinstall women in the exclusive role of homemaker. In the context of what is now labeled the postwar baby boom,

Calderone and PPFA were swimming against a strong pronatalist tide. A decade later, however, when the AMA announced its support of family planning as a part of "responsible medical practice," the world of American gender roles was beginning to shift on its axis, albeit imperceptibly at first. Even Friedan's groundbreaking book, *The Feminine Mystique* (1963), could cite a flurry of magazine articles from the early 1960s describing the malaise of American middle-class wives and mothers.[83] The introduction in 1960 of the birth control pill made contraception a topic of discussion for millions of women (and men). Calderone began to feel that her days as a pioneer on behalf of Planned Parenthood were drawing to a close. She also became, for the first time in her life, conscious of sex discrimination in the workplace when she concluded that she was being paid on a part-time basis for full-time work only because she was a married woman and "didn't need" more money.[84] In 1964 she resigned from Planned Parenthood.

Education for healthy sexuality, rather than merely the practices associated with reproduction, now claimed her interest. She cofounded and became executive director of the Sex Information and Education Council of the United States, to "establish man's sexuality as a healthy entity." Having become convinced that responsible reproduction was just one part of responsible sexuality and, indeed, humanity, Calderone decided that "handing out contraceptives was not enough," and she began concentrating on the sexual education of adults. As she told a reporter in 1967, "Our motive in sex instruction should not be just to prevent illegitimacy or venereal disease. It should also be the development of the personality of the individual, an integral part of which is sex."[85] For her efforts, *Newsday* dubbed her "The Grandmother of Modern Sex Education." But in what was a backhanded tribute to her effectiveness and a bellwether of the emerging radical right in American politics, Calderone and SIECUS were branded by such organizations as the John Birch Society and the Christian Crusade as communists and pornographers, and as "totally unsuitable to have any dealings whatsoever" with the education of children. Despite rebuttals from New York Republican senator Jacob Javits in the *Congressional Record* in 1969 and supportive coverage in the *New York Times*, as late as 1981 *Time* magazine published what the National News Council concluded was an unfair article on children's sex education, featuring distorted references both to Mary Calderone and to SIECUS.[86]

The Beginnings of a New Medical Feminism

In 1968 Carol Lopate published *Women in Medicine*, an analysis of the discriminatory culture in which women were discouraged from pursuing medicine as a career, based on the 1966 Macy Foundation Conference on Women for Medicine. By 1968 AMWA, too, had begun addressing the question of gender discrimination directly. For one thing, the age structure of the association's membership was beginning to shift. In 1969, medical students accounted for 46 percent of the members, and although they could not vote, their presence was beginning to be felt.[87] More important, as a result of the civil rights, antiwar, and still-embryonic women's movements, a full-scale reassessment of the meaning of race, social class, and, finally, gender, was beginning to register in the national consciousness. American society was confronting the question of how to reconcile "difference" and "equality."

As an example of this, Lopate's book directly challenged the adequacy of AMWA's position on women physicians' minority status in medicine. Although Lopate believed that AMWA could be justly proud of its history, she also believed that the association's leadership was partly responsible for its fairly low membership. In Lopate's view, women physicians as individuals were unwilling to risk the privileges of their minority status by confronting it directly. She faulted AMWA and its journal, *JAMWA*, for being just a "special interest group" rather than "an intellectually vital center for medical ideas" that could attract the interest of more women. If a separate association for women physicians was still professionally essential—and her interviews with women doctors strongly suggested that it was—her book seemed to ask, Why not make a virtue of that very necessity?[88]

AMWA's leaders, largely educated before World War II, were not yet ready to abandon an older conception of women physicians as a minority in medicine. They continued to search for ways to accommodate the new generation of women but found it difficult to imagine structural changes in society and in the profession that might move women out of their minority niche. Moreover, they saw women's problems as matters for women themselves to overcome. In 1967, for example, AMWA decided to sponsor a research survey on "household help," a problem they believed to be at the core of the professional difficulties of medical women. A discussion in *The Woman Physician*, "Household Help—

The Woman Doctor's Gordian Knot," revealed the unexpectedly wide-spread response that journal had received to a questionnaire about household help and child care. The article began with this anonymous quotation: "Household help is, in my opinion, a woman physician's greatest problem. All the rest are minuscule by comparison."[89] Interestingly, on July 17, 1970, the husband of the woman so quoted wrote privately to the editor to declare his disagreement and profound frustration with the article's analysis.

> It is my impression that the opinions set forth in your article (my wife's included) are essentially erroneous . . . That there is a problem for the woman intending to become a physician I would be the last to deny. But . . . [t]he problem that women physicians face is essentially the problem that all women suffer from: the masculine orientation of our society. If society could be made to recognize the right of women to a career, provision on a general scale would be made for the daily care of children in tax-subsidized centers— the children of housekeepers together with the children of physicians . . . [I]t is most curious that women physicians are not in the vanguard . . . storming the barricades of male privilege. They are grousing about household help, like so many suburban matrons.[90]

What was needed from a new generation of leaders was a response to the structural inequities of a workplace in which women were expected to carry the full burden of both a career and a family and in which the professions themselves remained largely impervious to the underutilization of women. In short, AMWA was being called on to be an active advocate of the positive advantages to society of making an equitable place for the woman physician.

In the late 1960s, slowly and not without strain, AMWA undertook this essential task of revision. At a 1968 conference titled "Meeting Medical Manpower Needs: The Fuller Utilization of the Woman Physician," sponsored by AMWA, the Women's Bureau, the President's Study Group on Careers for Women, and the U.S. Department of Labor, AMWA president Alice Chenoweth responded to the canard that educating women physicians is a "waste" of national resources. Challenging the statement that women physicians work only half as much as men, she asked, "Are there adequate data to support such a statement?" and "Can the amount of time spent be equated with the physician's effectiveness? What is the true measure of a physician's worth to society?" The conference concluded with this direct challenge to the prevailing definition of "physician productivity" and insisted that "talented women are needed in medicine, that their contribution is valuable and unique."[91]

The beginnings of AMWA's newly liberalized political agenda could also be seen in the resolutions adopted at its 1968 annual meeting. The membership acknowledged its common cause with other American women, first by supporting sex education as integral to a healthy society and second by urging federal tax credits for the domestic and child care expenses of all working women, not just women physicians. A second Josiah Macy, Jr. Foundation conference, The Future of Women in Medicine, also held in 1968, again addressed the need for more women physicians. The AAMC annual meeting that year focused on widening access to medical care by increasing the enrollment of minority, low-income, and women medical students.[92] But a full-scale attack on discrimination against women in medicine, like the campaign for women's medical education of a century earlier, required the energy of a general movement for women's rights to carry it forward. This was not long in coming.

8

Medicine and the New Women's Movement

> The fact is that in the field of medicine, the significant battle, the one that counts above others—equal opportunity to get into medical school—has been won.
>
> *E. Grey Dimond, M.D., 1983*

> However, men who have worked to eliminate discrimination cannot assume that their job is finished.
>
> *Marilyn Heins, M.D., 1983*

IF CAROL LOPATE'S 1968 study, *Women in Medicine*, conveys the liminal quality of the 1960s for women in medicine—a threshold reached but not yet crossed—then Mary Roth Walsh's book *Doctors Wanted: No Women Need Apply*, published nine years later, reflects the profound change in attitude and expectations brought about by the rebirth of feminism during the intervening years.[1] Passage in 1971 and 1972 of the "equal opportunity" amendments to the 1964 Civil Rights Act proved to be the catalyst women needed to increase their representation in many professional fields. Indeed, nothing better illustrates how closely the status of women physicians mirrored that of all white-collar American women than the impact of equal opportunity legislation on the medical profession. The rapid alteration of medicine's demographic profile reflected the effects of the sudden transformation of gender politics.

The Equal Opportunity Era

Titles VI and VII of the 1964 Civil Rights Act banned discrimination on the basis of race and sex but exempted colleges and universities from

216

compliance. Title VII was viewed even by the Equal Employment Opportunity Commission as unenforceable. Even after President Lyndon Johnson promulgated Executive Orders 11246 (1964–1965) and 11375 (1966–1970) banning sex discrimination in federal employment and among contractors to the federal government, the numbers of women hired for faculty positions or matriculating at medical schools remained modest. In response to calls by the American Medical Women's Association for tougher oversight, Dr. Donald Pitcairn, director of physician education for the Division of Health Manpower of the U.S. Department of Health, Education and Welfare, wrote, "There are no legal requirements for the admission of students in any category into the educational program of any medical school. Nor do we favor the establishment of such."[2]

Against this background of underenforcement, on January 31, 1970, the Women's Equity Action League (WEAL) filed a class action lawsuit against the University of Maryland and 250 other colleges and universities for race and sex discrimination in programs, personnel, and admissions policies. In June, WEAL presented extensive testimony concerning "discrimination against women in medical schools" to Representative Edith Green's Special Subcommittee on Education. Key testimony was offered by Dr. Frances S. Norris, representing AMWA. She argued, first, that "the rarity of women doctors . . . has limited the freedom of women to choose doctors of their own sex, and has deprived women of the benefits of research into diseases peculiar to women." Second, she noted that although the number of women applicants to medical schools had risen by more than 300 percent since the 1930s (compared with an increase of 29 percent for men), the rate of acceptance of women remained at a steady 50 percent of applicants per year. This, she charged, was the result of "an *arbitrary grouping of applicants by gender.*"[3]

On October 5, WEAL notified Elliot Richardson, secretary of Health, Education and Welfare, that it had filed formal charges of sex discrimination against every medical school in the United States under Amended Executive Order 11246. By then Congress was convinced of the need to strengthen Title VII. In October 1971 it passed the Public Health Service Act, including a provision barring employment discrimination in medical and other health professions schools. The Equal Employment Opportunity Act of 1972 overturned Title VII's exemption of institutions of higher learning. Finally, in June 1972, Representative Green shepherded the passage of Title IX of the Higher Education Act, banning discriminatory policies, such as in admissions and salaries, in any school receiving federal funds. As a result of this legislation, institutions receiving federal funding were required to submit to HEW an "affirmative action plan" (AAP) detailing the discrimination that presently existed and what steps were planned to reduce it. As of 1973, no medical school had submitted an AAP that had been approved.[4]

Such regulations were essential to level the playing field. Some writers have denied that overt discrimination against the admission of women and minorities was practiced, noting that men and women applicants were accepted to medical schools on average in about equal proportion to their percentage of the applicant pool. Yet Mary Walsh has noted that from at least 1959 to 1971 the Association of American Medical Colleges published a handbook on medical school admissions for prospective applicants that explicitly listed each school's preferences (if any) for its student body composition. Factors listed included sex, race, and age. As late as 1963, 15 percent of medical schools expressed an unabashed preference for male applicants; in 1969 four schools continued to do so. In fact, as Table 8.1 shows, women were accepted to medical school at a slightly lower rate than men from 1940 through 1960, despite a steady increase in the number of women applicants. The percentage of women accepted slightly exceeded the rate for men from 1964. From 1970 to 1974, in the immediate aftermath of the congressional hearings and passage of the equal opportunity amendments, the acceptance rate of women exceeded that of men more than at any other time (see Table 8.1). But during the preceding decades, the effect of a prejudicially chilly climate on the numbers of women who applied to medical schools seems incontrovertible. As one woman medical student wrote in 1973 for an anonymous questionnaire from Dr. Mary Howell, associate dean of students at Harvard Medical School, "On the subject of undergraduate support for women in pre-med—let's face it—there is damned near none and even none would be better than the generally negative advising that goes on."[5]

Given the long and complex history of gender disparity in the medical profession, few would have expected the discriminatory climate to disappear abruptly. Still, between 1970 and 1974, two years after pas-

Table 8.1. Medical school applicants and acceptances, by gender, for selected years, 1935–1995

Year	Men		Women	
	Applicants	Acceptances (%)	Applicants	Acceptances (%)
1935–36	12,051	6,521 (54.1)	689	379 (55.0)
1940–41	11,269	6,025 (53.5)	585	303 (51.8)
1950–51	21,049	6,869 (32.6)	1,231	385 (31.3)
1955–56	13,935	7,465 (53.6)	1,002	504 (50.3)
1960–61	13,353	7,960 (59.6)	1,044	600 (57.5)
1964–65	17,437	8,219 (47.1)	1,731	824 (47.6)
1970–71	22,253	10,214 (45.9)	2,734	1,295 (47.4)
1974–75	33,912	11,674 (34.4)	8,712	3,392 (38.9)
1979–80	25,919	12,156 (46.9)	10,222	4,730 (46.3)
1984–85	23,468	11,463 (48.8)	12,476	5,731 (49.9)
1989–90	16,369	10,522 (64.3)	10,546	6,453 (61.2)
1994–95	26,397	10,062 (38.1)	18,968	7,255 (38.2)

Sources: Figures for 1935–1965 are based on Carol Lopate, *Women in Medicine*, (Baltimore: Johns Hopkins University Press, 1968), Appendix 2, p. 194; for 1975–1990, American Medical Association, *Women in Medicine in America: In the Mainstream* (Chicago: AMA, 1991); Leah J. Dickstein, Daniel P. Dickstein, and Carol C. Nadelson, "The Status of Women Physicians in the Workforce," typescript prepared for Council on Graduate Medical Education, April 8, 1994, Table 5.1, p. 33. My calculations for 1970 are based on figures published in *JAMA* 218 (1971): 1217; for 1994–95, Janet Bickel et al., *Women in Academic Medicine Statistics, 1997* (Washington, D.C.: AAMC, 1997), Table 2.

sage of Title IX, the number of women applicants more than tripled. What might account for this? The rebirth of political feminism was one of several significant factors in the changed gender landscape of American medicine. Now that discrimination in admissions was proscribed by law, the percentage of women in the applicant pool, which began rising in 1969, became the fastest growing segment of the applicant population. Perhaps, as Walsh argues, "when women perceived an improvement in their chances of being accepted, they increased their applications."[6] Significantly, the scores of women students on the Medical College Admission Test (MCAT) continued to lag slightly behind those of their male peers. Yet admissions committees, perhaps in the context of increasing political awareness, were becoming convinced of the negligible effectiveness of these differences in predicting the success of medical students. As Robert F. Jones of the AAMC wrote in 1984, these differences seem to "underestimate the success of women medical students . . . medical schools appear to have compensated for this disparity by accepting women with lower average MCAT scores than men. Nothing in the data we have analyzed suggests that this practice is unsound."[7] This author's experience as a member of the admissions committee of a state-supported medical school in the 1990s, which paralleled reports from the University of California, Davis, and the Medical University of South Carolina, demonstrated that individual institutions can create sound admissions criteria that incorporate factors other than standardized test scores in admissions decisions.[8] When the choice was made to value gender diversity in the medical school population, it was not difficult to find qualified and competent students to fulfill that mission. The definition of what "qualifies" a student for admission is fundamentally a social decision, responding to whatever society values in its medical professionals at a given period.[9]

Increases in the proportion of women students actually began during the 1960s and accelerated with the passage of anti–sex discrimination statutes.[10] Chapter 7 described the positive impact of newly established medical schools on the percentage of women medical students in the 1960s. During the following decade, women occupied more than 65 percent of the newly created first-year class positions. Several factors besides feminist lobbying accelerated and broadened that trend, which soon included long-established schools as well as newer ones. A fear of a shortage of primary care physicians dominated national health care policy during the 1970s, encouraging a continued overall increase in student enrollments through capitated federal funding for medical schools. Existing schools increased the number of all students admitted, especially the proportion of women, which rose from 9.6 percent in 1971 to 26.5 percent in 1981, an increase of 176 percent. The overall number of medical students in those years rose from around 11,500 to around 17,000 in 1981, an increase of 48 percent. Between 1970 and 1980, the percentage of medical graduates who were women nearly tripled, reaching 23 percent in 1980 (see Table 8.2). Finally, in 1976 Congress mandated a linkage between a medical school's funding and the percentage of its residents training as primary care doctors, historically a field of greater interest to women. In 1970, for example, more

Table 8.2. Women medical school graduates for selected years, 1950–1997

Year	Number	% of graduating classes
1949–50	595	10.7%
1959–60	405	5.7%
1964–65	503	6.8%
1969–70	700	8.4%
1974–75	1,706	13.4%
1979–80	3,497	23.1%
1984–85	4,898	30.0%
1989–90	5,197	33.9%
1994–95	6,216	39.1%
1996–97	6,614	40.0%

Sources: JAMA "education numbers" for corresponding years and Janet Bickel et al., *Women in U.S. Academic Medicine Statistics, 1997* (Washington, D.C.: AAMC, 1997), Table 1. My thanks to Sara Clausen for research assistance in obtaining these figures.

than two-thirds of women residents specialized in pediatrics, internal medicine, psychiatry, or obstetrics-gynecology. In 1980, with the addition of family practice, these five specialties accounted for more than 70 percent of women residents; in 1995 they accounted for 68 percent.[11]

The increased presence of women students created opportunities and, of course, challenges to the existing institutional approaches to curriculum, mentoring, and professional development. Women medical students of the early 1970s were greeted by widely divergent expectations. Unlike earlier generations, who were expected to adapt as best they could to the conditions prevailing in coeducational schools, the "equal opportunity generation" of the 1970s was shadowed by largely unspoken questions about its power—and willingness—to resist business as usual. In 1977, even before the women of this cohort had completed their residencies, the National Center for Health Statistics compiled a survey of physician practice characteristics in ambulatory care. For the first time, it compared data for men and women, because of "mounting interest in the performance and productivity of women in traditionally male-dominated professions" like medicine. The survey concluded that "women and men in general and family practice and internal medicine spent about the same average number of hours per week in direct patient care," although women spent more time with each and saw fewer patients overall.[12] Some medical faculty expressed concern that women students would experience "conflict between defining and accepting their femininity and also functioning effectively, and therefore at times aggressively, in active competition with men." But the dearth of women faculty members as role models and mentors, and the underlying culture of "institutional sexism," soon were identified as the real obstacles to change.[13]

On many medical school campuses, especially where senior administrators were uncomfortable with the new influx of women students, the responsibility for responding to their needs fell to the office of the dean of students or academic affairs.[14] The administrators in these offices conveyed their impressions to the senior deans, who began discussing the issue with leaders of the AAMC. Thus, early in the 1973–74 academic year, the AAMC turned its attention to enhancing the professional development of women in medicine, according to Janet Bickel, associate vice president for institutional planning and development and director of women's programs at the AAMC. The organization began addressing these matters in response to a concern expressed by several medical school administrators—all of them women—that many women students had needs not effectively addressed at most medical campuses. For several years a number of these administrators, including Drs. Marilyn Heins, Norma Wagoner, and Carol Rosenberg, had hosted a women's breakfast at meetings of the AAMC's Group on Student Affairs. In 1974, responding to member institutions' new recognition of the concerns of women medical students, AAMC staff member Judy Braslow, followed by Kathleen Turner, began devoting 30 percent of her time to women's issues. A Women in Medicine Planning Committee was established, and from 1976 the AAMC published a yearly statistical compendium, *Participation of Women and Minorities on U.S. Medical School Faculties*, denoting the rank and departmental distribution of women faculty. Starting in 1978, most of the AAMC's member schools began appointing women's liaison officers (WLOs), followed a few years later by affiliated societies and hospitals. A representative committee of WLOs meets semiannually with AAMC representatives to guide programming. In 1987, the Women in Medicine program was expanded and relocated within the Division of Institutional Planning and Development, and Janet Bickel was hired to direct it. Since then

Bickel's program has published a quarterly newsletter, created semiannual professional development seminars for junior faculty and, more recently, senior faculty women, and published three extensive resource manuals, *Medicine and Parenting* (1991), *Building a Stronger Women's Program* (1993), and *Enhancing the Environment for Women in Academic Medicine* (1996). One indication of the AAMC's responsiveness to change is a six-page section of its application handbook directed at women and, more briefly, minority medical school applicants. This section now begins, "The AAMC works toward the removal of any barriers which stand in the way of women having successful careers in medicine." By 1993 close to half of the 140 medical schools in the United States and Canada had created a program, an office, or a committee on women in medicine. In 1996, the AAMC Project Committee on Increasing Women's Leadership in Academic Medicine published its findings and detailed a list of recommendations for the AAMC, medical school deans, and other administrators for facilitating the entry of more women into leadership positions in medicine.[15]

Following the AAMC's lead (and bowing to the decisive demographic shift in the profession), other institutions and organizations have created programs to support women in medicine. The AMA, for example, established a Department of Women in Medicine in 1989. It is directed by Phyllis Kopriva, who had been with the AMA for nearly a decade before taking this position. According to Kopriva, around 1979 the AMA board of trustees appointed an ad hoc committee on women in medicine consisting of women physicians. It was later made an advisory panel to the board, and in conjunction with Kopriva, the panel submits reports and policy recommendations to the AMA House of Delegates. The committee's underlying objective was to learn more about women in medicine and attract women members to the AMA. As Kopriva noted in 1990, women were distinctly underrepresented in AMA membership rolls, as active participants, and as leaders in the AMA. Indeed, the first woman on the AMA board of trustees, Dr. Nancy Dickey, a family practitioner in College Station, Texas, was elected only in 1989, after serving for about a decade on the AMA Council of Ethical and Judicial Affairs. In 1993 she became chair of the board. In 1997, at the age of forty-six, she was elected the AMA's first woman president. Dickey's career is emblematic of many of the recent changes in medicine for women. She is married and the mother of three children, she specializes in primary care, and she is an associate professor at Texas A & M University College of Medicine. Regarding the status of women in the AMA, she has said, "Our organization is perceived as an old boys' network with the emphasis on *old* and *boys*. I hope that simply by walking to the podium I negate both of those [perceptions]."[16]

In recent years the AMA has sponsored surveys on maternity leave policies for medical women, child care, part-time residencies, sexual harassment, gender bias in medical liability insurance rates, and the creation of liability insurance for part-time practitioners. Unlike the AAMC, however, which has acted directly to effect change, the primary mission of the AMA's Women in Medicine Department has been to gather data for consideration by its members. Kopriva commented, "What we are trying to do with our somewhat limited resources is to take these issues one by one and . . . collect as much information as possible, and then analyze it and . . . keep on top of it, keep monitoring it on an ongoing basis." A survey of women AMA members and nonmembers conducted by the Ad Hoc Committee in 1980 found three main reasons for women's disproportionately low interest in "organized medicine," including local and state medical societies and the national association. The reason cited most often was lack of time due to conflicting family obligations; second was women's perception of not being welcome, whether because of their having chosen to combine medicine and motherhood, or because of perceived sex discrimination, or because of "lack of confidence" in their organizational and political skills. (Two-thirds of the members sampled were over the age of thirty-six.) By shifting the age and gender composition of the House of Delegates, it was hoped, AMA policies would be more likely to attract women members and reflect the needs of younger physicians, many of whom are women. By 1993 women represented about 14 percent of the membership, as compared with less than 7 percent in 1980, indicating some success in closing the membership gap. But according to an AMA study, even in 1993 more than 50 percent of members of the House of Delegates were sixty-one years of age or older, although the median age of members was forty-seven. Still, in 1993 the AMA House of Delegates adopted recommendations by its Council on Ethical and Judicial Affairs that "immediate" steps be taken to redress the shortfall in the salaries and leadership roles of women in the profession.[17]

Even more than the AMA, AMWA was well placed to benefit from the rising number of women in the profession. Between 1970 and 1980, the number of female first-year students nearly quadrupled, rising to 28 percent of all students. Between 1970 and 1988 the number of women in practice reflected this increase, rising from 25,401 to 98,446 (see Table 8.3). More than 75 percent of these were primarily occupied in office-, HMO-, or hospital-based clinical care.[18] Women as a percentage of medical school faculties increased more slowly, from 14.5 percent in 1975 to 20.1 percent in 1989. In 1995 women comprised 25 percent of all full-time faculty but only 10 percent of full professors. In 1998 they represented 26 percent of full-time faculty and 11 percent of all full professors.[19] In 1974 AMWA member Marlys Witte and the staff of the AMWA Professional Resources Center at the University of Arizona Health Sciences Center had already conducted one of the first research studies to track the careers of medical women in academia. As she and her colleagues discovered, women were on the ground floor. In an article published in 1976, Witte and her colleagues wrote, "The figures for M.D. female Deans and Vice-Presidents for Health Affairs are simple: there are *none.*" Women constituted only 2.9 percent of full professors.[20] AMWA organized conferences to address the career development needs of women faculty, such as the 1979 AMWA Regional Conference and Workshop on Women in Medicine, with the intent of "outlining the goals and delineating the constraints which limit the expansion of the role of women in medicine [as well as] plans and strategies to overcome the restraints."[21]

Not only women faculty but also women students, weary of being mere observers, were beginning to make themselves heard in the stu-

Table 8.3. Physicians in the American workforce, by gender

	1960	1970	1980	1990	1995
Total	229,590	334,028	467,679	615,421	720,325
Men	213,918 (93.2%)	308,627 (92.4%)	413,395 (88.4%)	511,227 (83.1%)	570,921 (79.3%)
Women	15,672 (6.8%)	25,401 (7.6%)	54,284 (11.6%)	104,194 (16.9%)	149,404 (20.7%)

Sources: Carol Lopate, *Women in Medicine* (Baltimore: Johns Hopkins University Press, 1968), Appendix 4, pp. 196–197; Janet Bickel and Phyllis R. Kopriva, "A Statistical Perspective on Gender in Medicine," *JAMWA* 48, no. 5 (1993): 141–144; American Medical Association, *Physician Characteristics and Distribution in the U.S., 1995–1996* (Chicago: AMA, 1996).

dent associations of the AAMC, the AMA, and AMWA. AMWA en-
hanced student involvement by amending its bylaws. After 1973 every
junior branch was entitled to send one delegate per eight members to
sit as a member of the House of Delegates. By 1979 students made up
48 percent of the membership.[22] Student interests were beginning to be
felt in ways that would eventually be directed upward, toward AMWA's
national leadership. Interviews with prominent participants in these
meetings during the 1970s in the Boston and the New York City
branches revealed that strong student concern about fairness in medical
education, sexual harassment, access to top residencies, abortion and
other women's health issues, and—particularly for residents—the availability of child care, was critical to AMWA's heightened political awareness. Interest in such issues brought women faculty and students together to reform their local medical communities. It also sparked a
move to increase student power within AMWA itself. For example,
students and their advocates successfully changed the bylaws so that,
after 1982, student representatives could actually vote. Oddly, AMWA's
lobbying activities during the "manpower shortage" deliberations of
the 1960s apparently were unknown to these younger women. And
undoubtedly the younger generation's more directly assertive political
style seemed painfully different from the genteel lobbying of their elders. One woman, now an orthopedic surgeon, active as a national student coordinator for AMWA in the early 1980s, recalled that it wasn't
easy to interest women students in AMWA because it "was seen almost like an old ladies' club . . . The older women were almost like a
women's auxiliary of the hospital: they were interested in fund-raising
for worthy charities and they were interested in some women's health
issues and children's health issues, but they were not very inspiring."
Still, some older AMWA members, such as Bertha Offenbach in Boston
and Rosa Lee Nemir in New York, successfully persuaded the younger
generation, physicians as well as students, to work for change through
AMWA.[23] Interviews conducted with physicians in Boston; New York;
Washington, D.C.; and Louisville, Kentucky, made clear how much
AMWA was the beneficiary of an essentially reciprocal process, the
mutual politicization of women students and physicians during these
years.[24]

Nevertheless, rapid changes are never easy to endure, especially in an
association whose leaders have served it faithfully for many, many years.

The rapid increase in the number and political awareness of medical women was both exhilarating and, at times, confusing. AMWA experienced growing pains as its younger members vied with more traditional leaders to set their stamp on the association—in substance as well as style. At the 1980 annual meeting in Cambridge, Massachusetts, for example, an alternate slate of candidates for association offices was offered from the floor, an unprecedented event. (Delegates received the list of new names on the tabs of Lipton tea bags—a reference to the Boston Tea Party and the American Revolution.) Although the original slate was duly elected, AMWA responded to the insurgents' demands for increased access to the leadership process and greater organizational visibility on important issues. In 1981, a major restructuring of the organization and a clarification of its goals were begun in earnest. The amended AMWA constitution included among its stated objectives, "to encourage women to study medicine" *and* "to ensure equal opportunity to do so."[25] By 1985, despite the ambivalence of some older members, AMWA had redirected its mission and strategic plan to include both women physicians' professional issues, conceptualized as "looking inward," and health policy issues, such as tobacco use among women, breast cancer screening and detection, abortion rights, and the inclusion of proportional numbers of women in clinical research trials, issues that were directed toward the population at large. In short, the organization returned to its roots, taking an interest in women's health policy as a way of highlighting AMWA's uniqueness among medical associations.[26]

Women physicians also organized associations within their specialties, such as the Association of Women Surgeons (AWS) and the Association of Women in Psychiatry (AWP). Women surgeons have particularly tried to address the needs of women residents in surgery, who in 1996 ranged from about 20 percent of general surgery residents and 16.4 percent of plastic surgery residents to only 7.1 percent of orthopedic surgery residents.[27] The AWS publishes a pocket-size paperback called the *Pocket Mentor,* which discusses questions that can bedevil women starting out in what is frequently referred to as medicine's most macho specialty. From the historian's perspective, it provides a picture of issues women surgeons consider important. In addition to chapters on "Learning to be a Surgeon" and "Getting the Work Done," the guide also includes less traditional topics, such as "Surgical Politics,"

"Gender Issues," and "Taking Care of Yourself." The latter chapter, for example, considers such questions as how to preserve a personal relationship during the intense years of a surgical residency, avoiding an unplanned pregnancy during residency, negotiating maternity leave, and other critical issues for residents who are young and ambitious but also at their peak years for starting a family. Most important, though, the *Pocket Mentor* is designed to serve as a stopgap "mentor" for career planning, offering a residency calendar that spells out major career goals for each of the five to seven years of postgraduate training. It also advises women to seek out one or more mentors as early as possible. Women psychiatrists, too, have addressed professional development issues such as leadership training, and have been especially active as lobbyists to remove scientifically questionable, gender-linked diagnostic categories, such as "self-defeating personality disorder," from the *Diagnostic and Statistical Manual* published by the American Psychiatric Association (APA). The APA elected its first woman president, Dr. Carol Nadelson, in 1985.[28]

Paradoxically, many senior women faculty reported during the early 1990s that just when medical schools, the AAMC, and women's specialty groups began grappling with residual gender discrimination and the career development needs of faculty women, women medical students seemed to be losing sight of the issue of gender inequity. Of course, this reflected a trend that was not exclusive to medicine. A recent story in the *New York Times Magazine* labeled college students of the 1990s the "no-complaints generation." Women students at the University of Wisconsin, Madison, reported that "they expect little or no discrimination once they land in the workplace. 'I feel I don't have to fight so much,'" said one young woman.[29] Speaking specifically about younger women physicians, AMWA's executive director Eileen McGrath commented in 1990, "I think younger members don't see the need for [organizations for women physicians] as much . . . and I think the reason for that is you have a built-in networking group in medical school. [I]f you are with women all the time, both in medical school and in a practice situation, why pay dues and go somewhere for something you think you have there?" But, she added, "I think what they don't realize is that it's still a man's world and medicine is still a man's profession." One of AMWA's past presidents put it this way: "[For women] in medical school, it's like a game, you know it's a social situation, all your classmates are your buddies, everything is A-OK, but you haven't gotten into that other, the real stuff, you know, where people are clawing around for promotions and they are clawing around for appointments ... That's when ugliness comes out. That's when you have to know how to play the game." A resident who attended medical school in the early 1980s agreed. She also observed that residency, in contrast to undergraduate medical school, is "a time when women who may not have ever faced discrimination based on their sex, may face it there."[30] Judith Lorber, a sociologist who frequently writes about women in medicine, draws an important conclusion—namely, that senior women can't afford to neglect mentoring their junior women colleagues; nor, Lorber believes, can junior women faculty afford to eschew senior women colleagues as potential sponsors. Increased numbers alone, without coordinated, collective action, will not create power for women in medicine.[31]

Gender Bias and Sexual Harassment

In 1991 the near resignation of Dr. Frances Conley, professor of neurosurgery at Stanford University, in response to a climate of persistent sexual harassment, put younger women physicians on notice that not all gender inequities had disappeared, even two decades after the introduction of equal opportunity legislation. Indeed, while prejudicial gender stereotypes constituted the subliminal status quo for many medical institutions prior to the 1970s, episodes of flagrant sexual harassment often flared up when that norm was threatened. Sexual harassment is defined as unwelcome conduct of a sexual nature, physical or verbal, that constitutes a condition (or implied condition) of future or current employment, evaluation, or advancement, or that unreasonably interferes with an individual's work or academic performance, or creates "an intimidating, hostile, or offensive" work or educational environment. A 1991 AMA survey of work satisfaction among second-year medical residents found that 63 percent of the women respondents had experienced sexual harassment during the previous year, almost four times more than men respondents. Incidents reported included, in decreasing order of frequency, "sexual slurs, favoritism, sexual advances, denied opportunities, and poor evaluations." A woman who graduated from Georgetown University School of Medicine in 1988, for example, recalled having had to enlist the aid of a senior woman professor before any action was taken to stop the obscene and threatening letters that were being written to the members of a student women's group.[32]

Dr. Conley, the sole woman neurosurgeon in her department, tendered her resignation after learning that Stanford intended to hire as permanent department chair a neurosurgeon who had, in her view, condoned and even committed acts of sexual harassment, thereby creating a hostile work environment. Conley is one of the mere 1 percent of American neurosurgeons who are women.[33] In the course of a twenty-year career at Stanford, she had been subjected to instances of unwanted fondling and frequent references to her as "honey" by male colleagues in the operating room. Her decision to act was made not only on her own behalf but also to ensure that future generations of women surgeons would not have to endure either flagrant sexual abuse or, more commonly, the subtle put-downs termed "micro-inequities."[34]

Although Stanford reconsidered its choice for the chair of neurosurgery, resulting in Conley's rescinding her resignation, questions of persistent gender bias and sexual harassment continue to roil the professional waters in medical schools and hospitals. Sexual harassment refers to a continuum of hostile behaviors ranging from pointed sexual remarks to outright sexual assault. Such behavior can become the basis for legal action. Micro-inequities range on a continuum from unconscious professional slights or "invisibility" to conscious put-downs, exploitative assignments, and inequitable allocation of resources. To call these subtler manifestations of harassment micro-inequities is not to suggest that their effects are trivial. While they usually are not actionable, they can substantially reduce professional performance, visibility, status, and, most insidiously, self-esteem. Coping with these accumulating stressors has been likened to lifting a "ton of feathers."[35]

Some predicted that when the percentage of women physicians reached a level proportional to their number in the population at large, gender-based stereotyping and discrimination would decline. Women would move freely throughout the profession and would occupy leadership positions in more than token numbers. But compared with male physicians, women in medicine today neither earn salaries nor hold leadership positions in proportion to their experience and number— even when controlling for choice of specialty, number of patients seen per week, and number of years in practice. Primary care specialties— with the exception of obstetrics-gynecology—do pay, on average, less than the more procedure-oriented specialties such as surgery and radi-

ology. Younger physicians, such as the majority of women now in practice, also earn less than senior colleagues. Yet women, compared with
men in the same specialty and with the same years of experience, earn
less per patient and less per annum than men.[36] They also are underrepresented in the senior ranks in health care institutions, accounting for
just 10.2 percent of tenured clinical faculty as of 1994. Between 1976
and 1991, moreover, more than twice the proportion of men as women
had attained the rank of full professor; by 1997 the ratio of men to
women at that rank was more than three to one. Among hospital CEOs,
women made up about 10 percent as of 1995.[37]

One frequently cited explanation for these discrepancies has been
called the theory of "cumulative career advantages or disadvantages" or,
alternatively, the theory of "limited differences," a model I referred to
in the introduction. It hypothesizes that the advantages or disadvantages accrued from the early years of a career onward will accumulate,
or cascade, in importance over time. For example, if two junior faculty
members at the same institution, one male and one female, receive
different starting salaries and differing levels of commitment to providing protected time and facilities for research at the beginning of their
career, within a few years they will probably begin to demonstrate differing levels of research productivity. That, in turn, will be reflected in
their success in obtaining external grant funding to continue that productivity and, eventually, to obtain promotion and tenure. If, at the
same time, one of these physicians attracts the support of a sponsor or
mentor, this too will enhance his or her chances for success. And if
female physicians choose to slow down their rate of research productivity for a few years early on, to better accommodate the needs of family,
the long-term effects may well be magnified when they are compared
with males who began their career at the same time. The differences in
the career paths of Drs. Ruth Lawrence and Mary Ellen Avery, described in Chapter 7, illustrate the effect of cumulative advantages and
disadvantages, as well as the recognition that this phenomenon can only
be linked to gender in the aggregate. Individuals may well advance far
in medicine, regardless of gender or racial prejudice. Moreover, different individuals will respond differently to negative social expectations
and other barriers to success.[38] Yet, on the whole, women appear to
experience the effects of cumulative disadvantages more uniformly than
men.[39]

The situation at Harvard University Medical School, one of the premier medical colleges in the country, illustrates that the accumulation of advantages or disadvantages can characterize the culture of an institution as well as the fortunes of individual professionals. Harvard did not admit women students until 1945. Even then, they were admitted only after "'scenes of disorder and confusion'" during faculty meetings and at least one change of heart by the university board. The decision might have been delayed even further had not a shortage of medical school applicants during World War II added a sense of urgency. Women were not hired as full-time faculty until 1947. In 1919 the prominent industrial toxicologist, Dr. Alice Hamilton, was offered an appointment at the rank of assistant professor—but only if she would agree not to walk in the graduation procession; she insisted instead on a half-time appointment, continuing her work as an industrial consultant from her home base at Hull House in Chicago. In 1947, Dr. Marian Wilkins Ropes was appointed instructor in medicine, thereby becoming the first woman to receive a full-time appointment to the faculty. But women students and faculty remained an isolated minority. In 1972, in response to increased and vocal pressure from women students and faculty, Dr. Mary Howell, head of the pediatric behavioral medicine unit at Massachusetts General Hospital and assistant professor of pediatrics at Harvard Medical School, was appointed associate dean for faculty affairs, the first woman to attain that administrative rank at Harvard. The following year, a Joint Committee on the Status of Women (JCSW) was created. Yet by 1991 the handful of senior women on the faculty concluded that additional leverage had to be brought to bear on the administration. Whereas in 1991 women constituted nearly 10 percent of full professors in American medical schools, at Harvard they represented only 6.7 percent; at the level of associate professor, the national average was 19.5 percent, while at Harvard it was 13.7 percent. As a group of five senior women wrote in a letter to Dean Daniel Tosteson, "We have examined the situation and are seriously concerned about the subtle systematic biases that are built into the Harvard system." In 1995 the JCSW of the medical and dental schools, along with Harvard University President Neil Rudenstine, held a dinner to commemorate the fiftieth anniversary of women's admission to the medical school. As Rudenstine commented at the dinner, "Unlike most birthdays or anniversaries, we look back on this one and wonder, 'Why aren't we older?'"[40]

Had Dr. Mary Howell been asked that same question, her experience as a dean from 1972 to 1975 could have thrown considerable light on the matter. After graduating from Radcliffe, she went on to receive an M.D. and a Ph.D. (in developmental psychology), both from the University of Minnesota), and a law degree from Harvard. Howell was one of the rare women in the 1950s who combined medical school and motherhood. Married and divorced twice, Howell eventually raised seven children, four of whom were adopted. She was also one of the first physicians to become active in the women's health movement, contributing to the first edition of *Our Bodies, Ourselves: A Book by and for Women* (1971), and helping to organize the 1975 Boston conference at which the idea for the National Women's Health Network was first conceived. She said about the problems faced by women medical students, "It is clear that there is a direct relationship between discrimination against women as medical students and as patients: the one supports the other." Moreover, she also wrote, "If women physicians do not concern themselves about health care for women, then health care for women is not likely to improve in the near future."[41]

In 1973, Howell published the scaldingly critical book *"Why Would a Girl Go into Medicine?" Medical Education in the United States: A Guide for Women*. Because of her administrative position at Harvard, however, she wrote it under the pseudonym Margaret A. Campbell, M.D. In 1974 she summarized her findings in an opinion piece for the *New England Journal of Medicine*. (Under pressure from feminist colleagues, the second edition of the book acknowledged her authorship, to the embarrassment of Harvard.) Based on survey responses from women who were students during the debates over the equal opportunity legislation, the book provides vivid testimony to the discriminatory climate encountered by women in most American medical schools. Howell's idea for the survey sprang from her own experiences during a career spent pioneering new roles for women physicians. As a student and mother-to-be at the University of Minnesota Medical School, she herself had experienced genuine hostility. "The more pregnant I got, the more invisible I got—because doctors can't be pregnant." Her sense of commitment to women's participation in medicine, her memories of her own traumatic years as a medical student, and the frustrations she faced as associate dean at Harvard Medical School informed her critical account of the plight of women in medical school.

Howell was particularly sensitive to the possibility of a backlash against women entering the profession, because at the time, their number was increasing with such unexpected rapidity. She was appointed associate dean in 1972 through the efforts of a medical student–faculty committee rather than, Howell implied, because of any consensus among the senior administrators. She always felt quite embattled. As Howell told an interviewer in 1978, when the numbers of women first increase beyond the level of tokenism, they're seen as "very threatening. So I think that the [increasing] numbers meant that first of all, people got used to seeing a lot more women around, but secondly they weren't treated as pets; as the numbers got larger they seemed quite threatening. It engendered a kind of increased conservatism . . . a dig-in-the-heels kind of conservatism." Much of this was manifested in what Howell termed "belittling" comments. One survey respondent wrote that the new dean of her medical school had publicly commented, "I don't think women belong in medicine anyway." Another dean was quoted as saying in a public lecture, "Lord knows there are already too many women in medicine." After a few years as associate dean, Howell decided to return to patient care because she felt like a "token." As she told an interviewer in 1979, "Harvard got tremendous mileage out of my appointment . . . The administration expected us to act as student pacifiers rather than student advocates."[42]

As the exchange quoted at the start of this chapter between Drs. Dimond and Heins reveals, overcoming the barriers to a sound medical education was not the only hurdle facing women in medicine. Twenty-five years after Howell's survey was published, the question facing the profession is not whether women will become full, practicing members of the profession. Rather, it is whether they are free to advance as far as they wish in their chosen profession. Even in specialties that have long attracted women, such as internal medicine and pediatrics, in 1995 women held department chairs in only two and ten departments, respectively. One explanation was offered by a woman in pediatrics described as a "rising star" in her field. Writing to a woman colleague—a department chair—she observed, "In my experience, women are not advancing academically because the men around them are not doing for them what they do for men: talking them and their work up to leaders and study section members in their fields, promoting their visibility by arranging for them to give important talks, putting them up for and pushing their membership in academic societies, and so forth. This is what a mentor does."[43]

Clearly, recognizing potential obstacles to advancement for women was an essential first step. Learning to respond effectively to a negative professional climate, or even a neutral one, is the crucial next step. In short, many women physicians in the past two decades have concluded that they cannot afford to wait for mentors to come to them; they have had to either go out and recruit them or, much more difficult, become their own "mentors." One of the earliest systematic efforts to provide what the AAMC's Janet Bickel calls "mentor replacement therapy" for academic medical women began in 1987 when the AAMC Women in Medicine Program began holding annual three-day professional development seminars for junior faculty women. In 1992 the AAMC also began offering a version designed for senior women faculty. Many medical schools, too, have developed similar programs for faculty development. In their original form, the AAMC seminars included sessions on developing professional networks, mentoring, obtaining grants, department finances, negotiation skills, presentation skills, attributes of leaders and managers, effective strategies for career building and "self-promotion," team building, and conflict management. The workshops initially focused almost exclusively on career development. It is suggestive that, after the first few years, optional focus groups were added to address what Bickel terms a "holistic" approach to professional and personal well-being. These include sessions on how to balance multiple roles, particularly professional and personal or family responsibilities. As this book's last chapter will discuss, the need—and desire—to reconcile professional and family life is the woman physician's most intractable obstacle to professional equality.

African American Women Physicians since Integration

For one group of women physicians, the persistence of negative stereotypes came as no surprise. Many African American women physicians—even in the late twentieth century—acknowledge that racial discrimination may have abated somewhat but has not disappeared. Moreover, the combined effects of race and gender frame a context that confounds optimism with its seemingly glacial rate of social change. Between 1970 and 1990, for example, the percentage of African American women among all physicians in the United States rose from only .4 percent to about 1.3 percent.[44] Medical school admission statistics for the 1990s suggest that black women have been catching up and may even surpass black men. As of 1992, the ratio of male to female African American physicians was approximately two to one, comparable to the ratio of male to female physicians overall. And among black medical graduates, women outnumbered men between 1990 and 1995.[45] Yet a recent decline in the number of African Americans applying to medical schools suggests that progress, for black women *and* men, has merely inched forward.[46] In 1975, African Americans comprised 5 percent of medical school graduates, Native Americans, .2 percent, Hispanics and Latinos, 1.1 percent, and underrepresented minorities as a whole, 6.3 percent of total graduates. Twenty years later, the percentage of black medical graduates had risen only to 6.6 percent, Native Americans, .5 percent, and Hispanics and Latinos, 6.1 percent, altogether making up 13.2 percent of medical graduates in 1995. About 20 percent of the African Americans graduated from the four predominantly black medical schools, Howard, Meharry, Morehouse, and Charles R. Drew Postgraduate Medical School.[47]

African Americans, who for most of their long history in this country were denied adequate provision for health care and adequate educational opportunity to provide it for themselves, nevertheless have responded with persistent and passionate dedication to supply the needs of the black community.[48] Beyond support for a black professional class and professional institutions such as medical schools, hospitals, medical journals, and medical societies (the National Medical Association, or NMA, was founded in 1895, the National Association of Colored Graduate Nurses in 1908), community service has long been a fundamental value among many African Americans. This was true prior to the civil rights era, when hospitals first were required to integrate and predominantly white medical schools began increasing their percentage of black students, and it has continued up to the present.[49] In addition, African American women physicians have frequently experienced the "double jeopardy" associated with discrimination on account of gender *and* race. Like other women physicians, many retain the cultural imprint of Progressive-era ideals of women as "social housekeepers." Conscious of the dearth of external support to the black community for much of this century, black women in medicine and health care have frequently made community service a cornerstone of their professional life.[50]

The career of Dr. Edith Irby Jones exemplifies many of these themes. A Houston internist and the first woman to serve as president of the NMA, her career spans the entirety of the civil rights era. She was born in 1927, and in 1949, eight years before the "Little Rock Nine" painfully integrated Central High School in Little Rock, Arkansas, Edith Irby became the first African American to be admitted to the medical school of the University of Arkansas. She graduated in 1952. Irby was admitted into the school, but she was not allowed to eat, room, or even use bathroom facilities with the other students. The daughter of a sharecropper and a domestic worker, Irby was used to hardship. As a scholarship student at Knoxville College, her many part-time jobs had paid only thirty-five cents an hour, yet she had managed to send money home. She also took large risks on behalf of social justice by—secretly—spending many of her nights traveling around the state with a team of organizers from the National Association for the Advancement of Colored People (NAACP). Fortunately, after her first year, most of her expenses were funded by members of Little Rock's black community and, at the urging of the publisher of the *Journal of the National Medical Association*, by the Jesse Smith Noyes Foundation. (The medical school's custodial staff supported her too, in more subtle ways, for example by providing daily a vase of fresh flowers for her table in the adjoining—and segregated—staff dining room.) Irby recalls that several of the white women students became her good friends, but her strongest support through medical school came from her husband, Professor James B. Jones, whom she met and married when she was a second-year student.[51]

After an internship at University Hospital in Little Rock, Dr. Irby Jones practiced medicine in her hometown of Hot Springs for six years. But by then the racial climate in the state had become so polarized that she and her husband decided to move to the somewhat less threatening environment of Houston. By this time they had had two of their four children, but Dr. Jones was determined to complete a residency in internal medicine at Baylor College of Medicine Affiliated Hospitals. Baylor itself seems to have been hospitable to her, but the hospitals to which she was assigned continued to segregate her and limit her patient roster. She completed the last months of her residency at Freedmen's

Hospital in Washington, D.C. In 1962, ten years after graduation from medical school, Jones set up a private practice in Houston that continues to thrive. She sees patients in her office, attends at the predominantly black Riverside Hospital, and supervises residents at Baylor and the University of Texas Health Science Center in Houston. Dr. Jones's commitment to social betterment has led her to maintain a wide variety of professional and voluntary associations. In 1985 she was elected president of the National Medical Association. She has also been active in AMWA and Planned Parenthood, on the Houston school board, and in many other groups. In 1991 she sponsored the establishment of a medical clinic in Haiti. The themes of her 1985 presidential address to the NMA offer a convenient summary of her social values. In Jones's view, the NMA is "anointed to be the conscience of all providers of health care. We are anointed to tell the news that all must have access and availability to quality health care . . . we are the advocates for the poor, the minorities, the aged, and the imprisoned." She also expressed alarm at the dwindling number of black applicants to medical school and noted possible reasons for the decline: "lack of availability of financial resources for potential physicians, the lack of counseling and motivating of competent students to pursue medicine as a career, the double standards for meeting requirements to be accepted into medical school, and the hostile environment for black students."[52]

Arkansas was also the home of the first African American woman to become surgeon general, Dr. Joycelyn Elders. In fact, a speech given by medical student Edith Irby Jones while Elders was in college sparked Elders's ambition to become a doctor. She joined the army to take advantage of the GI Bill to finance her medical education. Elders attended the University of Arkansas School of Medicine during the worst of the Little Rock integration crisis, yet she recalls rooming with the two other women students in her class, both of whom were white. Like Jones, she also gained crucial moral support from her husband, whom she married during her final year of medical school. Both women suffered from segregation as students, and both are strongly committed to social justice. Yet Elders's career differs significantly from Jones's. Unlike her predecessor, Elders was encouraged to become an academic physician and a subspecialist in pediatric endocrinology—the first to be board certified in Arkansas. In 1976 she became a full professor at the University of Arkansas. The women's similarities, however, far overshadow their differences. Both strongly believed in using their education and opportunities to benefit the community. When Bill Clinton, as governor of Arkansas, persuaded her to take a leave of absence from the university in 1987 to become director of the state's Health Department, her premier goal was to reduce teenage pregnancies through a program that emphasized "early childhood education, comprehensive health education from [grades] K through twelve, parenting education, teaching young men to be responsible fathers . . . making services available in school . . . and giving hope." Unlike Jones, Elders is no diplomat, and her bluntness has produced some bumps along an otherwise smooth career path. When she and Governor Clinton announced her teen pregnancy prevention program for the state, for example, the press asked whether she was going to give out condoms in the schools. Elders replied, "We're not going to put them on the lunch trays, but, yes." After she moved to Washington, D.C., in 1993, her term as surgeon general was cut short, in fact, because Elders was unwilling to sacrifice policy for political finesse.[53]

It isn't hard to see how the legacy of slavery, legalized segregation, and other persistent forms of racism distinguish African American women physicians from the general population of American women in medicine. Less obvious are the ways in which commonalities among women doctors—regardless of race or ethnicity—often overshadow their differences. This issue was raised in a series of nine interviews conducted with African American women physicians in 1992 and 1993 by Professor Marian Gray Secundy of Howard University. Among other questions, she wondered "what they might tell me about the experience of being black women professionals and to hear whether they viewed themselves as similar or dissimilar to their white female counterparts." Among the most common reason these physicians gave for entering the profession was "a desire to be of service . . . a desire to help others." As Secundy noted, in this they resembled their professional forebears, the early generations of women physicians—white and black. Several sounded a distinctive note, however, when they invoked the power of a spirituality nurtured by their families and churches from a very young age. And almost all acknowledged a "commitment to their heritage." Finally, all pointed to the effects of racism encountered in their own professional lives. Perhaps for this reason, as Secundy commented, not one "made any distinction between white men and white women" when discussing their relationships with white physicians. She concluded, they "do not appear to have consciously struggled with gender issues as much as they have with issues of race and oppression."[54]

Exploring these questions in a comparative context and among a group of more than two hundred women physicians helped clarify the interconnectedness of race and gender for black women doctors. Statistician Marilyn Greer and I conducted an anonymous survey of African American and non–African American women physicians in the Houston-Galveston region, which has a population of more than 4 million.[55] We matched a population of 109 African American women physicians in this region with an equal number of randomly selected other women physicians to find out whether the two groups would differ significantly with regard to their backgrounds; major problems faced during their years as students, residents, and practitioners; and their motivation for their choice of specialty. We had a total response rate of 55.4 percent, 51 percent from the African American cohort and 59.8 percent from the non–African Americans, of whom 4.7 percent were Hispanic and 12.1 percent Asian. Our results did not contradict Secundy's but did put them in perspective. We found, as one might have predicted, that the African American respondents cited racial discrimination as a major obstacle during medical school, residency, and in practice. But we also found that black women perceived *gender* discrimination to be a greater obstacle than did the non–African American survey participants, both in medical school and in residency. They also experienced more perceived sexual harassment during residency. In their experience, the notion of "double jeopardy" was not an abstraction.[56]

In recent years American medical schools have reduced gender-specific obstacles to the success of women and minority students. But residency is still perceived to be a war zone. Being a resident, of course, is notoriously stressful for both male and female doctors.[57] But over and above the usual stressors—chiefly the lack of sleep and the "sink or swim" quality of many residency programs—we found that gender discrimination during residency was reported by many of our respondents, and with greater frequency by African Americans (32.7 percent vs. 19 percent). In addition, nearly 40 percent of the African American women experienced racial discrimination during residency, whether from fellow staff members or from patients. A memoir by African American physician and historian Vanessa Gamble recounts a story from her own residency in Philadelphia, when a (white) patient persistently mistook her for a maid. Her discomfiture was dramatically heightened by the evident amusement expressed by her fellow residents. No doubt experiences like Dr. Gamble's have not been uncommon.[58] Black women residents also have experienced a greater lack of mentors and role models than nonblack women. As one African American physician commented, "It's a difficult task to be in such a high-powered position, and no one around looks like you." Another wrote, "As a minority female physician I think you have to prove yourself in your profession. You are not treated equally even though the other physicians or administrators act like it." At the same time, many more of the black women in the survey than the others received support during all phases of their careers from their peers and from nurses, secretaries, and even the custodial staff. Study participants who were not African American tended to emphasize financial hardship as a significant obstacle during medical school. Still, they were three times more likely to receive encouragement or mentoring from faculty, both as students and as residents. But for both groups, residency seems to have been "quite a shock," as one respondent put it. More than any other stage of their career, residency presented the most difficult challenge.

Specialty choice was a subject of wide agreement among black and nonblack respondents. It is sometimes said that American society's deeply rooted gender norms subtly influence women physicians' specialty choices, a generalization that obscures the role of personal preference. Women doctors do consistently favor primary care over all other fields. As of 1995, five specialties accounted for more than 60 percent of women physicians' specialty choices: internal medicine (especially general internal medicine), pediatrics, family practice, obstetrics-gynecology, and psychiatry.[59] Our study, however, offered respondents the opportunity to list "personal preference/interest" as a reason for their specialty affiliation. Nearly three-fourths of both African American and non–African American physicians responding to the study listed "personal preference/interest" as either the first or second reason for their choice of specialty (57 percent of both groups listed it first). Twenty percent of non–African Americans, but only 10 percent of black women physicians, gave "quality of life issues" as the reason for their choice. No respondent listed gender discrimination as her most important reason for choosing her specialty; only two respondents cited it at all.

Reflecting national trends, two-thirds of those surveyed chose primary care specialties, including internal medicine (20 percent), pediatrics (16 percent), family practice (15 percent), and obstetrics (13 percent); 14 percent chose psychiatry. With the exception of pediatrics, which was the specialty choice of twice as many black women physicians as others surveyed, few differences were evident in their preferences. The preference of women physicians for primary care specialties is a significant—and long-standing—commonality between black and nonblack women physicians, reflecting both perceived opportunities and personal choice.[60]

Further, there were no striking differences between African American and all other women physicians in the study in the extent to which "community needs" influenced their choice of specialty or practice setting. It is true that black and Hispanic physicians of either sex have been shown to practice in communities of underserved black and Hispanic Americans more than white physicians of either sex.[61] In our study, furthermore, twice the number of African American women physicians reported spending time each week in church-related or community activities. In the general population of American physicians, however, white women physicians resemble underrepresented minority physicians more than they do white males in the proportion of poor and underserved patients in their practice populations, regardless of the race or ethnicity of those populations.[62]

The career and life choices of Dr. Judith Martin Cadore, a family physician now in practice with Dr. Edith Irby Jones, may clarify the comparison between African American and other women doctors. Dr. Cadore—a wife, a mother, and one of the study respondents—was brought up in the small Texas town of Bay City, about eighty miles south of Houston. The first African American to be valedictorian of her high school, she entered MIT as a National Merit Scholar, graduating in 1980; from 1984 to 1990 she attended the University of Texas Medical Branch. As she recalled, during her childhood "there was segregation, and so we had our own little world there . . . I felt very supported by my family, people in the neighborhood. There were very close links between the schools and the church . . . just a feeling of community, people moving up, and people wanting to see you achieve things that they hadn't been able to achieve." Her father was a science teacher, her mother a homemaker. But racial calm was a thin veneer in the Martin

Dr. Judith Martin Cadore (b. 1957), family practitioner, Houston. Undated photograph, ca. 1995. (Photograph courtesy of Dr. Cadore.)

family's part of the world. Dr. Cadore described her grandparents, children of ex-slaves, as pillars of the black community. Her grandmother, known as "Miss Virginia," wore "little gloves" and "dainty lace" when going out. But Miss Virginia, whose husband often worked out of town, kept a loaded shotgun by her bed. "They used to have night riders and just horrible things would happen. They'd burn people's houses, like Klansmen—that type of thing . . . So there was always this loaded shotgun next to this lace bed."[63]

Leaving Bay City for college in Boston was a wrenching change. (Cadore had had only enough money for a single ten-dollar college application fee. She had applied to MIT because she had heard it was "into science . . . It was so different from Bay City. Just being on the East Coast and being that far away from home.") Discrimination intruded occasionally, although Cadore remembers only one incident, when a professor asked her, "Why don't you go to a place where people like you [should] be?" By her senior year, Cadore was a double major in chemistry and creative writing and was attracting attention for her work as a folklorist, recording the oral traditions of African Americans in rural Texas. At the suggestion of MIT's president, Polaroid founder Dr. Edwin Land gave her an early model video camera to record her interviews the summer after graduation.

Cadore was accepted to medical school in Texas, but decided instead to work as a research chemist and after a few years perhaps, to attend medical school in Boston. Her family gently prodded her to return home, however, lest she be tempted to marry and settle down with someone from "up there." Besides, she herself felt that her life would be spent, ultimately, in Texas, practicing medicine nearer her family. The prospect of facing racism once again made her decision to return a difficult one. What she found, though, was a combination of subtle racism and blatant sexism. According to Cadore, just under 6 percent of the students at the University of Texas Medical Branch were African American during her years there, of whom half were women. Overall, women constituted about one-fourth of the class. Yet, she remembered, anatomy lectures still included "surprise" slides such as a "flaming red, swollen penis," described by the professor as the sequella to a medical student fraternity party.

At the start of her third year, when medical students generally begin their clinical rotations on the wards, Cadore—like her peers—was both excited and apprehensive. "I guess for me (being a black person), there was a lot of encouragement from the janitors (and they were mostly black) and the cashiers (and they were mostly black), and they were very, very proud of you and they were looking out for you." They, and the few black women residents, were her "support system," giving her a perspective shared by few other students. In fact, Dr. Cadore's perspective on medicine reverberates with the dualistic view of the insider/outsider. She remembers, as a nine-year-old, waiting with her family outside the operating room double doors in a recently desegregated Houston hospital, only to be told that her aunt, a patient "on the *other* side of those doors," had died. "That was when I decided I wanted to be on the other side of that door . . . It was just that feeling of being on the outside . . . So for me, when I finally went on the [hospital] floor and I was finally on the other side of that double door, that was very exciting for me. I was very proud, I was very happy to be there. I tried my best, when I interacted with patients, to make them feel comfortable and not make them feel they were on the outside . . . I still do that."

During her third year she married Michael Cadore, a professional caterer and chef, whom she had met while in Boston. The first of their three children arrived during her final year, a common time among married women medical students. It was during her internship at UTMB, in the Department of Family Medicine, that Cadore had their second child, which, she said, she "would not advise *anyone* to do." Their youngest was born during Cadore's last year of residency. Despite hostile comments from some male residents in the program (many of whom had wives who were also having babies), she managed to complete the residency on time—a tribute, in part, to the dedicated help of her husband. She insists, "Even though it was difficult . . . to me, there's nothing like being a mother: when you go home you are the greatest person in the whole world."

Dr. Cadore's choice of specialty reflects her interest in patients as individuals with histories, stories, families. "Most of my patients—I can tell you a little about them. If I don't know their name, I know their story." Her choice of practice setting, a private group practice in a predominantly blue-collar community rather than a faculty position at a medical center, resulted from the influence of her supervising physician during part of her residency. Dr. Katie Youngblood had been in private practice for years before returning to the medical school to teach. "She was just like fresh air coming through," said Cadore, someone who knew the "real world" of practice, someone with "horse sense." More important, from Cadore's viewpoint, her mentor valued motherhood and had a realistic sense of what it took to make career and family work well together. As Cadore heard it, Dr. Youngblood's message was not one that many young women doctors readily accept: "She said there have to be sacrifices as a woman physician in a family." Even more, "You have to decide . . . where your priorities are . . . I thought she had a point there. So, I made a decision that was very difficult." She decided to leave academic medicine and join an established, all-white, all-male, formerly segregated family practice in predominantly blue-collar Texas City.

The real world, however, proved more real than Cadore had bargained for. Her suite of rooms, for example, was located in what had been the office's "colored section," something she learned from one of her patients. In fact, people would occasionally come into the office, see her, and just stare. One of the older physicians seemed uncomfortable, she recalled, but for the most part, "we always functioned as a group." The (female) staff, too, were supportive. Many patients "were really excited, because I was a female or, some of them, because I was black, and some of them, because I was a black female." Men would bashfully admit to never having undressed before any woman but their wife. But they never switched over to one of "the guys."

For three years Cadore "had a lot of patients and was doing well"— until the practice was sold to a large, Nashville-based HMO trying to break into the Houston-Galveston health care market. At that point all the physicians in the practice were required to sign a contract with a "noncompetition" clause, agreeing that if they left the practice, they would not practice for at least two years anywhere within a large, stipu-lated radius of any clinic owned by the company and, more crucial, would attend no patient previously in their care. Cadore refused to sign, accepting that she would leave the practice. In the week following her departure, she saw several long-time patients who needed her atten-tion, working temporarily out of a friend's office. But a week later she was handed a restraining order forbidding her to practice anywhere in Galveston or Harris County, and claiming she had violated a contract she had never signed. "So, I ended up going to court. You know, I had never been to court in my life!"

But Cadore was well represented, well supported by her family and patients, and well within her rights. Her opponent, the HMO, claimed that she was infringing on its corporate "goodwill," that her patients were in fact consumers of the HMO's health care, not patients engaged in an ongoing doctor-patient relationship. As she tells the story,

> Two of my patients took a day off from work and they went down to the court [at] the request of the attorneys. The other side had several doctors, some of their business people, they had a much bigger group . . . I was sitting there thinking, "I can't believe that I am in court. I didn't do anything wrong." I went to court because when I left, they [claimed] I took my patients with me. So one of the issues . . . which the patients expressed very eloquently, was, "I am not a thing, I am not a book, I am not a desk, I am a person and this is my doctor and I am going to go wherever she is, I do not care what kind of sign is on the door. This is my doctor!"

In her own cross-examination, which lasted two hours, Cadore made the same point as her patients: "Whose patient are you? Are you a patient of a corporation or is there an individual relationship there?" In the end, the judge took "about two minutes" to decide in her favor. She also won on appeal.

The lawsuit reinvigorated Cadore's original sense of her place in medicine. After taking some time off to be with her family, she began working with Dr. Edith Irby Jones in Houston. As of this writing, Dr. Jones is in her early seventies and "as sharp as anybody [just] out of residency . . . She just has this incredible energy." Although Dr. Jones is an internist, she handles her "overloaded" practice as a family practice office with, of late, assistance from Cadore. Cadore also works with Jones at Riverside Hospital in Houston, where they both put in long hours. Riverside, founded in 1924 as the Houston Hospital for Negroes, is one of the few remaining black hospitals in the country. After many years in decline, the hospital began focusing on psychiatric and substance-abuse cases. Today it is undergoing a renaissance, in part due to the efforts of Dr. Jones. A new unit will house psychiatric patients, while the main hospital will build up its medical and surgical services with help both from older physicians like Jones and from young black physicians, like Cadore, in an attempt to revive the tradition of neighborhood health care. As Cadore sees her, "Dr. Jones doesn't focus on the negative . . . She reminds me of my grandmother. [It's] like going home again."

Women physicians like Mary Howell and Judy Cadore, who have practiced medicine during the past three decades, have witnessed many changes in the professional landscape. Equal opportunity legislation, heightened awareness of race and sex discrimination by the public as well as the profession—all this has opened many doors and widened those openings that already existed. However, it seems unlikely that a simple increase in sheer numbers of women will be sufficient to bring true equality to the profession. We must now consider what remains to be done.

Conclusion: Reconciling Equality and Difference

> The reasons women in academic medicine are not succeeding at the same pace as men involve a complex combination of isolation, cultural stereotypes, and sexism, and difficulties of combining family responsibilities with professional demands.
>
> *AAMC Project Committee on Increasing Women's Leadership in Academic Medicine, 1996*

> I am relieved if rather than sex bias, the reason more women are not breaking through the glass ceiling of academic medicine is because their children are hanging on the tails of their white coats. Most of us are happy to have them there, and academic medicine offers a level of professional fulfillment, financial stability, and geographic flexibility that is well worth the juggle.
>
> *Christine Laine, M.D., Ph.D.*, Annals of Internal Medicine, *1998*

SEVERAL YEARS before passage of the Family and Medical Leave Act in 1993, a colleague described to me a cartoon that had appeared in the *Houston Post*. It showed a barnyard filled with hens, chicks, and roosters. Amid all the bustle, encircled by admiring poultry, stood a hen skillfully juggling eggs. At the edge of the circle, one of the onlookers whispered, "I just love the way she juggles family and career."

Unquestionably, the greatest obstacle still facing women practitioners is the need to accommodate the demands of childbearing and child rearing. Women in medicine in America have accomplished a great deal in a relatively short time. Nevertheless, a century and a half has been barely enough time to overturn nearly a millennium of discouragement and negative expectations. Given women's numerical standing in the profession until thirty years ago—given, in Rosabeth Moss Kanter's terms, their *skewed* minority of less than 8 percent—women have been a remarkably visible force both for better health care and for fairer treatment of women. Through such activities as founding the American Medical Women's Association and the American Women's Hospitals service, leading the U.S. Children's Bureau and Planned Parenthood, spearheading the Mississippi Health Project, making breakthroughs in medical science, and establishing the NIH Women's Health Research initiative, women physicians are fulfilling the promise of their diverse forebears. Still, Sarah Dolley's goal of "restoring the balance" remains elusive, as difficult to achieve as its corollary—reconciling professional equality and gender difference.

Like the cartoonist's feathered heroine, women physicians have become experts at juggling the personal and professional aspects of their lives. They have had little choice. In actuality, they are attempting to solve the dilemma of difference. What may be changing, unacknowledged by most policymakers in the medical profession, is the willingness of at least some men to participate in resolving the dilemma, to be more than part of the crowd of onlookers. Certainly if women physicians are to achieve any sense of balance in their own life—much less act as a model for their patients—it is essential that we find ways to reconcile their professional skills and aspirations with their personal and social responsibilities to family and community. The only way to do that is to acknowledge that this is a shared problem, not theirs alone.

As Martha Minow has shown, the "difference dilemma" arises whenever acknowledging differences—for example, women's childbearing potential or the responsibilities of parents—results in stigmatizing these very differences.[1] This dilemma can be resolved only by transcending "difference" through the recognition of commonality and common cause, something our society does not accomplish often—especially in regard to workers' family obligations. The vast majority of physicians under the age of fifty, male and female, are parents. Yet we still look only to *women* physicians to solve "their" career problem. Pathways to professional success are conceived as if they never intersect with, nor are even bordered by, the joys and responsibilities of family life.

We will not be able to take women's advancement in the medical profession for granted until we resolve the parental dilemma of difference. It continues to shadow their expectations for themselves and society's expectations for them. A 1992 study, for example, found that women were significantly less likely than men to envision careers in academic medicine, largely because of anxieties associated with the "competition between academia and . . . family commitments." A 1995 study strongly suggests that these concerns are valid. Comparing the careers of female medical faculty who have children with those of either female faculty without children or male faculty (with or without children), faculty mothers were found to face many more obstacles to academic success, including less institutional support for research and a slower rate of peer-reviewed publishing. No such differences were found when comparing the careers of the men with the women who did not have children. As an editorial accompanying the study emphasized, "Mothers, fathers, and others would benefit if academic medicine built some flexibility." As the study's authors suggested, "Some of these obstacles can be easily modified (for example, by eliminating after-hours meetings and creating part-time career tracks)." More important, however, academia must destigmatize those who choose these options, so that the "parent track" doesn't just run in circles. For example, Sonnert and Holton showed that women scientists tend to publish fewer but more frequently cited articles. Institutional promotion-and-tenure committees could choose to accept such publication patterns as a sign of high quality, not low productivity. But relatively few promotions committees adopt this position when weighing evidence for academic advancement.[2]

Can one be a successful physician (academic or otherwise) and still be a "good" parent? To some extent the answer depends on the values of the institution where one works. For a brief period during the 1980s the medical profession seemed ready to engage in this discussion. About the time of the debate over the Family and Medical Leave Act, which was finally passed on February 5, 1993, U.S. physician labor force projections foresaw a massive glut of doctors. These forecasts, ironically, also foresaw women physicians easing the projected glut by being *less* productive than men, as measured in hours worked and patients seen per week.[3] (The implication was that it would take more physicians to cover patient demand if many women, as then predicted, worked fewer than the statistically arbitrary norm of about sixty hours per week.[4]) In that context, and given the rapidly rising percentage of women doctors, there seemed to be little to lose from an open discussion of issues like parental leave, flex-time scheduling, or even job sharing for physicians.

(Pregnancy leave, too, may have become slightly less stigmatized within residency programs, but it has never been good news to tell a program director that you are—unexpectedly—pregnant). Thus the AMA, which during the 1980s called for the establishment of maternity leave for residents, began in 1991 to advocate "parental" leave: "The dramatic increase in the number and influence of women in medicine, and their relatively younger age, requires that the medical profession pay greater attention to their pregnancy and maternity leave needs . . . Young male physicians also want the opportunity to fully share in . . . child care and family responsibilities."[5]

Yet as a culture we do not seem to have internalized the message that men are parents, too. As one male resident wrote, "The prevailing attitude is that child-rearing is woman's work. If I ask to leave rounds to attend a sick child or cover for an absent baby-sitter, I run the risk of appearing as though I have a lackadaisical attitude or my work comes second . . . The first step is to try to understand that the enormous responsibility we have as parents is a shared one—not borne by the mother alone."[6] A 1996 position paper from the Association of American Medical Colleges Project Committee on Increasing Women's Leadership in Academic Medicine pointed out the greater "likelihood that physicians with a reasonable balance in their lives [would] be able to maintain enthusiasm for their work and to inspire patients toward good self-care."[7] Indeed, "reasonable balance" may well be the bare minimum requirement for the well-being of young physicians.

For example, rates of depression for physicians are relatively high overall, but for women they are above the norm both for the population at large and for male physicians. When the figures are adjusted for age, it now does not appear to be true (as was once believed) that the suicide rate of women physicians exceeds that of men; the rates are about equal. Nevertheless, because the rate for all men is usually higher than for women, this constitutes a relatively high suicide rate for women who are doctors. The suicide rate among women in medicine is in fact "highly elevated" above that of the general female population in Europe and North America, more so than the rate for male physicians compared with men in the general population.[8] These figures should remind us that heavy responsibility without commensurate authority can be a recipe for despair.

It is not yet clear, however, that medical centers, hospitals, and

HMOs can incorporate an ideal of "balance" into their group portraits of potential professional leaders. Physician-mothers, in particular, test the limits of medicine's capacity to reshape its model of career development for leadership. Overall, women physicians consider how their choice of specialty and practice setting will affect the balance between career and personal life—especially family life—to a significantly greater extent than men do.[9] For that reason, and perhaps because of increasingly restricted opportunities in regions where HMOs dominate health care employment, fewer women than men tend to practice in office-based settings, and more women, conversely, practice in outpatient clinics and health centers.[10] Although these differences have diminished recently, they have not disappeared. In 1989, twice as many women as men physicians were likely to be employees, even in group practice settings; the ratio of male to female physicians who were self-employed was three to two. By 1995 women were still more likely to be employees than were men physicians, although the significance of this difference had declined as general employment trends for physicians showed an overall movement toward employee status for both men and women. In 1995, 42 percent of all physicians were employees. Employee physicians worked about five hours less per week—a trade-off, perhaps, of lower income for better quality of life. The most lucrative form of practice continued to be self-employed group practice, a setting in which there were nearly twice the number of men to women physicians.[11] One respondent to the More and Greer survey (described in Chapter 8), a radiologist, commented that "[the] old boy network continues in residency and recruitment for private practice. [There have been] very few women in private practice then and now, unless on the 'mommy track' with no possibility of partnership." Women also may be more reluctant than many men to move from one geographic region to another in search of rapid career progress. Moreover, as of 1988, between 50 and 70 percent of women physicians are married to physicians, a trend that was expected to continue, and this heightens the difficulty of sharing child rearing responsibilities.[12]

In truth, it is not easy to disentangle the effects of choice and necessity in the career patterns of women physicians. The weight of cultural tradition in several specialties (surgery is a prime example[13]), the subtle effects of gender stereotyping, quality-of-life considerations such as a desire for a shorter residency and predictable hours after residency, and—not least—the personal interests and values of women physicians all contribute to decisions regarding specialty, practice setting, geographic location, and preferred patient population. Anthropologist Joan Cassell describes how these factors combine to encourage some women surgeons to choose—with enthusiasm—the "relatively low-prestige" subspecialty of breast surgery. As another recent study concluded, "The association between gender and career choice noted in many previous studies may in fact be a proxy for the effects of other closely related factors [such as prior positive clinical experiences, concern over job opportunities, and concern for societal needs]."[14] A major consideration for women professionals, including physicians, is the need to reconcile relationship building and childbearing with the demands of a developing career. Although men choosing a residency today are more likely than in the past to consider lifestyle and family issues in making their choices, for women the demands of childbearing heighten the importance of the choice of specialty and practice setting.[15] For many women physicians, the conflict between their personal and professional development is at its most intense, and least forgiving, when they are in their twenties to midthirties, when fertility has not yet begun to decline. Career clock and biological clock, in the words of one commentator, "tick in synchrony."[16]

These concerns were evident at a 1991 AAMC Professional Development Seminar for Junior Faculty Women.[17] Many of the women in attendance expressed concern over the difficulty of achieving genuine "balance"—that is, a balance that would compromise neither their career nor their commitment to their family, for many an essential value. One woman commented, "I have chosen the role of not being in the very, very top . . . I want a full, rounded life." Another emphasized, "There are many visions of 'success.'" Yet these sentiments do not, and should not be taken to, represent the choices of *all* women in medicine. Women doctors are as notable for their diversity as for their commonalities. At the same AAMC workshop, for example, another physician exulted in her ability to "get over her guilt" toward her husband and children when her new job as a hospital CEO demanded much more time and energy than anything she had ever done before. She found the challenges to be both "pleasurable" and "exciting."

Some of the advice given seminar attendees, however, may have given pause to those in attendance who still cherished the belief that developing a successful career was entirely in their own hands. Here are some of the "laws" of successful academic careers presented by Clyde Evans, associate dean for clinical affairs at Harvard Medical School, intended for both men and women: first, "If you don't know where you're going, any road will do"; second, "If you don't play, you can't win"; third, "A faculty appointment is only a hunting license—whatever you bag, you can keep"; fourth, "It's not what you know but who you know"; fifth, "The playing field isn't level—it's tilted in favor of those who follow these rules." Evans also knew that women and men do not face the same obstacles. To illustrate this, he referred to a classic *New Yorker* cartoon that shows a committee consisting of half a dozen men and one woman. The chairman is saying, "That was an excellent suggestion, Ms. McCarthy. Perhaps one of the men would like to make it."[18] In his presentation Evans provided a few corollaries specifically targeted to women: first, women need to pay careful attention to gaining appropriate credit for the work they do; second, women need mentors more than men do (but are less likely to find them); third, women have to either be twice as good as their male colleagues or work twice as hard to make sure their real worth finally registers. Thus the question remains: how can women accomplish all this—assuming they want to—and salvage some semblance of a "balanced" life? Moreover, should *women* be solely responsible for solving this dilemma?[19]

In 1991, nearly two decades after a feminist tide began swelling the numbers of American women attending medical schools, sociologist Judith Lorber explicitly asked the question implied by Evans's "laws." The author of a longitudinal comparison of male and female physicians' career development that spanned a full generation, Lorber now wondered, "Can women physicians ever be true equals in the American medical profession?" Probably not, she asserted, and for two main reasons: the demands of family time continue to conflict with the long hours necessary to supplement patient care with research and administrative duties, and continued "informal discriminatory practices" often keep women from reaching leadership positions even if they have the time to do the job. As a result, she predicted, women physicians either would be held at arm's length from leadership roles, or they would voluntarily choose lower-visibility slots as nontenure track clinicians in HMOs or academic health science centers to better control the demands on their time. Lorber believes that many male physicians still think that the "feminization" of the medical profession would be tantamount to a significant decline in its prestige and status.[20]

Ironically, an alternative motive for continued male bias is a fear of women doctors' potential advantage in the competition for patients, because "patients feel they offer a different [preferable] practice style."[21] This is especially important in managed care settings because, in a capitated payment system, "panel size [the number of patients enrolled under the care of an individual doctor] most clearly reflects the value of the physician to the HMO."[22] Recent studies seem to demonstrate small but significant differences in women doctors' practice styles that may enhance patient satisfaction. Specifically, women physicians engaged in greater numbers of "preventive services" and spent a greater proportion of time in discussions of patients' "family medical and social information," while men devoted more time to history taking related to patients' "current complaint or prior illnesses."[23] Although it is not clear that in academic clinical settings such factors have a positive impact on career development, since controlling clinical time in favor of research and administration often is more essential to academic advancement, Lorber noted in her study that women physicians do "successfully compete for institutional and federal funding of clinical research; they do not turn down the chance to head [clinical] services." All in all, Lorber, like Mary Howell, believes that because women physicians are no longer exotics, tokens, or mascots but rather have become viable professional colleagues and competitors, they will have to fight for every advantage and, at the same time, neutralize lingering negative stereotypes and subtle exclusionary tactics. She is not optimistic that women can overcome this "damned if they do, damned if they don't" double bind.[24]

Even the AAMC sometimes—inadvertently—perpetuates the old stereotypes: woman as underachieving dilettante, woman as overachieving supermom. In 1998 the AAMC's journal, *Academic Physician and Scientist*, published an article entitled "Academic Careers Need Not Compete with Family Life," on options for physician-parents trying to stay on the academic track. Unfortunately, as an irate reader noted, the article made reference only to *women's* careers. "Certainly," she wrote, "in the case of pregnancy, this is solely a woman's issue. However, once a child is born, until the father takes equal responsibility [for] child care women will continue to bear the brunt of professional liabilities." Ironically, the offending article appeared in the same issue as a profile of Dr. Deborah E. Powell, physician, wife, mother of four, and executive dean of the University of Kansas School of Medicine—a classic example of *over*achievement.[25]

Despite the obvious difficulties ahead, it would be inaccurate to conclude on a pessimistic note. The gender landscape of medicine has changed markedly over the past thirty years and is unlikely ever to return to the uniformity of its earlier decades. Women have become deans (eight women, as of this writing, are deans of American medical schools), hospital CEOs, department chairs, division chiefs, and principal investigators of large grant-funded studies in many fields at many medical schools. In 1991 Congress empowered the NIH to found an Office of Research on Women's Health, directed by Dr. Vivian Pinn. Since then, the Women's Health Research Initiative at NIH has helped generate basic and epidemiological research on breast and endometrial cancer, coronary artery disease, hormone replacement therapy, osteoporosis, and AIDS, conditions in which the cellular mechanisms, treatment interventions, or societal impact of disease may differ according to sex. As one researcher commented, women's health research is "moving out of the pelvis."[26] Some medical faculty have also taken the first steps toward developing an interdisciplinary medical curriculum and, eventually, a board-certified specialty in women's health by founding the American College of Women's Health Physicians and the National Academy on Women's Health Medical Education.

Women's health has attracted the interest of women physicians as a way to address women patients' often underserved health care needs, to provide a base for research, and to institutionalize at least one specialty that will positively value the perspective of women physicians and patients. Opponents of a separate specialty in women's health fear that, partly for lack of resources, it could become a marginalized site for low-tech, low-status health care—a feminine ghetto. It is still too early in the debate to assess the validity of the two positions, but the development of subspecialty tracks in women's health within generalist residencies in internal medicine and family medicine suggests that a viable compromise will emerge.[27]

It is also encouraging that many residency programs, HMOs, and faculty contracts now include provisions for parental leave, flex time, and tenure-clock "time-outs" to enable physician-parents to fulfill their dual responsibilities.[28] Men's changing expectations regarding greater equality in child rearing and household responsibilities encourage some optimism for the generation now in residency.[29] And a vanguard of elite medical schools is instituting internal measures to ensure that women's careers are not programmed to fail. At Johns Hopkins, for example, a report from the university provost's Committee on the Status of Women in 1989 demonstrated lower faculty salaries and slower rates of promotion for women faculty. In response, the Department of Medicine began systematic efforts to reduce "gender-based career obstacles for women." After internal departmental surveys corroborated perceptions of manifold, if subtle, career disadvantages for women, the department chair and an appointed task force began implementing a fifteen-year plan to ameliorate these conditions. Possibly the most critical intervention was the active leadership of the department administration in legitimizing the concerns of women on the faculty. In addition, the departmental leadership addressed questions of career development, fair apportionment of departmental resources, and increased inclusiveness in divisional and departmental decision making. It also sponsored leadership training for selected women faculty and colloquia on gender-related issues and effective mentoring. Finally, the time limits for tenure-track faculty at each rank were expanded to allow for greater career flexibility when desired. As a result, between 1990 and 1995 the proportion of female tenured associate professors increased from 7 percent to 37 percent, many times more than the increase in the number of women in the department during the same period.[30]

The questions that still dog the path of women physicians do not hinge on their ability or even their willingness to succeed in the practice of medicine. Rather, they arise from the medical profession's own crisis of professional values in an era of increasingly market-driven health care. To excel in the profession previous generations of women physicians were asked to organize their priorities in much the same way as their male colleagues. Today, women in medicine ask that American health care institutions modify their values to accommodate many definitions of success and diverse models of leadership. Some of their male colleagues, experiencing for the first time a loss of authority and control within medicine's increasingly corporate structure, are beginning to share these goals. As one male physician recently wrote, "Clearly we impoverish not only our emotional lives but our spiritual and intellectual ones as well by our inordinate adulation of work." Or, as a recent article put it, "Physicians often complain of having 'unbalanced' lives, with too much time devoted to work . . . Imbalance in life contributes to physician stress."[31] To return to the question with which this book began, can women physicians "restore the balance" in their profession, in their own lives, and perhaps even in the lives of their patients? It seems clear that the answer is yes, but they can not, should not—and may no longer have to—do it alone.

Notes

Index

Notes

Abbreviations

The following abbreviations are used for archival sources cited in the notes.

AMWA/NYH	American Medical Women's Association Archives, New York Hospital-Cornell Medical Center, New York
AWH/MCP	American Women's Hospitals Collection, Archives and Special Collections on Women in Medicine, MCP Hahnemann University, Philadelphia
BCA/RGH	Baker-Cederberg Archives, Rochester General Hospital, Rochester, New York
Disp./UR	Provident Dispensary Association, Edward G. Miner Library, University of Rochester School of Medicine and Dentistry, Rochester, New York
Door of Hope/RRL	Papers of the Door of Hope Home/North Haven, Inc., Department of Rare Books and Special Collections, Rush Rhees Library, University of Rochester, Rochester, New York.
DRB/RRL	Department of Rare Books and Special Collections, Rush Rhees Library, University of Rochester, Rochester, New York
GMB *Minutes*/NARC	General Medical Board, *Minutes*, National Archives Research Center, Suitland, Maryland
LHD/RPL	George W. Goler Collection, Local History Division, Rochester Public Library, Rochester, New York

MCP/MCP	Marion Craig Potter Papers, Archives and Special Collections on Women in Medicine, MCP Hahnemann University, Philadelphia
MCP/UR	Marion Craig Potter Collection, Edward G. Miner Library, University of Rochester School of Medicine and Dentistry, Rochester, New York
MME/SL	Martha May Eliot Papers, Schlesinger Library, Radcliffe College, Cambridge, Massachusetts
MSC/SL	Mary Steichen Calderone Papers, Schlesinger Library, Radcliffe College, Cambridge, Massachusetts
Obits./AMA	Mortuary Card file, Library of the American Medical Association, Chicago
Pract./BMS *Minutes*	Practitioners'/Blackwell Medical Society, *Minutes,* Edward G. Miner Library, University of Rochester School of Medicine and Dentistry, Rochester, New York
SAD/MCP	Sarah Read Adamson Dolley Collection, Archives and Special Collections on Women in Medicine, MCP Hahnemann University, Philadelphia
SAD/UR	Sarah Read Adamson Dolley Collection, Edward G. Miner Library, University of Rochester School of Medicine and Dentistry, Rochester, New York

Introduction

1. In 1991 Dr. Healy became the first woman to head the National Institutes of Health (NIH); in 1991 Dr. Frances Conley briefly resigned as professor of neurosurgery at Stanford to protest administrative indifference to pervasive sexual harassment of women physicians; Drs. Novello and Elders were, respectively, the first woman and the first African American woman surgeon general of the U.S. Public Health Service; Dr. Kirschstein is director of the NIH Institute of General Medical Sciences, and with Dr. Vivian Pinn, director of the NIH Office of Research on Women's Health, she headed up the women's health research initiative at the NIH; Dr. Susan Love is a surgeon specializing in breast surgery and a crusader for better research and treatment for breast cancer; Dr. Joyce Wallace was profiled by Barbara Goldsmith in the April 26, 1993, issue of *The New Yorker* (pp. 64–81) for her work with the Foundation for Research on Sexually Transmitted Diseases—specifically, for her campaign of prevention among New York City prostitutes; in 1997 Dr. Nancy Dickey became the first woman president of the American Medical Association. Also see Jennifer Steinhauer, "For Women in Medicine, a Road to Compromise, not Perks," *New York Times,* March 1, 1999, pp. 1–5 online. Available URL: http://www.nytimes.com/learning.

2. Martha Minow, *Making All the Difference: Inclusion, Exclusion, and American Law* (Ithaca, N.Y.: Cornell University Press, 1990), p. 21.

3. Perri Klass, *Other Women's Children* (New York: Random House, 1990), p. 14.

4. Iris Marion Young, "The Ideal of Community and the Politics of Difference," in Linda J. Nicholson, ed., *Feminism/Postmodernism* (New York: Routledge, 1990), pp. 300–323, quotation on p. 301.

5. Holt N. Parker, "Women Doctors in Greece, Rome, and the Byzantine

Empire," in Lilian R. Furst, ed., *Women Healers and Physicians: Climbing a Long Hill* (Lexington: University Press of Kentucky, 1997), pp. 131–150. But cf. Helen King, *Hippocrates' Woman* (London: Routledge, 1998), pp. 168–187. Doreen A. Evenden, "Gender Differences in Licensing and Practice of Female and Male Surgeons in Early Modern England," *Medical History* 42 (1998): 194–216; Monica H. Green, "Documenting Medieval Women's Medical Practice," in Luis Garcia-Ballester et al., eds., *Practical Medicine from Salerno to the Black Death* (Cambridge: Cambridge University Press, 1994), pp. 222–252.

6. Thomas Neville Bonner, *To the Ends of the Earth: Women's Search for Education in Medicine* (Cambridge, Mass.: Harvard University Press, 1992).

7. Two books are the starting point for any history of women doctors in the United States. Regina Markell Morantz-Sanchez's *Sympathy and Science: Women Physicians and American Medicine* (New York: Oxford University Press, 1985) remains the best history of women's medical education and careers from the 1830s to the early twentieth century. Mary Roth Walsh, in *"Doctors Wanted: No Women Need Apply": Sexual Barriers in the Medical Profession, 1835–1975* (New Haven, Conn.: Yale University Press, 1977), vividly portrays the discrimination confronting women physicians, especially in their fight for educational opportunity.

8. This figure is somewhat lower than was previously believed. See Tables 4.1 and 4.2 and the graph in Chapter 4 for a full explanation of the figures.

9. Herbert M. Morais, *The History of the Negro in Medicine* (New York: Association for the Study of Negro Life and History, 1967), pp. 10–31, quotation p. 16, Todd L. Savitt, *Medicine and Slavery: The Diseases and Health Care of Blacks in Antebellum Virginia* (Urbana: University of Illinois Press, 1978).

10. Dr. Peck was the first African American medical graduate; he was followed two years later by two graduates of Bowdoin in Maine. Morais, ibid., pp. 20, 30–31; Darlene Clark Hine, "Co-Laborers in the Work of the Lord: Nineteenth-Century Black Women Physicians," in Ruth J. Abram, ed., *"Send Us a Lady Physician": Women Doctors in America, 1835–1920* (New York: W. W. Norton, 1985), pp. 107–109.

11. Darlene Clark Hine, "Physicians, Nineteenth Century," in *Black Women in America: An Historical Encyclopedia*, vol. 2, ed. Darlene Clark Hine, Elsa Barkley Brown, and Rosalyn Terborg-Penn (Bloomington: Indiana University Press, 1993), pp. 923–926.

12. Hine, "Co-Laborers in the Work," pp. 107–109; Bonner, *To the Ends of the Earth*. Gloria Moldow gives the following statistics: in 1923, 91 black women had graduated from medical school, including 39 from Meharry, 25 from Howard, and 12 from integrated schools. Gloria Moldow, *Women Doctors in Gilded-Age Washington: Race, Gender, and Professionalization* (Urbana: University of Illinois Press, 1987), p. 175 n. 21.

13. *Physician Characteristics and Distribution in the U.S., 1995–1996* (Chicago: AMA, 1996); *Physician Characteristics and Distribution in the U.S., 1993–1994* (Chicago: AMA, 1994), cited in Leah J. Dickstein, Daniel P. Dickstein, and Carol C. Nadelson, "The Status of Women Physicians in the Workforce," 1994, typescript, Table 2, p. 21; Janet Bickel and Phyllis Kopriva, "A Statistical Perspective on Gender in Medicine," *Journal of the American Medical Women's Association* (hereafter *JAMWA*) 48 (1993): 141–144.

14. Vanessa Northington Gamble, "Taking a History: The Life of Dr. Virginia Alexander," Fielding H. Garrison Lecture, American Association for the History of

Medicine, May 8, 1998. Many thanks to Dr. Gamble for providing the typescript of her lecture. Bonner, *To the Ends of the Earth*, pp. 14–16; Hine, "Co-Laborers in the Work," p. 111; Moldow, *Women Doctors*, pp. 129–132.

15. Debra L. Roter and Judith A. Hall, *Doctors Talking with Patients/Patients Talking with Doctors* (Westport, Conn.: Auburn House, 1993), pp. 59–64, quotation on pp. 59–60.

16. Dalia Ducker, "Research on Women Physicians with Multiple Roles: A Feminist Perspective," *JAMWA* 49 (1994): 78–84, esp. 79. On women physicians in the present-day United Kingdom and Australia, see Rosemary Pringle, *Sex and Medicine: Gender, Power and Authority in the Medical Profession* (Cambridge: Cambridge University Press, 1998).

17. Bickel and Kopriva, "A Statistical Perspective," p. 141.

18. Rosabeth Moss Kanter, "Some Effects of Proportions on Group Life: Skewed Sex Ratios and Responses to Token Women," *American Journal of Sociology* 82 (1977): 965–990.

19. Cole, Burton, and Zuckerman were, in turn, influenced by the earlier work of sociologist Robert K. Merton. See Gerhard Sonnert and Gerald Holton, "Career Patterns of Women and Men in the Sciences," *American Scientist* 84 (1996): 63–71; Harriet Zuckerman, "Accumulation of Advantage and Disadvantage: The Theory and Its Intellectual Biography," in C. Mongardini and S. Tabboni, eds., *L'opera di R. K. Merton e la sociologia contemporeana* (Genoa: Edizioni Culturali Internationali Genova, 1989), cited in Sonnert and Holton; Margaret W. Rossiter, *Women Scientists in America: Struggles and Strategies to 1940* (Baltimore: Johns Hopkins University Press, 1982); Margaret W. Rossiter, *Women Scientists in America: Before Affirmative Action* (Baltimore: Johns Hopkins University Press, 1995); Morantz-Sanchez, *Sympathy and Science*. For an additional discussion of these ideas, see Chapter 8.

20. Judith Walzer Leavitt, "'A Worrying Profession': The Domestic Environment of Medical Practice in Mid-Nineteenth Century America," *Bulletin of the History of Medicine* 69 (1995): 1–29, quotation on 12, 13.

21. The term *civic professionalism* was first used by historian Thomas Bender. For its application to medicine, see Ellen More, "The Blackwell Medical Society and the Professionalization of Women Physicians," *Bulletin of the History of Medicine*, 61 (1987): 603–628, esp. 605–606.

22. Morantz-Sanchez, *Sympathy and Science*, esp. chap. 7; Mary Putnam Jacobi, *Pathfinder in Medicine* (New York: Women's Medical Society of New York State, 1925).

23. Klass, *Other Women's Children*, p. 234.

24. Morantz-Sanchez estimated that only about 25–35 percent of women physicians in the nineteenth century ever married. Of these, a significant number never had children. My figures for Rochester, New York, are slightly higher (see Chapter 2). She estimates that for the first half of the twentieth century, the percentage rises to 30–40 percent. However, many unmarried women doctors did create the conditions of family life by establishing homosocial households with other women, sometimes even adopting and raising children. Morantz-Sanchez, *Sympathy and Science*, p. 136.

25. Aristotle, *Nichomachean Ethics*, trans. Martin Ostwald (New York: Macmillan, 1962; 1986), p. 312; Martha C. Nussbaum, *The Fragility of Goodness: Luck and Ethics in Greek Tragedy and Philosophy* (New York: Cambridge University Press,

1986), pp. 299–300, 302. For these ideas I am indebted to my discussions with Professor Dawson S. Schultz.

26. Barbara Gutmann Rosenkrantz, "The Search for Professional Order in 19th Century Medicine," in Judith Walzer Leavitt and Ronald L. Numbers, eds., *Sickness and Health in America: Readings in the History of Medicine and Public Health*, 2nd ed. rev. (Madison: University of Wisconsin Press, 1985), pp. 219–232, esp. p. 229.

27. In addition to Morantz-Sanchez, *Sympathy and Science*, Bonner, *To the Ends of the Earth*, and Walsh, *Doctors Wanted*, see Regina Morantz-Sanchez, *Conduct Unbecoming a Woman: Medicine on Trial in Nineteenth-Century Brooklyn* (New York: Oxford University Press, 1999), on the revealingly atypical career of surgeon Mary Dixon Jones; Virginia Drachman, *Hospital with a Heart* (Ithaca, N.Y.: Cornell University Press, 1984), and forthcoming studies by Professor Arleen Tuchman of Dr. Marie Zakrzewska and Dr. Steven J. Peitzman of Woman's Medical College of Pennsylvania.

28. They continue: "Maternalism always operated on two levels: it extolled the private virtues of domesticity while simultaneously legitimating women's public relationship to politics and the state, to community, workplace, and marketplace. In practice, maternalist ideologies often challenged the constructed boundaries between public and private, women and men, state and civil society." See Seth Koven and Sonya Michel, "Womanly Duties: Maternalist Politics and the Origins of Welfare States in France, Germany, Great Britain, and the United States, 1880–1920," *American Historical Review* 95 (1990): 1076–1108, quotation on 1079.

29. Cf. Delese Wear, ed., *Women in Medical Education: An Anthology of Experience* (Albany: State University of New York Press, 1996), which came to my attention too late to be consulted.

1. The Professionalism of Sarah Dolley, M.D.

1. Typescript of letter from Sarah Read Adamson to Elijah F. Pennypacker, from Central Medical College, Syracuse, New York, February 3, 1850, SAD/MCP.

2. Letter from Sarah Adamson Dolley to Charles Sumner Dolley, November 22, 1896, SAD/UR. The Dolley Collection includes Boxes 1–12. Letters in Boxes 1–11 are in chronologic order and are cited by date only.

3. For a fuller discussion of the attitudes of physicians today, see Chapter 8. Also see Sarah E. Brotherton and Susan A. LeBailly, "The Effect of Family on the Work Lives of Married Physicians: What If the Spouse Is a Physician, Too," *JAMWA* 48 (1993): 175–181.

4. She interned at Philadelphia General Hospital (Blockley) until several weeks before her marriage in June 1851.

5. Mary P. Ryan, *Cradle of the Middle Class: The Family in Oneida County, New York, 1790–1865* (New York: Cambridge University Press, 1981), pp. 12, 15.

6. Judith Walzer Leavitt, "'A Worrying Profession': The Domestic Environment of Medical Practice in Mid-Nineteenth Century America," *Bulletin of the History of Medicine* 69 (1995): 1–29.

7. Although the majority of Victorian women physicians never married (see "The Question of Marriage" in this chapter), many unmarried women doctors established homosocial households with other women; some even adopted children.

8. For a careful analysis of Blackwell's belief that the pursuit of bench science would compromise physicians' clinical skills and even their regard for the humanity of the patient, see Regina Morantz-Sanchez, "Female Science and Medical Reform: A Path Not Taken," in Ronald G. Walters, ed., *Scientific Authority and Twentieth-Century America* (Baltimore: Johns Hopkins University Press, 1997), pp. 99–115; idem, "The Gendering of Empathic Expertise: How Women Physicians Became More Empathic than Men," in Ellen Singer More and Maureen A. Milligan, eds., *The Empathic Practitioner: Empathy, Gender, and Medicine* (New Brunswick, N.J.: Rutgers University Press, 1994), pp. 40–58, esp. pp. 42–44 and p. 56 n. 6–12.

9. William Coleman, "Health and Hygiene in the *Encyclopédie:* A Medical Doctrine for the Bourgeoisie," *Journal of the History of Medicine and Allied Sciences* 29 (1974): 399–421, esp. 400–402, 406–408, 411, 412.

10. Charles E. Rosenberg, "The Therapeutic Revolution: Medicine, Meaning, and Social Change in Nineteenth-Century America," in Morris J. Vogel and Charles E. Rosenberg, eds., *The Therapeutic Revolution: Essays in the Social History of American Medicine* (Philadelphia: University of Pennsylvania Press, 1979), pp. 3–26, esp. pp. 5–7. Coleman, "Health and Hygiene," p. 412, discerns this tradition in the *Encyclopédie.*

11. Ellen S. More, "'Empathy' Enters the Profession of Medicine," in More and Milligan, eds., *The Empathic Practitioner,* pp. 19–39.

12. Anson Rabinbach, *The Human Motor: Energy, Fatigue, and the Origins of Modernity* (New York: Basic Books, 1990), pp. 54, 70, 71.

13. Aristotle wrote that an individual "fulfills his proper function only by way of practical wisdom and moral excellence or virtue: virtue makes us aim at the right target, and practical wisdom makes us use the right means." Aristotle, *Nichomachean Ethics,* trans. Martin Ostwald (New York: Macmillan, 1962; 1986), pp. 169–170, 312–315. This discussion is indebted to a lecture by Professor Dawson Schultz, delivered at the Institute for the Medical Humanities, University of Texas Medical Branch at Galveston, April 23, 1991.

14. Aristotle, *Nichomachean Ethics,* p. 312. But there is no practical wisdom without virtue; only "cleverness." Also see Martha C. Nussbaum, *The Fragility of Goodness: Luck and Ethics in Greek Tragedy and Philosophy* (New York: Cambridge University Press, 1986), pp. 299–300, 302. The "virtues" are "characteristics," i.e., "fixed capacities for action, acquired by habit" and "united with right reason." But "right reason" in moral matters "is practical wisdom." Thus, "no choice will be right without practical wisdom and virtue." See Aristotle, *Nichomachean Ethics,* pp. 169–172; Helen King, *Hippocrates' Woman* (London: Routledge, 1998), pp. 171.

15. Samuel Haber, *The Quest for Authority and Honor in the American Professions, 1750–1900* (Chicago: University of Chicago Press, 1994), pp. 9–14; Martin S. Pernick, "Medical Professionalism," in Warren T. Reich, ed., *Encyclopedia of Bioethics* (New York: Macmillan, 1978), pp. 1028–1034.

16. Barbara Gutmann. Rosenkrantz, "The Search for Professional Order in 19th Century Medicine," in Judith W. Leavitt and Ronald L. Numbers, eds., *Sickness and Health in America: Readings in the History of Medicine and Public Health,* 2nd ed. rev. (Madison: University of Wisconsin Press, 1985), pp. 219–232, esp. p. 229.

17. John Harley Warner, *The Therapeutic Perspective: Medical Practice, Knowledge, and Identity in America, 1820–1885* (Cambridge, Mass.: Harvard University Press, 1986), esp. pp. 11, 13, 58–65, skillfully elucidates these ideas. See also Martin

Pernick, *A Calculus of Suffering* (New York: Oxford University Press, 1985), for the related doctrine of "conservative" medicine.

18. Burton J. Bledstein, *The Culture of Professionalism: The Middle Class and the Development of Higher Education in America* (New York: W. W. Norton, 1976), p. 134.

19. Warner, *The Therapeutic Perspective*, pp. 8, 14–17; Genevieve Miller, "Dolley, Sarah Read Adamson," in Edward T. James, Janet Wilson James, and Paul S. Boyer, eds., *Notable American Women, 1607–1950* (Cambridge, Mass.: Harvard University Press, 1971), pp. 497–499.

20. See Regina Markell Morantz-Sanchez, *Sympathy and Science: Women Physicians in American Medicine* (New York: Oxford University Press, 1985); Mary Roth Walsh, *Doctors Wanted: No Women Need Apply* (New Haven, Conn.: Yale University Press, 1977); Ruth J. Abram, ed., *"Send Us a Lady Physician": Women Doctors in America* (New York: W. W. Norton, 1985); Gloria Moldow, *Women Doctors in Gilded-Age Washington* (Urbana: University of Illinois Press, 1987); and the forthcoming history of Woman's Medical College of Pennsylvania and Medical College of Pennsylvania by Steven J. Peitzman. On religious nonconformity and the first generation of women doctors, see Edward Atwater, "Women Who Became Doctors before the Civil War," paper given as the Radbill Lecture, College of Physicians of Philadelphia, April 18, 1989. My thanks to Dr. Atwater for providing me with the manuscript. One of Adamson's teachers, Dr. Bartholemew Fussel, was a charter member of the American Anti-Slavery Society. He, his son, and Elijah Pennypacker were all "conductors" on the Underground Railroad who harbored fugitives from slavery. Bartholemew Fussel taught private classes for women in physiology and hygiene. See Deborah Jean Warner, *Graceanna Lewis: Scientist and Humanitarian* (Washington, D.C.: Smithsonian Institution Press, 1979), p. 95.

21. "Aunt Gracie," as Dolley called her, was eight years her senior. Warner, *Graceanna Lewis*, pp. 16–23, 33–36, 39, 95–96; Margaret Rossiter, *Women Scientists in America* (Baltimore: Johns Hopkins University Press, 1982); Mary A. E. Wager, "Women as Physicians," *The Galaxy* 6 (December 1868): 780; Miller, "Dolley," *Notable American Women*.

22. Unless otherwise noted, quotations followed by dates in parentheses are taken from the letters of Sarah Read Adamson Dolley to her son, Dr. Charles Sumner Dolley, Boxes 1–11, SAD/UR.

23. Gulielma Fell Alsop, *History of the Woman's Medical College, 1850–1950* (Philadelphia: J. B. Lippincott Company, 1950), pp. 8–10; Morantz-Sanchez, *Sympathy and Science*, pp. 76, 397 n. 33; Martin Kaufman, "The Admission of Women to Nineteenth-Century American Medical Societies," *Bulletin of the History of Medicine* 50 (1976): 251–260.

24. As quoted in Wager, "Women as Physicians," p. 780.

25. See Morantz-Sanchez, *Sympathy and Science*, p. 49.

26. Cleveland Medical College was the one exception. Between 1852 and 1856 it graduated six women; no other women graduated from regular coeducational medical colleges until 1869. Linda Lehmann Goldstein, "'Without Compromising in Any Particular': The Success of Medical Coeducation in Cleveland, 1850–1856," *Caduceus* 10 (1994): 101–115.

27. Thomas Bonner estimates the total number of women medical graduates before 1915 to be about 3,400. Thomas Neville Bonner, *To the Ends of the Earth:*

Women's Search for Education in Medicine (Cambridge, Mass.: Harvard University Press, 1992), pp. 14–16, 142–155; Atwater, "Women Who Became Doctors," pp. 1–21. A total of seventeen women's medical colleges, regular and sectarian, were founded between 1848 and 1895. But by 1918 all but the Woman's Medical College of Pennsylvania (WMCP) had closed. WMCP retained its tradition of all-women's education until 1971, when it became coeducational. Morantz-Sanchez, *Sympathy and Science*, pp. 80–81, 246–248, 350; Walsh, *Doctors Wanted*, Table 2, p. 180; Abram, *"Send Us a Lady Physician,"* p. 75.

28. Stephen J. Peitzman, "Medical and Collegiate: Student Life in the Golden Age of the Woman's Medical College of Pennsylvania," paper given at the seventy-first annual meeting of the American Association for the History of Medicine, Toronto, May 9, 1998.

29. Frederick C. Waite, "Dr. Lydia Folger Fowler," *Annals of Medical History* 4 (1932): 290–297.

30. Ibid., pp. 290–291. The fourth woman in their class, Mrs. Margaretta B. Gleason, was the wife of a Philadelphia physician. It is not known if her husband was related to the husband of Rachel Gleason. See also Susan Cayleff, *Wash and Be Healed: The Water-Cure Movement and Women's Health* (Philadelphia: Temple University Press, 1987), pp. 25, 26, 111.

31. Obituary for Lester Clinton Dolley, M.D., *Rochester (N.Y.) Evening Express*, April 6, 1872, typescript, Box 10, folder 25, SAD/MCP; William G. Rothstein, *American Physicians in the 19th Century: From Sects to Science* (Baltimore: Johns Hopkins University Press, 1972), pp. 218–219.

32. John S. Haller, Jr., *Medical Protestants: The Eclectics in American Medicine, 1825–1939* (Carbondale: Southern Illinois University Press, 1994), pp. 32–37; Warner, *The Therapeutic Perspective*, pp. 91–92.

33. Beach's approach was sketched out in a series of widely reprinted volumes, beginning with his *The American Practice of Medicine* (1833). Cf. Wooster Beach, *Beach's Family Physician and Guide to Home Health and Happiness* (Cincinnati: Moore, Wilstach, Keys and Co., 1859), introduction, pp. 229–239, 254, 360–366; Rothstein, *American Physicians*, pp. 217–219; Warner, *The Therapeutic Perspective*, pp. 91–92. Also see Norman Gevitz, "Three Perspectives on Unorthodox Medicine," pp. 1–28, esp. p. 19, and William G. Rothstein, "The Botanical Movements and Orthodox Medicine," pp. 29–51, both in Norman Gevitz, ed., *Other Healers: Unorthodox Medicine in America* (Baltimore: Johns Hopkins University Press, 1988).

34. Bonner, *To the Ends of the Earth*, pp. 14–16.

35. Naomi Rogers, "Women and Sectarian Medicine," in Rima D. Apple, ed., *Women, Health and Medicine in America: A Historical Handbook* (New Brunswick, N.J.: Rutgers University Press, 1990), pp. 273–302.

36. Bonner, *To the Ends of the Earth*, chaps. 2 and 8; also see Atwater, "Women Who Became Doctors," pp. 1–21. Both Bonner and Atwater essentially confirm the claim made by John B. Blake, "Women and Medicine in Ante-Bellum America," *Bulletin of the History of Medicine* 39 (1965): 99–123, that the majority of the early women graduates were products of sectarian schools. As I noted in the Introduction, the first two African American women medical graduates, Drs. Rebecca Lee and Rebecca J. Cole, graduated from regular women's medical schools. For the vast majority of nineteenth-century black women medical graduates, the breakdown

between regulars and sectarians is not yet known. See also Cayleff, *Wash and Be Healed*, pp. 68–70.

37. Bonner, *To the Ends of the Earth*, pp. 142, 156–57.

38. Rothstein, "The Botanical Movements"; Atwater, "Women Who Became Doctors."

39. Rothstein, "The Botanical Movements"; Cayleff, *Wash and Be Healed*, p. 6.

40. Fewer than one-fourth of the medical schools founded between 1880 and 1906, the peak years for proliferation of American medical colleges, were sectarian schools. The percentage of sectarian graduates decreased from less than 25 percent to less than 10 percent during the same period. Rothstein, *American Physicians*, pp. 217–228, 287, Table XV.1. Cf. Atwater, "Women Who Became Doctors," p. 2.

41. Letter from Sarah Adamson Dolley to Charles Sumner Dolley, October 17, 1904, SAD/UR. Full accounts of the medical curricula of reputable nineteenth-century medical colleges may be found in Kenneth M. Ludmerer, *Learning to Heal: The Development of American Medical Education* (New York: Basic Books, 1985), and William G. Rothstein, *American Medical Schools and the Practice of Medicine* (New York: Oxford University Press, 1987).

42. Letters from Sarah Adamson Dolley to Charles Sumner Dolley, February 11, 1874; August 29, 1879; September 7, 1892; January 8, 1895; October 17, 1904, all in SAD/UR; letter to Elijah F. Pennypacker, February 3, 1850, SAD/MCP; Wager, "Women as Physicians," pp. 781, 782.

43. For example, in the 1896 letter to her son already cited, Dolley invoked the teachings of Dr. Orin Davis, an eclectic surgeon from Attica, New York, when she asserted, "In all departures from health of body, mind, or spirit, I believe there is a loss of balance . . . [T]he principle of restoration must be to restore that balance." Dolley to Charles Sumner Dolley, November 22, 1896, SAD/UR.

44. Waite, "American Sectarian Medical Colleges," pp. 148–166; idem., "Lydia Folger Fowler," pp. 290–292; Blake, "Women and Medicine," p. 114.

45. I am indebted to Dr. Edward Atwater for finding Rachel Gleason's "fading diploma" at the Arnot Ogden Hospital in Elmira, New York. See Edward C. Atwater, "The Area's First Women Doctors," *The Bulletin of the Monroe County Medical Society* 1991 (September/October): 13–14.

46. This is carefully described in John Harley Warner, *Against the Spirit of System: The French Impulse in Nineteenth-Century American Medicine* (Princeton: Princeton University Press, 1998), esp. pp. 104–108.

47. Alsop, *History*, p. 9.

48. Letter from Charles Adamson and Mary C. Adamson to Dr. L. C. Dolly [*sic*], July 8, 1851, collection of Dr. Edward Atwater, Rochester, New York. My gratitude to Dr. Atwater for sending me the typescript. Also see Charles E. Rosenberg, *The Care of Strangers* (New York: Basic Books, 1987), pp. 175–184; Janet Golden and Charles E. Rosenberg, *Pictures of Health: A Photographic History of Health Care in Philadelphia* (Philadelphia: University of Pennsylvania Press, 1991), p. 77.

49. Letter from Sarah Adamson Dolley to Charles Sumner Dolley, October 17, 1904, SAD/UR; John Welsh Croskey, *History of Blockley: A History of Philadelphia General Hospital from Its Inception, 1731–1928* (Philadelphia: F. A. Davis Company, 1929), pp. 32–33, 99, 258–259. Adamson had a far easier time of it at the hospital

than Blackwell, from whom the house officers deliberately hid their patient notes. Warner, *Against the Spirit*, pp. 49–50; Abram, *"Send Us a Lady Physician,"* p. 75. Much earlier, in 1810, the Philadelphia Almshouse, Blockley's predecessor institution, had accepted a Mrs. Lavender as an assistant midwife in order to "perfect" her education. Croskey, *History of Blockley*, pp. 384–385.

50. Letter from Charles Adamson and Mary C. Adamson to Dr. L. C. Dolly [*sic*], July 8, 1851; New York State Medical License No. 376 for Lester C. Dolley and Sarah R. A. Dolley, 1862; both in the personal collection of Dr. Edward Atwater. Dolley Family Genealogy, typescript, p. 16, Box 10, folder 25, SAD/UR.

51. Gertrude Stuart Baillie, "Should Professional Women Marry?" *Woman's Medical Journal* 2 (1894): 33–34.

52. Morantz-Sanchez, *Sympathy and Science*, pp. 136–138. Ellen More, "The Blackwell Medical Society and the Professionalization of Women Physicians," *Bulletin of the History of Medicine* 61 (1987): 603–628.

53. Morantz-Sanchez, *Sympathy and Science*, p. 136.

54. Goldstein, "Without Compromising," pp. 111–112. Although Bertha Van Hoosen never married, her household included her mother, sister, and niece, for whom she was principal caregiver for many years. Bertha Van Hoosen, *Petticoat Surgeon* (Chicago: Pellegrini and Cudahy, 1947), p. 53. On Mary Putnam Jacobi's volatile marriage to Abraham Jacobi, see Joy Harvey, "Clanging Eagles: The Marriage and Collaboration between Two Nineteenth-Century Physicians, Mary Putnam Jacobi and Abraham Jacobi," in Helena M. Pycior, Nancy G. Slack, and Pnina G. Abir-Am, eds., *Creative Couples in the Sciences* (New Brunswick, N.J.: Rutgers University Press, 1996), pp. 185–195.

55. Sarah R. A. Dolley, Journal, 1860, from the collection of Dr. Edward Atwater, Rochester, New York, to whom I am most grateful.

56. The phrase is taken from David McDonald, "Organizing Womanhood: Women's Culture and the Politics of Woman Suffrage in New York State, 1865–1917," Ph.D. diss., State University of New York at Stony Brook, 1987, pp. 7, 8. I am grateful to Professor McDonald for sending me the chapters pertaining to Rochester's women activists.

57. Blake McKelvey, *Rochester: The Flower City, 1855–1890* (Cambridge, Mass.: Harvard University Press, 1949), pp. 5, 301–302, 385, 386.

58. Rochester City Hospital, like St. Mary's a private foundation, was chartered in 1851 but did not open until 1864. Ibid., pp. 33–36, 79–81.

59. Martin Kaufman, "Edward Mott Moore," *Dictionary of American Medical Biography*, pp. 534–535; Croskey, *History of Blockley*, pp. 444–445.

60. "April 8, 1860 . . . Yesterday I sent in my bill to Mr. Waring. He sent word he wanted to see Doctor [Lester]. I have felt very much worried about it as I did not want to send in an unreasonable bill, and for the benefit Mrs. W has received I thought it not too much." Sarah R. A. Dolley, Journal. Cf. a letter of March 9, 1866, from Sarah's uncle, Dr. Hiram Corson, to Lester ("My Dear Nephew"), in which Corson directed medical discussions individually to each of them. He asked Lester about Rochester's sanitary regulations and mentioned an operation to remove an ovarian tumor; with Sarah, he discussed a case of "glossitis," an inflammation of the tongue and palate. Collection of Dr. Edward Atwater, Rochester, New York.

61. Biographical details contained in the following paragraphs about the lives of Sarah Dolley and her family are drawn from the following sources: letters of Sarah

Read Adamson Dolley and Charles Sumner Dolley, SAD/UR; letters from Charles Sumner Dolley to Frederick C. Waite, August 24, 1930, and May 26, 1939; letter from Frederick C. Waite to Charles Sumner Dolley, September 15, 1930; and letter from Sarah Read Adamson to Elijah Pennypacker, February 3, 1850, SAD/MCP; biographies of the antecedents and descendants of Dr. Lester C. Dolley, typescript, n.d., all in Box 10, folders 22, 25, SAD/UR. Miller, "Dolley"; Waite, "Lydia Folger Fowler"; Phebe A. Hanaford, *Women of the Century* (Augusta, Me.: True and Co., 1883), pp. 547–550; Hiram Corson, *The Corson Family: A History of the Descendants of Benjamin Corson* (Philadelphia: Henry Lawrence Everrt, 1906), pp. 82–85; Mary A. E. Wager, "Women as Physicians," *The Galaxy* 6 (December 1868): 774–789; J. M. Anders, "Hiram Corson . . . Clinician Pioneer and Reformer," *The General Magazine and Historical Chronicle* 1932: 86–94; Jane Marsh Parker, *Rochester: A Story Historical* (Rochester, N.Y., 1884), pp. 264–266; J. Smith Futhey and Gilbert Cope, *History of Chester County, Pennsylvania* (Philadelphia: Henry Lawrence Everrt, 1881), p. 461.

62. E. Anthony Rotundo, *American Manhood: Transformations in Masculinity from the Revolution to the Modern Era* (New York: Basic Books, 1993), pp. 26–30, 42–51. Also see Ryan, *Cradle*, pp. 160–185; Steven Mintz, *A Prison of Expectations: The Family in Victorian Culture* (New York: New York University Press, 1983), p. 84; T. Walter Herbert, *Dearest Beloved: The Hawthornes and the Making of the Middle-Class Family* (Berkeley: University of California Press, 1993), p. 6; Nancy M. Theriot, *The Biosocial Construction of Femininity* (New York: Greenwood, 1988).

63. See Shirley Roberts, *Sophia Jex-Blake* (London: Routledge, 1993), pp. 125–137.

64. Only two years earlier the young Mary Putnam had written Elizabeth Blackwell of her "thorough contempt" for the slowness of her students at the Woman's Medical College of the New York Infirmary. "Be very prudent and patient!" reads Blackwell's response, written on December 31, 1871. Mary Putnam Jacobi, *Life and Letters*, p. 307.

65. For a detailed account of Charles Sumner Dolley's career, see Ellen S. More, "Doctors or Professors? Late Victorian Physicians and the Culture(s) of Professionalism," *Canadian Review of American Studies*, 23 (1993): 125–148.

66. Letters from Sarah Dolley to Charles Sumner Dolley, January 14, 26, 1885, and December 8, 1890. See also letters of November 1876, n.d.; October 25, 1877; March 24, 1880; August 8, 1881; September 27, 1881; December 30, 1886; and March 15, 1887, all in SAD/UR. Dolley traveled to Paris for postgraduate clinical training in 1869 and again in 1875. In 1875 she also attended clinics in Prague and Vienna. Parker, *Rochester: A Story Historical*, p. 265, handwritten extract in Box 10, folder 22, SAD/UR; Miller, "Dolley," p. 498.

67. More, "Doctors or Professors?" pp. 125–148.

68. Cf. Janet Oppenheim, *Shattered Nerves: Doctors, Patients, and Depression in Victorian England* (New York: Oxford University Press, 1991), pp. 240–243.

69. Letter from Charles Sumner Dolley, to Sarah Adamson Dolley, January 31, 1892, SAD/UR.

70. "Obituary: Charles Sumner Dolley, M.D.," *New York Times*, July 11, 1948. The author wishes to thank the archivists at the University of Pennsylvania Rare Books and Special Collections Archive, and especially Sandra Markham, for their assistance in gaining access to the trustees' records of these events.

71. Sumner's image of a "hot box" refers to the problem of designing railroad car axles that would tolerate intense friction without heating up and locking the wheels.

72. Letter from Sarah Adamson Dolley to Charles Sumner Dolley, Box 1, folder 9, n.d., SAD/UR.

73. Ibid., December 10, 1879, and June 15, 1885.

74. The Fortnightly Ignorance Club met regularly to discuss literary and political questions. On January 17, 1881, they discussed Horatio Charles Wood, "The Value of Vivisection," *Scribner's Monthly* 20 (1880): 766–770.

75. Because Dolley believed in a somatic basis for neurasthenia resulting from irritation of the nerve sheaths by poorly digested foods (ptomaines), a treatment she recommended for Sumner in 1888 included an antacid, a diuretic-diaphoretic, and even the heroic mercurial agent calomel to induce sweating and stimulate his metabolism. Yet her letters clearly indicate her simultaneous belief that emotional factors also played a role in the condition. Letters from Sarah Adamson Dolley to Charles Sumner Dolley, February 9, 1888, and February 16, 1888, SAD/UR; Warner, *The Therapeutic Perspective*, pp. 85–90, 101, 102, 149–159; *Dorland's American Illustrated Medical Dictionary*, 12th ed., rev. (Philadelphia: W. B. Saunders Company, 1924).

76. Letters from Sarah Adamson Dolley to Charles Sumner Dolley, October 30, 1891; April 30, 1888; October 8, 1888; and October 31, 1891, SAD/UR. On Mary Putnam Jacobi's ambivalence regarding specialism for nineteenth-century women doctors, see Morantz-Sanchez, *Sympathy and Science*, pp. 198–199.

77. Letter from Sarah Adamson Dolley to Charles Sumner Dolley, November 15, 1885, SAD/UR. Also see Leslie J. Reagan, "Linking Midwives and Abortion in the Progressive Era," *Bulletin of the History of Medicine* 69 (1995): 569–598, esp. 575–576; idem., *When Abortion Was a Crime: Women, Medicine and Law in the United States, 1867–1973* (Berkeley: University of California Press, 1997), pp. 11–12, 57–58.

78. Dolley's cash flow troubles persisted, however. On October 27, 1886, she noted that so many debts were coming due by the middle of November that she might have to take out a short-term loan. Including her loan payments, taxes, and insurance policies for herself, for Sumner, and for his wife, Elizabeth, she calculated her indebtedness at $217.11. She solved the problem with a $30.00 short-term loan from the businessman husband of her colleague Dr. Mary Slaight (December 22, 1886).

79. Letter from Sarah Adamson Dolley to Charles Sumner Dolley, February 1, 1885, SAD/UR; Bonner, *To the Ends of the Earth*, p. 28; Morantz-Sanchez, *Sympathy and Science*, p. 92; Walsh, *Doctors Wanted*, p. 186.

80. See her letter to Elijah Pennypacker, February 3, 1850, typescript, SAD/MCP.

81. Letters from Sarah Adamson Dolley to Charles Sumner Dolley, July 27, 1882, and May 21, 1885, SAD/UR.

82. Ida Husted Harper, *The Life and Work of Susan B. Anthony*, vol. 3 (Indianapolis: The Hollenbeck Press, 1908), p. 1204; Nancy A. Hewitt, *Women's Activism and Social Change: Rochester, New York, 1822–1872* (Ithaca, N.Y.: Cornell University Press, 1984), pp. 24–25, 231–232, 253–254. Also see Paul E. Johnson, *A Shopkeeper's Millennium* (New York: Hill and Wang, 1978).

83. Hewitt, *Women's Activism*, pp. 211, 213; also see Harper, *Susan B. Anthony*, vol. 1., p. 424.

84. Letters from Sarah Adamson Dolley to Charles Sumner Dolley, October 11, 1887, and April 11, 1896, SAD/UR.

85. Ibid., April 30, 1893, and January 11, 1894.

86. Letter from Dr. Emily Blackwell to Dr. S R A [sic] Dolley, February 26, 1867, in the collection of Dr. Edward Atwater, Rochester, New York.

87. "Death Takes Woman Long Doctor Here," *Democrat and Chronicle* (February 18, 1934), p. 1; letters from Sarah Adamson Dolley to Charles Sumner Dolley, March 18, 1886, and December 30, 1895, SAD/UR.

88. Letter from Sarah Adamson Dolley to Charles Sumner Dolley, September 14, 1898, SAD/UR.

89. See Chapter 2 for a fuller discussion of these institutions.

90. "Sarah Read Adamson Dolley, M.D.," *JAMA* 54 (1910): 152–153; "Fortieth Annual Meeting of the American Medical Association," ibid., 13 (1889): 28; "Report of Section on State Medicine endorsing the implementation of a State Board of Medical Licensure in Each State: Report Adopted," ibid., 102–103.

91. In 1902 the AMA reconstituted its House of Delegates with a fixed limit of 150 members, or a ratio of approximately 1 : 15. By 1915, when the first woman was elected as a delegate under the new rules, the ratio stood at 1 : 700 and AMA membership had climbed to more than 70,000. As for leadership positions for women, see Chapter 5. Paul Starr, *The Social Transformation of American Medicine* (New York: Basic Books, 1982), pp. 109, 110, 461 n. 78; Walsh, *Doctors Wanted*, p. 155 n. 13; "Minutes of the House of Delegates, Fifth Meeting—Thursday, June 25," *JAMA* 63 (1914): 108–109; Bertha Van Hoosen, "Shall Medical Women Hold Official Positions in the A.M.A.?" *Medical Woman's Journal* 34 (1927): 287–288; *Digest of Official Actions, 1846–1958*, vol. 1 (Chicago: American Medical Association, 1959), pp. 158, 744–747.

92. Dolley's opinion that homeopathy was a "fraud" did not stop her from attending an occasional patient at one of Rochester's homeopathic hospitals. Letters from Sarah Adamson Dolley to Charles Sumner Dolley, July 30, 1890, and January 3, 1895, SAD/UR.

93. Ibid., June 4, 1888; John Harley Warner, "Ideals of Science and Their Discontents in Late Nineteenth-Century Medicine," *Isis* 82 (1991): 454–478, esp. 470–477.

94. Sarah Dolley, "President's Address," Women's Medical Society of New York State, Rochester, March 11, 1908; reprinted in *Woman's Medical Journal* 18, no. 4 (April 1908): 63–65. The last sentence quoted here appeared in both her 1874 address in Philadelphia and in a letter she wrote her son May 21, 1893, SAD/UR.

95. Letter from Sarah Adamson Dolley to Charles Sumner Dolley, August 19, 1909, SAD/UR. The committee was the idea of Dr. Rosalie Slaughter Morton, who was appointed its first chair, but the idea of naming Dolley honorary chair may have been suggested by Morton's friend Dr. Marion Craig Potter. See Chapter 3.

96. Last Will and Testament of Sarah R. A. Dolley, October 30, 1909; Appraiser's Report, January 7, 1911; Reports of Kathleen Buck, Executrix, January 13, 1911, and April 22, 1913. At the time of her death, Dolley owned one valuable piece of property running from East Avenue to Main Street in Rochester's rapidly appreciating business district and three others of lesser value. Together they were valued

at $62,300. Over the next two years Buck sold them for $73,000. After deducting for mortgages and other debts, Dolley's son received approximately $30,000 from the estate. His mother also left smaller bequests to St. Mary's Hospital, to St. Anne's Home for the Aged, and to her brother, grandchildren, and a niece. Buck received an executor's commission of $900. My thanks for research assistance from Corinne Sutter-Brown in locating these documents in the archives of the Surrogate's Court of Rochester, New York.

2. Gendered Practices

1. Part of this chapter appeared as "The Blackwell Medical Society and the Professionalization of Women Physicians," *Bulletin of the History of Medicine* 61 (1987): 603–628. It was originally presented at the fifty-eighth annual meeting of the American Association for the History of Medicine in Durham and Chapel Hill, North Carolina, May 2, 1985. Many thanks are due to the past and present history of medicine librarians at the Edward G. Miner Library of the University of Rochester School of Medicine and Dentistry, Janet B. Joy and Christopher Hoolihan, and to Sandra Chaff, Margaret Jerrido, and Jill Gates Smith, formerly of the Medical College of Pennsylvania's Archives and Special Collections on Women and Medicine.

2. The concept of "separate spheres" is not unproblematic. Nevertheless, as with any group pursuing a new avenue of social change, women physicians were significantly restricted in their movement into the profession for at least 75 years after first gaining entry to American medical education, both by dominant gender norms and by their own desire to adhere to them. The separatist institutions described in this chapter, however, were the prelude to the development of a concept of "maternalist" politics (addressed in Chapter 3) that presupposed the desirability of carrying so-called maternal values into the wider domain of public policy. See Linda K. Kerber, "Separate Spheres, Female Worlds, Woman's Place: The Rhetoric of Women's History," *Journal of American History*, 75 (1988): 9–39; and Jill K. Conway, Susan C. Bourque, and Joan W. Scott, "Introduction: The Concept of Gender," *Daedalus* 116 (1987): 21–30.

3. Rebecca J. Tannenbaum, "Earnestness, Temperance, Industry: The Definition and Uses of Professional Character among Nineteenth-Century American Physicians," *Journal of the History of Medicine and Allied Sciences* 49 (1994): 251–283.

4. Estelle Freedman, "Separatism as Strategy: Female Institution Building and American Feminism," *Feminist Studies* 5 (1979): 512–529; Cora Bagley Marrett, "On the Evolution of Women's Medical Societies," *Bulletin of the History of Medicine* 53 (1979): 434–438. Cf. Virginia G. Drachman, *Hospital with a Heart: Women Doctors and the Paradox of Separatism at the New England Hospital, 1862–1969* (Ithaca, N.Y.: Cornell University Press, 1984), esp. pp. 14, 15.

5. Thomas Bender, "The Cultures of Intellectual Life: The Cities and the Professions," in John Higham and Paul J. Conkin, eds., *New Directions in American Intellectual History* (Baltimore: Johns Hopkins University Press, 1979), pp. 190, 191. Bender distinguishes between antebellum "civic professionalism" and late nineteenth-century "disciplinary professionalism." Cf. Robert J. Wiebe, *The Search for Order, 1877–1922* (New York: Hill and Wang, 1967). For varying assessments of professionalization, see Talcott Parsons, "Profession," in David Sills, ed., *Interna-*

tional Encyclopedia of the Social Sciences (New York, 1968), pp. 536–546, cited in Gerald L. Geison, ed., *Professions and Professional Ideology in America* (Chapel Hill: University of North Carolina Press, 1983), Introduction; Christopher Lasch, *Haven in a Heartless World: The Family Besieged* (New York: Basic Books, 1977), pp. xiv, xv; Burton J. Bledstein, *The Culture of Professionalism: The Middle Class and the Development of Higher Education in America* (New York: W. W. Norton, 1976), chap. 1; Eliot Freidson, *Profession of Medicine: A Study of the Sociology of Applied Knowledge* (New York: Harper and Row, 1970), pp. 22–33; Magali Sarfatti Larson, *The Rise of Professionalism* (Berkeley: University of California Press, 1977), chaps. 4 and 5; Martin Pernick, *A Calculus of Suffering* (New York: Oxford University Press, 1985), esp. chaps. 2 and 11 and the afterword; Samuel Haber, *The Quest for Authority and Honor in the American Professions, 1750–1900* (Chicago: University of Chicago Press, 1991); Andrew Abbott, *The System of Professions: An Essay on the Division of Expert Labor* (Chicago: University of Chicago Press, 1988). Also see John C. Burnham, *How the Idea of Profession Changed the Writing of Medical History*, Medical History, Supplement No. 18 (London: Wellcome Institute, 1998). On medical societies in this period, see Charles E. Rosenberg, "Doctors and Credentials: The Roots of Uncertainty," *Transactions and Studies of the College of Physicians of Philadelphia* 6 (1984): 295–302.

6. Kenneth M. Ludmerer, *Learning to Heal* (New York: Basic Books, 1985).

7. Mary Ryan, *Womanhood in America*, 2nd ed. (1975; New York: Franklin Watts, 1979), pp. 135–145, among others, uses the term *social housekeeping* to suggest the continuity between the Victorian ideology of motherhood and the "extra-familial" activities of social feminists. Besides the works of Morantz-Sanchez, Walsh, and Drachman cited above, also see Gloria Moldow, *Women Doctors in Gilded-Age Washington: Race, Gender, and Professionalization* (Chicago: University of Illinois Press, 1987).

8. Sociologist Theda Skocpol writes that late nineteenth-century women's groups "were often 'maternalist' in orientation—that is, devoted to extending domestic ideals into public life—and they demanded benefits and services especially targeted on mothers and children." Theda Skocpol, *Protecting Soldiers and Mothers: The Political Origins of Social Policy in the United States* (Cambridge, Mass.: The Belknap Press of Harvard University Press, 1992), p. 36.

9. Like William L. O'Neill, *Everyone Was Brave: A History of Feminism in America*, 2nd ed. (Chicago: Quadrangle Books, 1971), and Karen Blair, *The Clubwoman as Feminist: The Woman's Culture Club Movement in the United States, 1868–1914* (Ann Arbor, Mich.: University Microfilms, 1976); I am using the term *social feminism*, as distinct from feminism, to suggest the broad coalition of nineteenth-century middle-class women, both volunteers and professionals, who under the banner of female moral superiority sought to bring about a wide variety of social reforms ranging from temperance and the support of foreign missions to social purity and settlement-house work. "Feminist," in my usage, refers specifically to the doctrine of female equality, and to support for women's rights, particularly female suffrage. Many social feminists supported suffrage, too, but only as one of a number of desirable reforms. Dolley, Dr. Marion Craig Potter, and their colleagues supported the broad goals of social feminism but did not subordinate the issue of suffrage to other social issues. See Nancy F. Cott, *The Grounding of Modern Feminism* (New Haven, Conn.: Yale University Press, 1987).

10. See Marrett "On the Evolution," pp. 439, 440, for statistics on male and female membership in state medical societies. There is some dispute about the identity of the earliest women's medical society. Walsh, *Doctors Wanted*, p. 104, names the New England Hospital Medical Society (est. 1878) as the first. Marrett (p. 445), cites the Alumnae Association of the Woman's Medical College of the New York Infirmary (est. 1873), since it was the direct forerunner of the Women's Medical Association of New York City. See also Martin Kaufman, "The Admission of Women to Nineteenth Century Medical Societies," *Bulletin of the History of Medicine* 50 (1976): 251–260.

11. They were, in alphabetical order: Drs. Sarah Adamson Dolley (Central Medical College, 1851); Frances Fidelia Hamilton (Woman's Medical College of the New York Infirmary, 1874); Sarah Perry (University of Buffalo, 1882); Marion Craig (Potter) (University of Michigan, 1884); Anna Searing (University of Michigan/homeopathic, 1868); Mary Slaight (University of Buffalo, 1880); Mary E. Stark (Woman's Medical College of the New York Infirmary, 1880); Harriet M. Turner (Woman's Medical College of Pennsylvania, 1886). University of Rochester School of Medicine and Dentistry, Edward G. Miner Library, Practitioners'/Blackwell Medical Society, *Minutes*, book 1, pp. 1–64; book 3, pp. 123–24 (hereafter, Pract./BMS *Minutes*).

12. See Table 4.2, Chap. 4, for percentages of American women physicians in 1915 and 1920. The overall figure for Rochester physicians is the result of a hand count comparing listings in the Rochester City Directory to those of the state medical directory every five years from 1910 to 1930, including only male and female regulars and homeopaths.

13. Frances Hurlbut-White also graduated from the University of Michigan's homeopathic department. I am indebted to Micaela Sullivan, then research associate at the Library of the American Medical Association, Chicago, for very kindly providing me with standard biographical information for many of the society's members (hereafter cited as Obits./AMA).

14. However, neither electrotherapeutics nor hydrotherapy was considered off limits to the regulars. Between 1890 and 1891 both the Practitioners' and the Pathological societies heard presentations, one of them in a gymnasium, on these therapeutic techniques. Pract./BMS *Minutes*, book 1, p. 57; book 2, pp. 7, 12–13, 112, 172–173, 216, 219; book 3, p. 35; Pathological Society, *Minutes, 1885–1895*, pp. 260–261, 281. *Woman's Medical Journal*, first published in 1893, commonly ran articles in the 1890s on the use of hydrotherapy, electrotherapeutics, and gymnastics: see indexes to vols. 1–3, 5, 6.

15. See Florence A. Cooksley, "A History of Medicine in the State of New York and the County of Monroe, Continued, Part II," *New York State Journal of Medicine* 36 (1936): 2003.

16. Obits./AMA.

17. Morantz-Sanchez, *Sympathy and Science*, Table 5.1, p. 137. Her calculations for the Woman's Medical College of Pennsylvania suggest a 35 percent marriage rate for graduates between 1852 and 1900, a figure she considers conservative (p. 407 n. 127).

18. Bertha Van Hoosen, "Notes and Comments," *Medical Woman's Journal* 36 (1929): 23. Women physicians' low representation in specialty societies such as the New York Obstetrical Society persisted into the 1930s. Charles R. King, "The New

York Maternal Mortality Study: A Conflict of Professionalization," *Bulletin of the History of Medicine* 65 (1991): 476–502, esp. 497.

19. Only one member of the Practitioners'/Blackwell Medical Society, Eveline Ballintine, was a member of a national specialty society (the American Medico-Psychological Association), as compared with thirty-one members, or 34 percent, of the Pathological Society in the sample year, 1910. The year 1910 was chosen for comparison because by then, in both medical societies, more than half the members had been in practice for at least twenty years. See also Rosemary Stevens, *American Medicine and the Public Interest* (New Haven, Conn.: Yale University Press, 1971), pp. 43–54, 77–78, 133–134; Ludmerer, *Learning to Heal*, pp. 87, 104.

20. Edward C. Atwater, "Medical Politics in Rochester, 1865–1925," *Bulletin of the Monroe County Medical Society* (1976): 115–122; Florence A. Cooksley, "A History of Medicine in the State of New York and the County of Monroe, Continued, Part II, Chapter IV," *New York State Journal of Medicine* 37 (1937): 88.

21. *Hospital Review* 34 (1898): 76.

22. Pract./BMS *Minutes*, book 1, p. 161; book 2, pp. 41, 83, 94, 140, 152, 171, 181; book 3, p. 61; Pathological Society, *Minutes, 1897–1904*, p. 261. Curt Proskauer, "Development and Use of the Rubber Glove in Surgery and Gynecology," *Journal of the History of Medicine* 13 (1958): pp. 373–381, reprinted in Gert H. Brieger, ed., *Theory and Practice in American Medicine: Historical Studies from the Journal of the History of Medicine and Allied Sciences* (New York: Science History Publications, 1976), pp. 203–211, traces the introduction of rubber gloves to about 1890 at Johns Hopkins.

23. See Morantz-Sanchez, *Sympathy and Science*, chap. 8 and p. 425 n. 55.

24. Compare the wording of the Pathological Society's statement of purpose: "social intercourse and mutual improvement in the medical and kindred sciences." See Cooksley, "History . . . Part II," (1936), p. 2003.

25. Ludmerer, *Learning to Heal*, p. 158.

26. See Kenneth M. Ludmerer, "The Plight of Clinical Teaching in America," *Bulletin of the History of Medicine* 57 (1983): 218–229; idem., *Learning to Heal*, pp. 74, 116, 154, 160–161. On competing concepts of science among nineteenth-century physicians, see John Harley Warner, "Science in Medicine," *Osiris*, 2nd series, vol. 1 (1985): 37–58. Cf. Bender, "Cultures," pp. 181–195.

27. Pract./BMS *Minutes*, book 1, pp. 118, 119.

28. Pract./BMS *Minutes*, book 2, p. 107.

29. Pathological Society, *Minutes, 1885–1895*, pp. 153, 154.

30. See Regina Morantz and Sue Zschoche, "Professionalism, Feminism, and Gender Roles: A Comparative Study of Nineteenth-Century Medical Therapeutics," *Journal of American History* 67 (1980): 568–588; and Pernick, *A Calculus*, chap. 11, for detailed analyses of the evidence for male and female physicians' therapeutics. Morantz-Sanchez concluded that although men's and women's therapeutic practices differed only slightly, the character of the doctor-patient relationship was significantly dependent on the gender of the physician. Pernick, however, found that, for the decision to administer anesthesia, physicians' gender differences may have been significant. The present study compares the opinions, not the actual therapeutics, of the Practitioners' and Pathological Societies, since comparable statistics for both societies' members could not be constructed.

31. Pathological Society, *Minutes, 1885–1895*, pp. 139, 332; Pract./BMS *Min-*

utes, book 2, pp. 160–161, 174–175. On the risks and benefits of chloroform, see Martin S. Pernick, *A Calculus,* and W. S. Playfair, *A Treatise of . . . Midwifery,* 5th American ed., ed. Robert P. Harris (Philadelphia: Henry C. Lea, 1889), pp. 299–301.

32. See Regina Morantz-Sanchez, "The Gendering of Empathic Expertise," in Ellen Singer More and Maureen A. Milligan, eds., *The Empathic Practitioner: Empathy, Gender, and Medicine* (New Brunswick, N.J.: Rutgers University Press, 1994), pp. 40–58; Ornella Moscucci, *The Science of Woman: Gynaecology and Gender in England, 1800–1929* (Cambridge: Cambridge University Press, 1990), p. 128, found "marked differences" among the surgical practices of male gynecologists in London's Middlesex Hospital in the late nineteenth century.

33. Pract./BMS *Minutes,* book 2, pp. 176, 183.

34. Pathological Society, *Minutes, 1885–1895,* pp. 119 ff.; *1897–1904,* p. 264. Pract./BMS *Minutes,* book 2, pp. 175, 183. Morantz-Sanchez, *Sympathy and Science,* p. 222, notes a trend in the 1890s toward surgical conservatism in leading male gynecologists such as Howard Kelly. Edna L. Manzer, "Woman's Doctors: the Development of Obstetrics and Gynecology in Boston, 1860–1930," Ph.D. diss. Indiana University, 1979, pp. 254–270, relates increased surgical conservatism to gynecologists' need to differentiate their specialty from general and abdominal surgery and legitimate their claims to greater expertise in the treatment of women's diseases. Also see Judith M. Roy, "Surgical Gynecology," in Rima D. Apple, ed., *Women, Health, and Medicine in America,* (New York: Garland Publishing, 1990), pp. 173–195.

35. Gerald N. Grob, *Mental Illness and American Society, 1875–1940* (Princeton, N.J.: Princeton University Press, 1982), p. 65, specifically names Pennsylvania, Massachusetts, and New York, and notes that by 1900 as many as two hundred women physicians had been so employed. Also see Constance McGovern, "'Doctors or Ladies?' Women Physicians in Psychiatric Institutions, 1872–1900," in Judith Walzer Leavitt, ed., *Women and Health in America* (Madison: University of Wisconsin Press, 1984), pp. 438–452.

36. Roy, "Surgical Gynecology," pp. 180–182; Edward C. Atwater, "'Making Fewer Mistakes': A History of Students and Patients," *Bulletin of the History of Medicine* 57 (1983): 175; Regina Markell Morantz, "Feminism, Professionalism, and Germs: The Thought of Mary Putnam Jacobi and Elizabeth Blackwell," *American Quarterly* 34 (1982): 474; Charles E. Rosenberg, "American Medicine in 1879," in Abram, ed., *"Send Us a Lady Physician,"* pp. 24, 25.

37. The same conclusion can be drawn for the homeopathic women physicians in Rochester at that time. Marcena Sherman-Ricker, for example, was Susan B. Anthony's physician in Anthony's later years. In 1906 she was listed as a gynecologist on the staff of the elite Rochester Homeopathic Hospital. At her death in 1933 she was credited with founding the Baptist Home in Fairport, New York, as well as the Door of Hope Home in Rochester. In 1931 she was also an international delegate for the Women's Christian Temperance Union. "In Memoriam," *Medical Woman's Journal* 40 (1933): 55.

38. Moldow correlated a decline in age at graduation with these trends. However, as the 1880s' graduates in Rochester were both young in age at graduation and among the physicians most committed to feminine social reform, age at graduation does not appear to be the salient factor. Gloria Melnick Moldow, "The Gilded Age,

Promise and Disillusionment: Women Doctors and the Emergence of the Professional Middle Class, Washington, D.C.," Ph.D. diss., University of Maryland, 1980.

39. Mary A. E. Wager, "Women as Physicians," *Galaxy* 6 (1868): 774–789. After completing medical school, Dr. Sarah Perry enrolled in the Nurse Training School of Rochester City Hospital for the sake of the clinical experience. Very little has been written about the culture of the nineteenth-century normal school. What has been written seems to suggest that the students, if not the faculty, were mostly female by 1890. On this subject I have profited from conversations with historians of American education Lynn Gordon and Harold Wechsler. Also see Myra H. Strober and Audri Gordon Lanford, "The Feminization of Public School Teaching: Cross-sectional Analysis, 1850–1880," *Signs* 11 (1986): 212–235.

40. Typescript of curriculum vitae for Marion Craig Potter, from the private collection of Dr. Marion Craig Potter, her granddaughter; I am most grateful to Dr. Potter. Provident Dispensary Association, *Minutes*, pp. 75–81; Pract./BMS *Minutes*, book 1, p. 131; book 2, p. 144; book 3, pp. 125 ff. Also see Blake McKelvey, *Rochester: The Quest for Quality* (Cambridge, Mass.: Harvard University Press, 1956), pp. 11–12, 194–195, 226.

41. Cott, *The Grounding of Modern Feminism*, pp. 13–50.

42. Letter from Sarah Dolley to Charles Sumner Dolley, February 15, 1909, SAD/UR; see also Marrett, "On the Evolution," p. 438.

43. Michael M. Davis, Jr., and Andrew R. Warner, *Dispensaries: Their Management and Development* (1918; New York: The Arno Press, 1977), pp. 1–41.

44. Charles E. Rosenberg, "Social Class and Medical Care in Nineteenth-Century America: The Rise and Fall of the Dispensary," in Charles E. Rosenberg, *Explaining Epidemics and Other Studies in the History of Medicine* (New York: Cambridge University Press, 1992), pp. 155–177.

45. Davis and Warner, *Dispensaries*, p. 10.

46. Morantz-Sanchez, *Sympathy and Science*, p. 73.

47. See Moldow, *Women Doctors in Gilded-Age Washington*, pp. 75–93, for a good description of the women's dispensaries of Washington, D.C.

48. Morantz-Sanchez, *Sympathy and Science*, p. 173.

49. *Handbook of Charity Organization* (1882), quoted in Davis and Warner, *Dispensaries*, p. 45.

50. Letter from E. V. Stoddard, vice president, New York State Board of Charities, to Dr. Mary Slaight, treasurer, Provident Dispensary Association, February 8, 1896, pp. 112–115, Disp./UR; Davis and Warner, *Dispensaries*, p. 45.

51. Davis and Warner, *Dispensaries*, pp. 13, 73, 78, 83.

52. Minutes of the Provident Dispensary Association, pp. 3–17, 33, Disp./UR.

53. Ibid., pp. 15–25, 45.

54. Ibid., pp. 55, 63.

55. Annual Report for 1890, p. 65, Disp./UR.

56. Minutes, Provident Dispensary Medical Staff, pp. 70–78, Disp./UR.

57. Ibid., pp. 105–127.

58. Pract./BMS *Minutes*, book 2, p. 193. Pathological Society, *Minutes, 1897–1904* (June 23, 1898), pp. 57–59; Dr. Soble, at the time of this meeting, was on the staff of the City Hospital Outpatient Department. In March of the following year, however, he submitted a courteous letter of resignation to the executive committee

of the medical staff. *Hospital Review* 54 (1898): 76; "Correspondence, 1893–1906," Rochester General Hospital, Medical Staff Collection, Box 2.

59. *Hospital Review* 57 (1902): 57; Rochester City Hospital, Medical Staff, *Minutes, 1873–1924*, p. 381; Pract./BMS *Minutes*, book 2, p. 201. For a full account of this, see Chapter 4.

60. Pract./BMS *Minutes*, book 2, pp. 179, 181, 186, 199–206, 210, 217–219; book 3, pp. 85–86, 99. They also encouraged "out-of-town" members to "draw to [them]selves desirable talent and freshen [their] work," as one of the younger members remarked in 1903 (ibid., p. 257). See Morantz, "Feminism, Professionalism, and Germs."

61. Eveline Ballintine, "Presidential Address" (January 11, 1906), Box 1, folder 9, MCP/UR. The Door of Hope and other such establishments are discussed in Chapter 3.

62. Walsh, *Doctors Wanted*, pp. 188–93; Drachman, *Hospital with a Heart*, pp. 151–157; and Morantz-Sanchez, *Sympathy and Science*, pp. 244–249, 255–261.

63. A handmade Valentine's Day card from Dr. Louise Hurrell to Dr. Marion Craig Potter read, "From Venus, Jupiter and Mars, From suns, and moons and distant stars, / We beg, we plead, we do beseech, / That Doctor Potter we may reach, / Who'll make a Constitution for us, / And all the Blackwell's will implore us, / To organize—to organize." MCP/UR.

64. As quoted in Pract./BMS *Minutes*, book 3, pp. 40, 113; also see book 2, p. 243. According to Marion Craig Potter, Rochester's women physicians would have renamed their society in 1907 after Dr. Dolley rather than Blackwell, but they "could not stand the approbrium [*sic*] of being called the 'dollies'." Marion Craig Potter, typescript autobiographical sketch, December 18, 1939, p. 2, MCP/UR.

65. Pract./BMS *Minutes*, book 3, pp. 264–289.

66. Evelyn Ballintine, Kathleen Buck, Sarah Dolley, and Marion Craig Potter all served as president of the Women's Medical Society of New York State; Evelyn Baldwin served as vice president.

67. Pract./BMS *Minutes*, book 2, p. 144, and book 3, p. 285; Minutes, pp. 64, 74–75, Disp./UR.

68. Sarah Dolley, "President's Address," *Woman's Medical Journal* 18 (1908): 65.

69. Morantz-Sanchez, *Sympathy and Science*, pp. 276, 277.

70. Box 1, folder 3, MCP/UR.

71. As early as 1921, Dr. Kate C. Mead, chair of the MWNA's Committee on Organization, reported that many women physicians responded negatively to her membership solicitations with the comment that they "did not believe in segregation from the men." *Medical Woman's Journal* 28 (1921): 152.

72. The comment was made to Dr. Louise Tayler-Jones, chair of the Committee on Organization and Membership of the MWNA, and repeated at its annual meeting in June 1930. *Medical Woman's Journal* 37 (1930): 6, 192.

73. Cf. the experience of women physicians in the Netherlands: Hilary Marland, "'Pioneer Work on All Sides': The First Generations of Women Physicians in the Netherlands," *Journal of the History of Medicine and Allied Sciences* 50 (1995): 441–477.

74. Interview with Dr. Mary Saxe, Rochester, New York, April 20, 1985.

3. Maternalist Medicine

1. See Chapter 4 for a full discussion of this problem.

2. Seth Koven and Sonya Michel, "Womanly Duties: Maternalist Politics and the Origins of Welfare States in France, Germany, Great Britain, and the United States, 1880–1920," *American Historical Review* 95 (October 1990): 1076–1108, quotation on p. 1077.

3. Seth Koven and Sonya Michel, "Introduction: 'Mother Worlds,'" in Seth Koven and Sonya Michel, eds., *Mothers of a New World: Maternalist Politics and the Origins of Welfare States* (New York: Routledge, 1993), pp. 10–11; Paula Baker, "The Domestication of Politics: Women and American Political Society, 1780–1920," in Linda Gordon, ed., *Women, the State, and Welfare* (Madison: University of Wisconsin Press, 1990), pp. 55–91 (originally published in *American Historical Review* 89 (1984): 620–647).

4. Barbara J. Berg, *The Remembered Gate: Origins of American Feminism* (New York: Oxford University Press, 1978), pp. 147–153, 177–183; also see Mary Ryan, *Cradle of the Middle Class: The Family in Oneida County, New York, 1790–1865* (New York: Cambridge University Press, 1981).

5. Theda Skocpol, *Protecting Soldiers and Mothers: The Political Origins of Social Policy in the United States* (Cambridge, Mass.: The Belknap Press of Harvard University Press, 1992), pp. 3, 34–36; Mary Ryan, in *Womanhood in America* (New York: Franklin Watts, 1975), esp. pp. 135–144, termed this phenomenon "social housekeeping." Also see Koven and Michel, "Womanly Duties."

6. On public health nurses, see Susan Reverby, *Ordered to Care: The Dilemma of American Nursing, 1850–1945* (New York: Cambridge University Press, 1987), and Barbara Melosh, *"The Physician's Hand": Work Culture and Conflict in American Nursing* (Philadelphia: Temple University Press, 1982).

7. Nancy S. Dye, "Introduction," in Noralee Frankel and Nancy S. Dye, eds., *Gender, Race, and Reform in the Progressive Era* (Lexington, Ky.: University of Lexington Press, 1991), pp. 1–4; Koven and Michel, "Introduction: 'Mother Worlds.'"

8. AMWA, Minutes of Council Meeting, April 27, 1920, Box 1, folder 3. AMWA/NYH.

9. Koven and Michel, "Introduction: 'Mother Worlds,'" pp. 6, 10–20; Molly Ladd-Taylor, *Mother-Work: Women, Child Welfare, and the State, 1890–1930* (Urbana: University of Illinois Press, 1994), p. 168 ff.; Dye, "Introduction," pp. 1–4; John Duffy, *The Sanitarians: A History of American Public Health* (Urbana: University of Illinois Press, 1990), p. 210.

10. Richard A. Meckel, *Save the Babies: American Public Health Reform and the Prevention of Infant Mortality, 1850–1929* (Baltimore: Johns Hopkins University Press, 1990), pp. 108–110; Steven J. Peitzman, "Forgotten Reformers: The American Academy of Medicine," *Bulletin of the History of Medicine* 58 (1984): pp. 516–528, esp. 517–524.

11. Peitzman, "Forgotten Reformers."

12. Joseph B. Chepaitis, "Federal Social Welfare Progressivism in the 1920s," *Social Services Review* 46 (1972): 213–229.

13. Dr. Josephine Baker (1873–1945), see S. Josephine Baker, *Fighting for Life* (New York: MacMillan, 1939), pp. 47, 54, 55, 83, 84; "Baker, Sara Josephine," in

Barbara Sicherman and Carol Hurd Green, eds., *Notable American Women: The Modern Period* (Cambridge, Mass.: Harvard University Press, 1980), pp. 85, 86.

14. By 1894 Strauss had begun pasteurizing the milk. Duffy, *The Sanitarians*, pp. 185, 213–215.

15. Baker, *Fighting for Life*, p. 145. Annie Goodrich had been codirector of the Henry Street Visiting Nurses Service and the dean of the Army School of Nursing. In 1923 she became the founding dean of the Yale University School of Nursing. Dorothy M. Brown, *Setting a Course: American Women in the 1920s* (Boston: Twayne, 1987), p. 155.

16. Baker, *Fighting for Life*, 236, 237.

17. Ibid., pp. 89, 90, 95, 107.

18. Rima D. Apple, *Mothers and Medicine: A Social History of Infant Feeding, 1890–1950* (Madison: University of Wisconsin Press, 1987), pp. 102–105.

19. Baker, *Fighting for Life*, p. 127.

20. Marion J. Morton, *And Sin No More: Social Policy and Unwed Mothers in Cleveland, 1855–1990* (Columbus: Ohio State University Press, 1993), pp. 106–109.

21. Molly Ladd-Taylor, "Hull House Goes to Washington: Women and the Children's Bureau," in Frankel and Dye, eds., *Gender, Race and Reform*, pp. 110–126; Molly Ladd-Taylor, "'My Work Came Out of Agony and Grief': Mothers and the Making of the Sheppard-Towner Act," in Koven and Michel, eds., *Mothers of a New World*, pp. 321–342; Lela B. Costin, *Two Sisters for Social Justice: A Biography of Grace and Edith Abbott* (Urbana: University of Illinois Press, 1983), pp. 130–132, 146–147, 170–171. Costin cites Grace L. Meigs, *Maternal Mortality from All Conditions Connected with Childbirth in the United States and Certain Other Countries* (Washington, D.C.: Children's Bureau, 1917), publication no. 19.

22. Skocpol, *Protecting Soldiers and Mothers*, pp. 372–422, 424–427.

23. Barbara J. Nelson, "The Origins of the Two-Channel Welfare State," in Frankel and Dye, eds., *Gender, Race and Reform*, pp. 138–139; Eileen Boris, "The Power of Motherhood: Black and White Activist Women Redefine the 'Political,'" in Koven and Michel, eds., *Mothers of a New World*, pp. 217–219. Also see Jane Lewis, "Women's Agency, Maternalism and Welfare," *Gender and History* 6, no. 1 (1994): 117–123. For a careful assessment of maternalist reformers' work, see Linda Gordon, *Pitied But Not Entitled: Single Mothers and the History of Welfare* (New York: Free Press, 1994).

24. Dye, "Introduction," p. 4; Lewis, "Women's Agency," pp. 117–123. Also see Ladd-Taylor, *Mother-Work*, "Conclusion."

25. Atina Grossmann, "German Women Doctors from Berlin to New York: Maternity and Modernity in Weimar and in Exile," *Feminist Studies* 19 (1993): 65–88, quotation on p. 65.

26. Regina Morantz-Sanchez, *Sympathy and Science: Women Physicians and American Medicine* (New York: Oxford University Press, 1985), chaps. 1 and 2; Gulielma Fell Alsop, *History of the Woman's Medical College, Philadelphia, Pennsylvania, 1850–1950* (Philadelphia: J. B. Lippincott Company, 1950), p. 76.

27. Bonnie Ellen Blustein, *Educating for Health and Prevention* (New York: Social Science History Publications, 1993), pp. 20–22.

28. American Medical Association Public Health Education Committee, report of meeting held July 20, 1909, "Resolutions and Outline for Work" (New York: AMA, n.d.), in Box 1, MCP/MCP.

29. Kate C. Mead, "The Duty of Medical Women for Public Health Education," *Woman's Medical Journal* 24 (1914): 89–92.

30. Apple, *Mothers and Medicine*, pp. 100–105, 130–131, 230 n. 8.

31. Mead, "The Duty of Medical Women," pp. 89–92.

32. Quoted in Barbara Sicherman, "Gender, Profession and Reform in the Career of Alice Hamilton," in Frankel and Dye, eds., *Gender, Race and Reform*, pp. 127–147, quotation on pp. 136–137.

33. Blustein, *Educating for Health*, pp. 9–22.

34. Penina Migdal Glazer and Miriam Slater, *Unequal Colleagues: The Entrance of Women into the Professions, 1890–1940* (New Brunswick, N.J.: Rutgers University Press, 1987), pp. 94–98.

35. Dorothy Porter, "Introduction," in Dorothy Porter, ed., *The History of Public Health and the Modern State* (Amsterdam and Atlanta: Editions Rodopi B.V., 1994), p. 15. Also see Elizabeth Fee, "Public Health and the State: The United States," in ibid., esp. pp. 234–256, and Barbara Rosenkrantz, "Cart before Horse: Theory, Practice and Professional Image in American Public Health," *Journal of the History of Medicine and Allied Sciences* 29 (1974): 55–73.

36. Unfortunately they sometimes lost most of what they gained when municipal governments changed hands. See Judith Walzer Leavitt, *The Healthiest City: Milwaukee and the Politics of Health Reform* (Princeton, N.J.: Princeton University Press, 1982).

37. Fee, "Public Health and the State," pp. 236–239; Duffy, *The Sanitarians*, pp. 194–196; George Rosen, *A History of Public Health* (New York: M.D. Publications, 1958; 1976), pp. 354–370; Morantz-Sanchez, *Sympathy and Science*, p. 160.

38. Porter, "Introduction," p. 15; Morantz-Sanchez, *Sympathy and Science*, chap. 10.

39. Two years later Anthony wrote Chief Physician Goler regarding her next-door neighbor's barn, complaining that it was "a very great nuisance to [her] olfactories." She added: "It is offensive enough in the winter, when frozen up, but now with the warm south and west winds it is simply intolerable." Susan B. Anthony to George W. Goler, March [22], 1897. In George W. Goler Collection, Local History Division, Rochester Public Library, Box 4, folder N. Also see Blake McKelvey, *Rochester: The Quest for Quality, 1890–1925* (Cambridge, Mass.: Harvard University Press, 1956), pp. 30–32; *Rochester City Directory*, 1890.

40. The six Blackwell Medical Society members who served as city physicians were Drs. Minerva Palmer, Harriet Turner, Cornelia White-Thomas, Lucy Baker Foster, Mary Saxe, and M. May Allen. The seventh, Dr. Grace Carter, was a homeopath. In 1900, part-time city physicians earned $480; Goler earned $2,600 as full-time director of the Health Bureau. See *Rochester City Directory*, 1900, 1922, 1932, 1959; Florence Cooksley, "A History of Medicine in the State of New York and the County of Monroe, Part II," *New York State Journal of Medicine* 36 (1936) and 37 (1937); Obituary, Dr. Harriet Turner, *Democrat and Chronicle*, February 20, 1934; Obituary, Dr. M. May Allen, *JAMA* 124 (1944): 664.

41. The philanthropist Nathan Strauss is credited with opening the first privately sponsored free milk stations in New York City, in 1893. Blake McKelvey, "The History of Public Health in Rochester, New York," *Rochester History* 18 (1956): 18; Duffy, *The Sanitarians*, pp. 185, 213–215.

42. *Rochester Women's Educational and Industrial Union Yearbook*, "Secretary's Re-

port," 1896–1897, in WEIU Collection, DRB/RRL; Rochester Board of Health, *Annual Report, 1900,* LHD/RPL; David B. Brady and Alfred D. Kaiser, *Fifty Years of Health in Rochester, New York, 1900–1950* (Rochester: Health Bureau of Rochester, 1950), pp. 5, 9–11; Blake McKelvey, "Historic Origins of Rochester's Social Welfare Agencies," *Rochester History* 9 (1947): 31. I am grateful to Corinne Sutter-Brown for skilled assistance in research on materials relating to George W. Goler.

43. Brown became the first head of the obstetrics inpatient service at Rochester General. See William M. Brown, "A Modification of the Technic of Abdominal Caesarian Section," *American Journal of Obstetrics* 72 (1915): 415–420; George W. Goler, "Medical School Inspection, Nursing Sick Poor," typescript, pp. 29, 61–65, Box Ma–Me, folder "Medical School Inspectors and Nurses, etc.," and George Goler, "Child Welfare," typescript, Box A–C, folder "Child Welfare," LHD/RPL; McKelvey, "The History of Public Health in Rochester," pp. 11, 61.

44. Waller is quoted in Morton, *And Sin No More,* pp. 57–64.

45. Ibid., pp. 22–32, 40–48, 60; Regina G. Kunzel, *Fallen Women, Problem Girls: Unmarried Mothers and the Professionalization of Social Work, 1890–1945* (New Haven, Conn.: Yale University Press, 1993), pp. 14–16; Joan Jacobs Brumberg, "'Ruined' Girls: Changing Community Responses to Illegitimacy in Upstate New York, 1890–1920," *Journal of Social History* 18 (1984): 247–272; Charles E. Rosenberg, *The Care of Strangers: The Rise of America's Hospital System* (New York: Basic Books, 1987), pp. 269–271.

46. Paula Giddings, *When and Where I Enter: The Impact of Black Women on Race and Sex in America* (New York: Bantam Books, 1988), chaps. 4, 6; Elisabeth Lasch-Quinn, *Black Neighbors: Race and the Limits of Reform in the American Settlement House Movement, 1890–1945* (Chapel Hill: University of North Carolina Press, 1993).

47. Morton, *And Sin No More,* pp. 88–98.

48. Giddings, *When and Where I Enter,* pp. 97–98.

49. On the growing tension between the ideals of maternalist evangelism and professional social work in maternity homes, see Regina G. Kunzel, "The Professionalization of Benevolence," *Journal of Social History* 22 (1988): 21–43.

50. Minutes, November 1, November 8, December 4, 1894, Door of Hope/RRL.

51. Minutes, March 7, 1901, July–December 1904, November 23, 1905, November 1, 1911, "A Study of Northaven, Inc.," typescript (May 1968), p. 1, all in Door of Hope/RRL.

52. Annual Report, March 1, 1897–March 1, 1898; Minutes, June 30, 1898, Door of Hope/RRL.

53. Cf. Kunzel, "The Professionalization of Benevolence."

54. The first medical staff consisted of six women, of whom three were homeopaths and three were regular physicians. Minutes, January 17, 1895, Door of Hope/RRL.

55. Ibid., January 24, 1895; Judith Walzer Leavitt, *Brought to Bed: Childbearing in America, 1750–1950* (New York: Oxford University Press, 1986).

56. Minutes, October 25, 1900, Door of Hope/RRL. See Morton, *And Sin No More,* pp. 22–32. Historian Marion Morton found that in another northern indus-

trial city, Cleveland, by 1904 most births to women residing in public institutions occurred in hospitals.

57. Medical Staff Minutes, May (n.d.) 1910, January 15, 1911, January 18 and April 12, 1912, Door of Hope/RRL.

58. Allen F. Davis, *Spearheads for Reform: The Social Settlements Movement and the Progressive Movement, 1890–1914* (1967; New Brunswick, N.J.: Rutgers University Press, 1984), pp. 3–12, 18–22, 37.

59. This is a conservative estimate. The *Handbook* recorded only those physicians who were on staff at the time the settlement sent in its report. Data from Rochester, moreover, show that these reports were not always complete. Robert A. Woods and Albert J. Kennedy, eds., *Handbook of Settlements* (New York: Charities Publications Committee of the Russell Sage Foundation, 1911), p. vi; Jeffrey Brosco, "Pride and Prevention: Child Health in 20th-Century Philadelphia," *Collections* 24 (January 1992): 1–5.

60. "The Baden Street Settlement, 1901–1909," typescript, n.d., pp. 1–9, DRB/RRL; Blake McKelvey, "Walter Rauschenberg's Rochester," *Rochester History* 14 (1952): 1–27; Lewis Street Center Monthly Report, January 1917, Box 2, folder 2, Lewis Street Center Papers, DRB/RRL.

61. *Medical Directory of New York, New Jersey, and Connecticut*, 1925.

62. Collection Prospectus, The Lewis Street Center Papers, 1907–1971, typescript, and "Lewis Street Center Fiftieth Anniversary, 1907–1957," n.p., Lewis Street Center Papers, DRB/RRL; McKelvey, "Walter Rauschenberg's Rochester," p. 20.

63. The eight physicians included Drs. Elsa Will Leveque, Dr. Lucy Baker Foster, Dr. Ida Porter, Dr. Gertrude McCann, Dr. Dorothy Worthington, Dr. Mary Jane Foley, Dr. Emma Gibbons, and Dr. Julia Russell. With the exception of Porter, all were relatively new physicians when they began their clinic work. See State of New York Board of Charities Application for Dispensary License in behalf of the Housekeeping Center of Rochester, 57 Lewis Street, November 23, 1917, Box 1, folder 8; Minutes of Executive Committee Meeting, June 1917, Box 2, folder 1; "Managers of the Lewis Street Center and Their Dates of Office," Box 1, folder 9, all in Lewis Street Center Papers, DRB/RRL.

64. Monthly Reports for November 1916 and May 1917, Box 3, folder 4, and Monthly Report for March 1924, Box 3, folder 5, Lewis Street Center Papers, DRB/RRL.

65. "History of Lewis Street Center Given at Thirtieth Anniversary Dinner by Mrs. Helen Rochester Rogers, November 16, 1937," Box 1, folder 1; Dispensary Monthly Reports, March and November 1924, November 1927, Box 3, folder 5, and June–December, 1931, Box 3, folder 6; Minutes of Board of Directors of Lewis Street Center and Dispensary, October 1931, Box 2, folder 2; Minutes of Dispensary Board of Managers, October–December 1930 and June, 1931, Box 2, folder 1, all in Lewis Street Center Papers, DRB/RRL.

4. Redefining the Margins

1. Minutes of the Annual Meeting of the Board of Directors, Rochester City Hospital, January 18, 1900; Medical Staff Minutes, May 27, 1900; letter from Dr.

Charles A. Dewey to Dr. Marion C. Potter, June 14, 1900; letter from Dr. F. W. Zimmer to Dr. Charles A. Dewey, March 4, 1902, all in BCA/RGH. My thanks to Mr. Philip Maples, hospital archivist, for his great help in using this excellent hospital collection. See below for a fuller discussion of this incident.

2. See Paul Starr, *The Social Transformation of American Medicine* (New York: Basic Books, 1982); Kenneth M. Ludmerer, *Learning to Heal: The Development of American Medical Education* (New York: Basic Books, 1985); William G. Rothstein, *American Medical Schools and the Practice of Medicine* (New York: Oxford University Press, 1987). Cf. Charles E. Rosenberg, *The Care of Strangers: The Rise of America's Hospital System* (New York: Basic Books, 1987); Rosemary Stevens, *In Sickness and in Wealth: American Hospitals in the Twentieth Century* (New York: Basic Books, 1989).

3. H. G. Weiskotten, "Present Tendencies in Medical Practice," *Bulletin of the Association of American Medical Colleges* 2 (1927): 29–40; Charles R. King, "The New York Maternal Mortality Study: A Conflict of Professionalization," *Bulletin of the History of Medicine* 65 (1991): esp. pp. 495–498.

4. Weiskotten, "Present Tendencies," pp. 29–40; Thomas Neville Bonner, *Medicine in Chicago, 1850–1950* (Madison, Wisc.: The American History Research Center, 1957), p. 102.

5. From its founding in 1888 until 1981, the American Association of Obstetricians and Gynecologists included only a handful of women members. See Edward Stewart Taylor, *History of the American Gynecological Society . . . and American Association of Obstetricians and Gynecologists* (St. Louis: C. V. Mosby, 1985), pp. 28, 112–128; Margaret Marsh and Wanda Ronner, *The Empty Cradle: Infertility in America from Colonial Times to the Present* (Baltimore: Johns Hopkins University Press, 1996), pp. 86–88; Rosemary Stevens, *American Medicine and the Public Interest* (New Haven, Conn.: Yale University Press, 1971), pp. 77–97; Regina Morantz-Sanchez, *Sympathy and Science: Women Physicians and American Medicine* (New York: Oxford University Press, 1985); and Mary Roth Walsh, *Doctors Wanted: No Women Need Apply* (New Haven, Conn.: Yale University Press, 1976), for women's exclusion from hospital staffs through exclusion from better internships. Monica Green and others note that as far back as the late Middle Ages, the rise of medical education as a university-trained profession was associated with the declining status of women practitioners. Monica H. Green, "Documenting Medieval Women's Medical Practice," in Luis Garcia-Ballester et al., eds., *Practical Medicine from Salerno to the Black Death* (Cambridge: Cambridge University Press, 1994), pp. 222–252.

6. Midwives were another target. Charlotte G. Borst, *Catching Babies: The Professionalization of Childbirth, 1870–1920* (Cambridge, Mass.: Harvard University Press, 1995).

7. Stevens, *In Sickness and in Wealth*, p. 54, Weiskotten, "Present Tendencies," pp. 44–45. Cf. Regina Morantz-Sanchez, "Making It in a Man's World: The Late-Nineteenth-Century Surgical Career of Mary Amanda Dixon Jones," *Bulletin of the History of Medicine* 69 (1995): 542–568; Dale C. Smith, "Appendicitis, Appendectomy, and the Surgeon," *Bulletin of the History of Medicine* 70 (1996): 414–441; Mary-Jo DelVecchio Good, *American Medicine: The Quest for Competence* (Berkeley: University of California Press, 1995), p. 94; Ornella Moscucci, *The Science of Woman: Gynecology and Gender in England, 1800–1929* (Cambridge: Cambridge University Press, 1990), pp. 165–206; Edna Louise Manzer, "Woman's Doctors:

The Development of Obstetrics and Gynecology in Boston," Ph.D. diss., Indiana University, 1979, pp. 285–302, 314–317.

8. Vanessa Northington Gamble, *The Black Community Hospital: Contemporary Dilemmas in Historical Perspective* (New York: Garland, 1989) and idem., *Making a Place for Ourselves: The Black Hospital Movement, 1920–1945* (New York: Oxford University Press, 1995); Edward C. Atwater, "Of Grandes Dames, Surgeons, and Hospitals: Batavia, New York, 1900–1940," *Journal of the History of Medicine and Allied Sciences* 45 (1990): 414–451; Starr, *The Social Transformation*, pp. 169–177.

9. Cf. Alice G. Bennett, "Would You Let Your Daughter Study Medicine?" *Woman's Medical Journal* 25 (1915): 128–129. In Rochester, for example, Marion Craig Potter consulted Dr. Sarah Dolley before attending medical school. Rochester physician M. May Allen and Louise Hurrell consulted Dr. Potter early in their training. Also see letters of appreciation to Potter on the occasion of the fiftieth anniversary of her graduation from medical school, "Regrets" folder, Envelope 4, MCP/MCP.

10. Recent studies demonstrate the same effects. A classic study is by Judith Lorber, *Women Physicians* (New York: Tavistock, 1984). For the sciences in general, see Gerhard Sonnert and Gerald Holton, "Career Patterns of Women and Men in the Sciences," *American Scientist* 84 (1996): 63. Morantz-Sanchez, *Sympathy and Science*, notes that women medical students encountered resistance on the wards of gender-integrated hospitals (p. 174).

11. Walsh, *Doctors Wanted*, pp. 185–186, argues for a dramatic flagging of the rate of increase of women physicians after 1910. Although I argue that the true number of women physicians was lower than she reported for the turn of the century, I agree that their continued low percentages in the early twentieth century requires explanation. Cf. Morantz-Sanchez, *Sympathy and Science*, for a multicausal analysis of the reasons for the slow rate of increase; also cf. Regina Markell Morantz, "Women in the Medical Profession: Why Were There So Few?" *Reviews in American History* (June 1978): 163–170. Bertha Van Hoosen, "Report of the National Committee on Medical Opportunities for Women," *Woman's Medical Journal* 37 (1930): 200–202; Ludmerer, *Learning to Heal*, p. 241.

12. See Walsh, *Doctors Wanted*, for an analysis emphasizing overt and deliberate discrimination in medical school admissions, and, Morantz-Sanchez, *Sympathy and Science*, for a multicausal analysis emphasizing both the realities of discrimination in certain schools plus the several factors causing some women to decide against careers in medicine in the first place.

13. By 1918 Woman's Medical College of Pennsylvania was the only women's school still in existence. Bonner, *Medicine in Chicago*, p. 154; Bertha Van Hoosen, "Report of the National Committee," pp. 200–202; Ludmerer, *Learning to Heal*, p. 241; Darlene Clark Hine, *Speak Truth to Power: The Black Professional Class in United States History* (New York: Carlson, 1996), pp. 183–186. According to Todd Savitt, by 1918 the only remaining African American medical schools, out of fourteen previously in existence, were Meharry Medical College of Nashville, Tennessee, and Howard University School of Medicine in Washington, D.C. Todd L. Savitt, "'A Journal of Our Own': *The Medical and Surgical Observer* at the Beginnings of an African-American Medical Profession in Late 19th-Century America," Part 1, *Journal of the National Medical Association* 88 (1996): 52–60, esp. the table on p. 54. My thanks to Dr. Savitt for making this article available to me.

14. The percentage of women students jumped temporarily to 5.8 percent as a result of openings made possible by World War I, but it declined to 4.4 percent by 1930. Carol Lopate, *Women in Medicine* (Baltimore: Johns Hopkins University Press, 1968), Appendix 1, p. 193; Morantz-Sanchez, *Sympathy and Science*, pp. 249, 314; Ludmerer, *Learning to Heal*, pp. 247–248; Walsh, *Doctors Wanted*, pp. 180, 186; clipping dated March 22, 1918, Box 1: Clippings 1906–1920, MCP/UR. Patricia Hummer, *Decade of Elusive Promise* (Ann Arbor, Mich.: University Microfilm, 1979), Table 2, pp. 143, 144, gives the following percentages for females among all enrolled medical students: 1920, 5.9 percent; 1921: 6.0; 1922, 6.3. In 1923 the number began to drop: 1923, 6.0 percent, 1924, 5.4; 1925, 5.0; 1926, 4.9; 1927, 4.9; 1928, 4.5; 1929, 4.4; 1930, 4.4.

15. Between 1904 and 1915 the percentage of women medical graduates declined from 3.4 percent to 2.6 percent, but it rose to 4.5 percent by 1920 as a result of openings created by World War I; it never again fell below 4.1 percent. In addition to sources cited above (n. 14), see "Medical Education," *JAMA* 105 (1935): 685. In earlier years, the AMA reported slightly higher figures, e.g. 4.0 percent for 1904 and 3.7 percent for 1915; "Medical Education Statistics," ibid. 65 (1915): 691.

16. Contrary to prior estimates, the percentage of practicing women physicians actually increased until 1920—albeit in very small increments. The U.S. Census for 1900 recorded 7,399 women physicians, or 5.6 percent of all physicians. But this figure likely included some who were not medical graduates or who were no longer in practice. As early as 1890 an overview of women physicians by Dr. Mary Putnam Jacobi expressed strong doubts about the numbers of women listed as "physicians" in the census. Thomas Bonner estimates the number of women in practice at the turn of the century at approximately 3,400, or 2.5 percent. Sources for women in practice from 1900 to 1950 include Thomas Neville Bonner, *To the Ends of the Earth: Women's Search for Education in Medicine* (Cambridge, Mass.: Harvard University Press, 1992), p. 203 n. 92; and Walsh, *Doctors Wanted*, p. 186. In 1916 Dr. Mary Sutton Macy conducted a survey for the *Woman's Medical Journal* yielding the estimate of 5,124 women in practice in 1916: about 3.5 percent of doctors. Dr. Macy noted that 34 women physicians were African Americans, which is probably an underestimate. She also calculated that only 7.36 percent of the active women physicians practiced sectarian medicine. Mary Sutton Macy, "The Field for Women of Today in Medicine," *Woman's Medical Journal* 27 (1917): 52–56. Also see *Woman's Medical Journal* 26 (1916): 97. In 1917–18 Dr. Marion Craig Potter conducted a census of women physicians for the General Medical Board and the Medical Women's National Association based on respondents' replies and the 1915 *American Medical Association Directory*. Potter's census identified a total of 5,322 women in active practice, or 3.6 percent (182 were unclassified because of no known address). This represents a modest increase in the percentage of women practitioners between 1900 and 1915, rather than a decline. Marion Craig Potter, *Census of Women Physicians*, in MCP/UR. See Table 4.2.

17. "Annual Report of the Committee on Opportunities for Medical Women," June 1940, typescript, Box 1, folder 17, AMWA/NYH. Also see Rothstein, *American Medical Schools*, pp. 131–138.

18. One study of medical school admissions found that "an astonishingly high percent of leakage which occurs between women accepted and those actually en-

rolled in these first-year classes [of 1927–1930] bears testimony that women are unwilling to assume the role of being the 'only woman' in the class." Van Hoosen, quoting from the doctoral thesis of Mrs. Lester Bartlett, in "Report of the National Committee," pp. 200–202. Cf. Morantz-Sanchez, *Sympathy and Science*, p. 174. Recent studies demonstrate the same effects. Cf. Lorber, *Women Physicians;* Sonnert and Holton, "Career Patterns of Women and Men in the Sciences," pp. 63.

19. Rosenberg, *The Care of Strangers*, pp. 262–285.

20. Dorothy Gies McGuigan, *A Dangerous Experiment: 100 Years of Women at the University of Michigan* (Ann Arbor, Mich.: Center for Continuing Education of Women, 1970), pp. 75–80; Horace W. Davenport, *Fifty Years of Medicine at the University of Michigan, 1891–1941* (Ann Arbor: University of Michigan Press, 1986), pp. 20–21.

21. Unsigned biographical sketch of Dr. Marion Craig Potter, personal collection of her granddaughter, Dr. Marion Craig Potter also of Rochester, New York. I am most grateful to Dr. Potter for allowing me access to these papers.

22. McGuigan, *A Dangerous Experiment*, pp. 75–80; Carol Marinch, M.D., lecture delivered at the annual meeting of the American Association for the History of Medicine, May 2, 1986; Joel D. Howell, "Introduction," in Joel D. Howell, ed., *Medical Lives and Scientific Medicine at Michigan, 1891–1969* (Ann Arbor: University of Michigan Press, 1993), p. 5 n.9; Kenneth M. Ludmerer, "The University of Michigan Medical School: A Tradition of Leadership," in Joel D. Howell, ibid., pp. 16–18. Also see William James Mayo, "Recollections of the University of Michigan Medical School, 1880–1883," in *Collected Papers of the Mayo Clinic*, vol. 29 (Philadelphia: W.B. Saunders, 1937), pp. 935–937; Davenport, *Fifty Years of Medicine at the University of Michigan*, pp. 20–21; Martin Kaufman, *Homeopathy in America: The Rise and Fall of a Medical Heresy* (Baltimore: Johns Hopkins University Press, 1971), pp. 93–109.

23. Morantz-Sanchez, *Sympathy and Science*, pp. 72–73, 165–166; Harry F. Dowling, *City Hospitals: the Undercare of the Underprivileged* (Cambridge, Mass.: Harvard University Press, 1982), pp. 61, 62.

24. Marion Craig Potter, autobiographical sketch, typescript, December 18, 1939, in MCP/UR.

25. Record Book of Dr. Marion Craig, Churchville, New York, August 3, 1884–November 1885, in Accession no. 138, Envelope 1, MCP/MCP.

26. Record Book, August 15–19, 1884, Accession no. 138, Envelope 1, MCP/MCP.

27. Record Book, December 25, 26, 28, 1884; February 3, 1885, Accession no. 138, Envelope 1, MCP/MCP.

28. Side A, tape 1, of interview by Dr. Edward C. Atwater of Dr. James Craig Potter (son of Dr. Marion Craig Potter), January 24, 1975. My thanks to Dr. Atwater for the loan of this tape.

29. Ellen More, "The Blackwell Medical Society and the Professionalization of Women Physicians," *Bulletin of the History of Medicine* 61 (1987): 603–628.

30. Marion Craig Potter, notes for autobiographical sketch, MCP/UR.

31. Interview of Dr. Marion Craig Potter (granddaughter), Rochester, New York, January 13, 1984; taped interview with Georgia Potter Gosnell, Rochester, New York, July 1987. My sincere thanks to Mrs. Ruth Atwater for arranging the interview with Mrs. Gosnell and to Dr. Marion Craig Potter and Mrs. Thomas

Gosnell, Dr. Potter's granddaughters, for graciously sharing documents and recollections related to their grandmother.

32. Taped interview of Dr. James Craig Potter.

33. The cages of white rats were intended for one of her son's research projects. Interview of Mrs. Thomas (Georgia) Gosnell; interview of Dr. James Craig Potter.

34. Personal communication from Potter's granddaughter Marion Craig Potter, April 13, 1998; interview of Mrs. Thomas (Georgia) Gosnell.

35. Gerald E. Markowitz and David Rosner, "Doctors in Crisis: Medical Education and Medical Reform during the Progressive Era, 1895–1915," in Susan Reverby and David Rosner, eds., *Health Care in America*, (Philadelphia: Temple University Press, 1979), esp. pp. 188–190; Morantz-Sanchez, *Sympathy and Science*, pp. 109, 250; Rosalie Slaughter Morton, *A Woman Surgeon* (New York: Frederick A. Stokes Company, 1937); Ruth J. Abram, ed., *"Send Us a Lady Physician": Women Doctors in America* (New York: W. W. Norton, 1985), pp. 155–162; Eve Fine, "Separate But Integrated: Bertha Van Hoosen and the Founding of AMWA," *JAMWA* 45, no. 5 (Sept./Oct. 1990): 181–190.

36. More, "The Blackwell Medical Society," pp. 611–612; Edward C. Atwater, "A Look at the 'Dozen and One Club,'" *Bulletin of the Monroe County Medical Society*, 1 (1982): 20–23.

37. Cf. J. J. Taylor, *The Physician as a Businessman* (Philadelphia: Medical World, 1892), cited in Robert L. Martensen, "Sundown Medical Education: Top-down Reform and Its Social Displacements," *JAMA*, 273, no. 4 (1995): 271. I am grateful to Dr. Martensen for this reference.

38. Margaret Elliott and Grace E. Manson, "Earnings of Women in Business and the Professions," *Michigan Business Studies* 3, no. 1 (1930): esp. 130. A small sample of women in practice between 1912 and 1921 taken by Martha Tracy found that 41.6 percent earned from $2,000 to $5,000; about 13 percent earned more. Martha Tracy, "Women Graduates in Medicine," *Bulletin of the Association of American Medical Colleges*, 2, no. 2 (1922): 21–28. Marion Craig Potter, M.D., Office Ledgers, January 1, 1890 to December 31, 1939, in Box RG-13, "Medical Staff," BCA/RGH. Last Will and Testament of Marion Craig Potter, filed for probate in Monroe County Surrogate's Court, April 7, 1943. My thanks to Ms. Corinne Sutter-Brown for the discovery of this document.

39. Fine, "Separate But Integrated;" Bertha Van Hoosen, *Petticoat Surgeon* (Chicago: Pellegrini and Cudahy, 1947).

40. These changes have been carefully studied recently: see especially Morris J. Vogel, *The Invention of the Modern Hospital: Boston, 1870–1930* (Chicago: University of Chicago Press, 1980); David Rosner, *A Once Charitable Enterprise: Hospitals and Health Care in Brooklyn and New York, 1885–1915* (Cambridge: Cambridge University Press, 1982); Dowling, *City Hospitals;* Starr, *The Social Transformation*, esp. pp. 145–179; Ludmerer, *Learning to Heal;* Rosenberg, *The Care of Strangers;* Stevens, *In Sickness and in Wealth;* Joel D. Howell, *Technology in the Hospital* (Baltimore: Johns Hopkins University Press, 1995); and the authors featured in Diana Elizabeth Long and Janet Golden, eds., *The American General Hospital: Communities and Social Contexts* (Ithaca, N.Y.: Cornell University Press, 1989).

41. Cf. Chester R. Burns, "The Development of Galveston's Hospitals during the Nineteenth Century," *Southwestern Historical Quarterly* 97 (1993): 238–263, esp. pp. 244, 246.

42. David Rosner, "Doing Well or Doing Good," in Long and Golden, eds., *The American General Hospital*, p. 158; Stevens, *In Sickness and in Wealth*, pp. 20–24.

43. Stevens, *In Sickness and in Wealth*, p. 68; also see Rosenberg, *The Care of Strangers*, esp. pp. 262–285.

44. Morris J. Vogel, "The Transformation of the American Hospital," in Reverby and Rosner, eds., *Health Care in America*, pp. 105–116; Rosenberg, *The Care of Strangers*," p. 9.

45. Howell notes that for the hospitals in his sample, the number of surgical procedures rose steeply between 1900 and 1925. Howell, *Technology in the Hospital*, pp. 61–68, 276–277 n. 85. Stevens, *In Sickness and in Wealth*, pp. 58–67. The AMA calculated that 900 to 1,000 internships remained unfilled. Homer F. Sanger, "Inquiry into the Demand and Supply of Interns," *JAMA* 82, no. 12 (1924): 992–994; "Hospital Service," *JAMA* 84, no. 13 (1925): 969; Ludmerer, *Learning to Heal* pp. 219–233.

46. Thomas N. Bonner, *American Doctors and German Universities* (Lincoln: University of Nebraska Press, 1963); Ludmerer, *Learning to Heal*, pp. 32, 33; Steven J. Peitzman, "'Thoroughly Practical': America's Polyclinic Medical Schools," *Bulletin of the History of Medicine* 54, no. 2 (1980): 166–187; Morantz-Sanchez, "Making It in a Man's World," pp. 552–553.

47. Manzer, "Woman's Doctors: The Development of Obstetrics and Gynecology in Boston," pp. 285–302; Lawrence D. Longo, "Obstetrics and Gynecology," in Ronald L. Numbers, ed., *The Education of American Physicians* (Berkeley: University of California Press, 1980), pp. 205–225.

48. Stevens, *In Sickness and in Wealth*, pp. 58–61, 65–67.

49. Rothstein, *American Medical Schools*, pp. 132–133.

50. See George Rosen, *Fees and Fee Bills, Bulletin of the History of Medicine Supplement No.6* (Baltimore: Johns Hopkins University Press, 1946), pp. 74, 89–91.

51. Stevens, *In Sickness and in Wealth*, p. 21; Walsh, *Doctors Wanted*, pp. 133–136; interview of Mrs. Thomas (Georgia) Gosnell.

52. Excluding interns and residents, the number was about 87,000. "Hospital Service in the United States," *JAMA* 94, no. 13 (1930): 992–994. Also see Edward C. Atwater, "'Making Fewer Mistakes': A History of Students and Patients," *Bulletin of the History of Medicine* 57, no. 1 (1983): 165–187.

53. Eve Fine, "Separate but Integrated." New York surgeon Mary Dixon Jones, like Van Hoosen in Chicago, solved the hospital problem by founding one of her own. Cf. Regina Morantz-Sanchez, "The Gendering of Empathic Expertise: How Women Physicians Became More Empathic Than Men," in Ellen Singer More and Maureen A. Milligan, eds., *The Empathic Practitioner: Empathy, Gender, and Medicine* (New Brunswick, N.J.: Rutgers University Press, 1994), pp. 40–58.

54. Richard C. Cabot, "Women in Medicine," *JAMA* 65, no. 11 (1915): 947–948; S. Adolphus Knopf, "The Woman Physician and Professor Cabot," *Woman's Medical Journal* 25, no. 7 (1915): 159–160. Also see Miriam Glazer and Penina Migdal Slater, *Unequal Colleagues* (New Brunswick, N.J.: Rutgers University Press, 1987). Margaret Rossiter, *Women Scientists in America: Struggles and Strategies to 1940* (Baltimore: Johns Hopkins University Press, 1985), describes the analogous struggles of early women scientists.

55. Stevens, *In Sickness and in Wealth*, pp. 65–67. Starr, *The Social Transformation*, p. 167, gives 1919 as the year that the AMA published its set of minimum standards

for internships, the same year the American College of Surgeons established minimum requirements for hospital standardization and surgeon certification.

56. See the series of articles published by Isabelle T. Smart, "Report on Internships for Women," *Woman's Medical Journal* 27, no. 2 (1917): 37, no. 3: 58–59, no. 4: 88–91, no. 6: 142–145, no. 7: 162–163. The estimated numbers of hospitals are derived from Joan E. Lynaugh, "From Respectable Domesticity to Medical Efficiency: The Changing Kansas City Hospital, 1875–1920," in Long and Golden, eds., *The American General Hospital*, p. 24.

57. Carol Lopate, *Women in Medicine* (Baltimore: Johns Hopkins University Press for the Josiah Macy, Jr., Foundation, 1968), Appendix 1, p. 193.

58. This was a decline from 1923, when 135 approved internship sites were open to women candidates. "Hospital Service," *JAMA*, 968.

59. "Medical Education," *JAMA* 105, no. 9 (1935): 685; Emily Dunning Barringer, *Bowery to Bellevue* (New York: W. W. Norton, 1950).

60. Smart, "Report on Internships for Women," *Woman's Medical Journal* 27, no. 6 (1917): 143; see also no. 2: 37, no. 3: 51–59, no. 4: 88–91, no. 6: 142–145, no. 7: 162–163.

61. Even by 1944, only 23 of the 110 African American–run hospitals were fully approved for internships by the American College of Surgeons. Gamble, *Making a Place for Ourselves*, p. 183; cf. Hine, *Speak Truth to Power*, pp. 187–189. Quotation is from Vanessa Northington Gamble, "Taking a History: The Life of Dr. Virginia Alexander," Fielding H. Garrison Lecture, American Association for the History of Medicine, May 8, 1998, typescript, p. 10. Many thanks to Dr. Gamble for providing me with her text.

62. Isabella Vandervall, "Some Problems of the Colored Woman Physician," *Woman's Medical Journal* 27, no. 7 (1917): 156–158; Emma Wheat Gillmore, "A Call to Arms," *Woman's Medical Journal* 27, no. 8 (1917): 183–184; Darlene Clark Hine, "'Co-Laborers in the Work of the Lord': Nineteenth-Century Black Women Physicians," in Abram, ed., *"Send Us a Lady Physician,"* pp. 107–120; Vanessa Gamble, "Physicians, Twentieth Century," in *Black Women in America: An Historical Encyclopedia*, vol. 2, ed. Darlene Clark Hine, Elsa Barkley Brown, and Rosalyn Terborg-Penn (Bloomington: Indiana University Press, 1993), pp. 926–928; Ellen Craft Dammond, "Interview with May Edward Chinn (June 27, July 13, September 12, 1979)," in *Black Women Oral History Project*, vol. 2, ed. Ruth Edmonds Hill (Westport, Conn.: Meckler, 1991), pp. 420–514; Sara Lawrence Lightfoot, *Balm in Gilead: Journey of a Healer* (Reading, Mass.: Addison-Wesley, 1988); Gamble, *Making a Place for Ourselves*, pp. 183–184; Bertha Van Hoosen, "The Woman Physician—Quo Vadis?" *Medical Woman's Journal* 36, no. 1 (1929): 1–4.

63. The actual percentages of all physicians who were board certified by that date were still low: 8 percent of men, 5 percent of women. "Hospital Service in the United States," *JAMA* 94, no. 13 (1930): 993; cf. Rothstein, *American Medical Schools*, pp. 136–137; Margaret D. Craighill, "An Analysis of Women in Medicine Today," 1940, typescript, Box 1, folder 18, AMWA/NYH.

64. "Hospital Service," *JAMA* 84, no. 13 (1925): 968; "Hospital Service," *JAMA*, pp. 992–993; Starr, *The Social Transformation*, p. 358; "Report of the Committee on Opportunities for Medical Women," *Women in Medicine* 66, no. 10 (1939): 20. Cf. the collection of letters of recommendation for internships and residencies in the papers of George Whipple, first dean of the University of Roch-

ester School of Medicine and Dentistry, George Whipple Collection, Edward G. Miner Library, University of Rochester School of Medicine and Dentistry; Rochester, New York: Merze Tate, "Interview with Lena Edwards, M.D. (November 13, 14, 1977)," in *Black Women Oral History Project*, vol. 3, ed. Hill, pp. 341–431, esp. p. 351.

65. Bertha Van Hoosen, "Opportunities for Medical Women Interns," *Medical Woman's Journal* 33 (1926): 341–343. See also Stevens, *In Sickness and in Wealth*, p. 118.

66. Craighill, "An Analysis," found that the most common specialties for women were pediatrics, psychiatry/neurology, and pathology; obstetrics-gynecology ranked sixth. Cf. Tracy, "Women Graduates," p. 26; Macy, "The Field for Women of Today in Medicine," 49–59; Van Hoosen, "Opportunities for Medical Women Interns," pp. 341–343; Van Hoosen, "Report of the Committee on Medical Opportunities for Women," *Medical Woman's Journal* 35, no. 7 (1928): 198–199; Morantz-Sanchez, *Sympathy and Science*, pp. 333–336, 442 n. 52; Starr, *The Social Transformation*, p. 358.

67. The APS membership was limited to seventy-five from 1912 to 1938, when numerical restrictions were eliminated. Thirty-five women became members between 1928 and 1965; some also were officers. Harold Kniest Faber and Rustin McIntosh, *History of the American Pediatric Society* (New York: McGraw-Hill, 1966). For Mary Putnam Jacobi's exclusion from the APS, cf. Joy Harvey, "Clanging Eagles: The Marriage and Collaboration between Two Nineteenth-Century Physicians, Mary Putnam Jacobi and Abraham Jacobi," in H. M. Pycior, N. G. Slack, and P. G. Abir-Am, eds., *Creative Couples in the Sciences* (New Brunswick, N.J.: Rutgers University Press, 1996), p. 193. At the local level, Bertha Van Hoosen never was admitted to the Chicago Gynecological and Obstetrical Society. Bertha Van Hoosen, "AMWA, Inc. Looks Backward," typescript, Bertha Van Hoosen Papers, folder 5, p. 5, Special Collections, University Library, University of Illinois at Chicago. I am grateful to Eve Fine for sharing this information.

68. Dorothy Reed Mendenhall, "Unpublished Memoir," in Jill Ker Conway, ed., *Written by Herself: Autobiographies of American Women* (New York: Vintage Books, 1992), p. 191; Donald Fleming, *William H. Welch and the Rise of Modern Medicine* (Baltimore: Johns Hopkins University Press, 1954), p. 133; Elizabeth Griego, "Clelia Duel Mosher: The Making of a 'Misfit,'" in Geraldine Joncich Clifford, ed., *Lone Voyagers: Academic Women in Coeducational Universities, 1870–1937* (New York: The Feminist Press, 1989), pp. 153–155.

69. Macy, "The Field for Women of Today in Medicine," pp. 49–59, esp. pp. 54, 55. Three women were assistant, 6 were associate, and 3 were full professors. These figures exclude the faculty of the Woman's Medical College of Pennsylvania. Among those who specialized, the largest number by far practiced gynecology or a combination of obstetrics and gynecology. The second highest number in a single specialty practice were in ophthalmology, followed by psychiatry, neurology, and pathology, in that order.

70. Bertha Van Hoosen, "Committee on Opportunities for Medical Women," *Bulletin of the Medical Women's National Association* no. 17 (July 1927): 21–23. Van Hoosen reported that of these 148, only 5 held research positions. The University of California had 22 women on the medical faculty, but none higher than clinical assistant professor. The University of Chicago had 15, with 3 at the level of clinical

assistant professor and none higher. Cornell and Yale each had 8, with 1 each at the assistant professor level. Loyola University Medical School, College of Medical Evangelists at Loma Linda, University of Illinois Medical School, Ohio State Medical School, and the University of Texas Medical Branch each had 1 female full professor, including Van Hoosen herself, who was also a department head. Woman's Medical College of Pennsylvania had a faculty of 81 that included 50 women; 10 were full professors, 3 were associate professors, and 37 were assistants or instructors; eleven of the school's twenty departments were headed by women.

71. Rosenberg, *The Care of Strangers;* Lynaugh, "From Respectable Domesticity to Medical Efficiency"; Atwater, "Women, Surgeons, and a Worthy Enterprise: The General Hospital Comes to Upper New York State"; and Rosner, "Doing Well or Doing Good."

72. *The Hospital Review* 37, no. 8 (April 15, 1901): 89. For information on Helen Gamwell, my appreciation to Philip G. Maples, curator of the Baker-Cederberg Museum and Archives, Rochester General Hospital. Also see Theresa K. Lehr and Philip G. Maples, *To Serve the Community: A Celebration of Rochester General Hospital* (Virginia Beach, Va.: Donning, 1997).

73. *Hospital Review* (November 1882–April 1889): April 9, 1888, pp. 169–171; June 4, 1888, p. 177; October 24, 1888, pp. 193, 194; Ladies' Board of Managers Minutes, April–November, 1892, BCA/RGH; Medical Staff Minutes, January 12– May 10, 1896, BCA/RGH; David Lovejoy, "The Hospital and Society: The Growth of Hospitals in Rochester, New York, in the Nineteenth Century," typescript paper, (winner of the Cushing Prize), 1975, Edward G. Miner Library, University of Rochester School of Medicine and Dentistry, Rochester, New York, p. 21, Appendices 1–3; Virginia Jeffrey Smith, *A Century of Service: Rochester General Hospital, 1847–1947* (Rochester, N.Y.: Rochester General Hospital, 1947), pp. 90–92.

74. Ladies' Board of Managers Minutes, vol. 5, February 3–April 6, 1896, pp. 106–112, BCA/RGH.

75. "Board of Directors Annual Report," *Hospital Review* 34, no. 5 (January 15, 1898); Medical Staff Minutes, May 10, 1896, and January 17–31, 1897; Board of Directors Minutes, January 21, 1897; Ladies' Board of Managers Minutes, July 6, 1896, all in BCA/RGH.

76. Rosenberg, *The Care of Strangers,* p. 334, makes this point with great force.

77. Letter from Mrs. Helen M. Craig to the Medical Staff of the Rochester City Hospital, December 14, 1897, in Medical Staff Collection, correspondence files, BCA/RGH; Minutes of Annual Meeting of Board of Directors, January 20, 1897 [*sic:* 1898], and January 19, 1899, BCA/RGH; "Tribute to Dr. Evelyn Baldwin," *Woman's Medical Journal,* 27, no. 4 (April 1917): 91. Cf. Ladies' Board of Managers Minutes for 1894–1898, BCA/RGH.

78. "Annual Report of the Rochester City Hospital," *Hospital Review* 35, no. 6 (January 16, 1899): 53–60.

79. Minutes of the Annual Meeting, Board of Directors, January 18, 1900, BCA/RGH.

80. Ibid.

81. Letter from Julius M. Wile to Dr. Charles A. Dewey, March 19, 1900, Medical Staff Collection, BCA/RGH.

82. Medical Staff Minutes, May 27, 1900, Medical Staff Collection, BCA/RGH. The male outpatient physicians were hardly indifferent to the outcome of the

dispute. Potter's chief rival for patient referrals, Dr. Lewis W. Rose, an 1887 graduate of New York University Medical College who in 1901 was named both secretary of the outpatient department and specialist in diseases of women. Rose wrote a letter to the president of the medical staff shortly after Wile's letter, resigning from the outpatient staff so as "not to embarrass" the staff any longer with his presence. Letter of Dr. Lewis W. Rose to Dr. John W. Whitbeck, April 27, 1900, Medical Staff Collection, BCA/RGH. In fact Dr. Rose remained on the staff until December 1911. *Hospital Review* 48, no. 3 (December 1911): 21.

83. Interview with Dr. James Craig Potter. Dr. Edward W. Mulligan, who had graduated from Rush Medical College in 1883 and then spent a year at Bellevue Medical College, was appointed to the outpatient department in 1889. In 1892, his offer to furnish a private men's ward to replace a public ward was accepted by the Ladies' Board, but the medical staff was allowed to make the final decision on this change. In 1896 he was promoted to the inpatient junior staff; by 1898 he was admitted to the surgical senior staff.

84. Letter from Dr. Charles A. Dewey to Dr. Marion C. Potter, June 14, 1900; letter from Dr. F. W. Zimmer to Dr. Charles Dewey, March 4, 1902, both in Medical Staff Collection, BCA/RGH.

85. Medical Staff Minutes, December 16, 1900, Medical Staff Collection, BCA/RGH. I have seen no direct evidence, however, that Palmer ever was asked to leave.

86. It was the hospital's staff surgeons, such as Mulligan, who represented the chief competition for all three. By the end of 1911 Dr. Rose was no longer listed on the City Hospital staff roster (see n. 82).

87. See Smith, "Appendicitis, Appendectomy, and the Surgeon," pp. 434 ff. (n. 7 above), for a discussion of the appendectomy craze during this period. Also see Rochester City Hospital, *Register*, vol. 7, 1902–1906, BCA/RGH. In the very few cases for which comparisons are possible, Dr. Mulligan kept his patients in the hospital much longer than did Dr. Potter, even when both doctors' patients were receiving free service. For example, in treating ovarian cysts, Potter's three patients averaged a 27-day hospitalization; Mulligan's one patient stayed for 36 days. Each physician treated three cases of lacerated cervix; Potter's averaged 22 days each, while Mulligan's averaged 31 days. With childbirth, Mulligan was only slightly more conservative than his women colleagues. Dr. Baldwin's eight cases averaged a 14-day stay, while Dr. Mulligan's three cases remained in hospital for 17 days. Cf. Howell, *Technology in the Hospital*, p. 61; Atwater, "Women, Surgeons, and a Worthy Enterprise," pp. 44–45.

88. Rochester City Hospital, *Register*, vol. 8, 1907, BCA/RGH.

89. Medical Staff Minutes, January 11, 1909, BCA/RGH. Potter was an active gynecological surgeon, at least in her earlier private practice. Cf. Mary E. Stark, "Clinical Studies in Alexander's Operation," *Woman's Medical Journal* 3, no. 3 (1894): 57–60.

90. The three physicians were Drs. Hastings, Young, and Little. Cf. "Rochester City Hospital Interns, 1864–1896" typescript, n.d., Medical Staff Collection, BCA/RGH; *Hospital Review* 36, no. 6 (February 1900): 61, and 46, no. 5 (February 1910): 43, BCA/RGH.

91. Stevens, *In Sickness and in Wealth*, pp. 82, 90, 93–5, 102, 114–118.

92. Chapter 5 discusses the overseas hospitals organized by women physicians

during World War I. John M. Swan and Mark Heath, *A History of United States Army Base Hospital No. 19* (Rochester, N.Y.: Wegman Publishing Co., 1922), p. 3.

93. Brig. Gen. J. R. Kean, "Note," in Maj. Gen. George de Tarnowsky, "Advance Surgical Formations: Modifications Which Seem Necessary in Light of the German Offensive of March–July, 1918," *The Military Surgeon* 43 no. 9 (March 1919): 256–257. On base hospitals, see Maj. Gen. M. W. Ireland, ed., *The Medical Department of the United States Army in the World War*, vol. 2: *Administration, American Expeditionary Forces*, by Col. Joseph H. Ford, M. C. (Washington, D.C.: U.S. Government Printing Office, 1927), p. 618.

94. See Paul B. Beeson and Russell C. Maulitz, "The Inner History of Internal Medicine," in Russell C. Maulitz and Diana E. Long, ed., *Grand Rounds: One Hundred Years of Internal Medicine* (Philadelphia: University of Pennsylvania Press, 1988), pp. 16, 25, 28. The authors trace the rising status of internal medicine, in part, to the effects of World War II.

95. Stevens, *In Sickness and in Wealth*, pp. 90–101, 104, 114–118, quotation on pp. 90–91.

96. Medical Staff Minutes, pp. 504–507, 514–516, 520–521, 526–527, 530–534, 558–559, 569–572, BCA/RGH. Up until 1930, several other women held brief junior staff affiliations, but they do not appear to have seen inpatients. Personal communication, Philip G. Maples, October 8, 1998. Many on the senior visiting staff were invited to assist the department heads at the new medical school of the University of Rochester, although none became a regular member of the faculty. Cf. George W. Corner, M.D., "Foundation and Earliest Years," in *To Each His Farthest Star: University of Rochester Medical Center, 1925–1975* (Rochester, N.Y.: University of Rochester Medical Center, 1975), pp. 50–51.

97. A profile honoring Potter at her retirement concluded: "Undoubtedly she considers the work she has done in periodic health examinations and social hygiene as her outstanding interests." "Marion Craig Potter," *Medical Woman's Journal* 45, no. 8 (1938): 247; Curriculum vitae, Marion Craig Potter, papers of Dr. Marion Craig Potter II; "Regrets" Folder, ca. November 1934, Envelope 4, MCP/MCP.

98. Starr, *The Social Transformation*, pp. 167–175, quotation on p. 167.

99. Morantz-Sanchez, *Sympathy and Science*, pp. 160–162.

5. Getting Organized

1. Portions of this chapter first appeared in Ellen S. More, "'A Certain Restless Ambition': Women Physicians and World War I," *American Quarterly* 41 (1989): 636–660, and "Rochester 'Over There': Gender and Medicine in World War I," *Rochester History* 51 (1989): 12–31.

2. Paul Starr, *The Social Transformation of American Medicine* (New York: Basic Books, 1982), chap. 3, esp. pp. 109–112.

3. *Woman's Medical Journal* 18, no. 5 (1908): 123–125.

4. Nancy Cott, *The Grounding of Modern Feminism* (New Haven, Conn.: Yale University Press, 1987), pp. 215–239; Margaret W. Rossiter, *Women Scientists in America: Struggles and Strategies to 1940* (Baltimore: Johns Hopkins University Press, 1982), pp. 51–99.

5. Cott, *The Grounding of Modern Feminism*, pp. 3–43; Peter G. Filene, *Him/Her/Self: Sex Roles in Modern America*, 2nd ed. (Baltimore: Johns Hopkins Uni-

versity Press, 1986), pp. 6–112; Carroll Smith-Rosenberg, "The New Woman as Androgyne," in Carroll Smith-Rosenberg, ed., *Disorderly Conduct* (New York: Oxford University Press, 1985), pp. 285–296; Rosalind Rosenberg, *Beyond Separate Spheres: Intellectual Roots of Modern Feminism* (New Haven, Conn.: Yale University Press, 1982), pp. 108–110; Dorothy M. Brown, *Setting a Course: American Women in the 1920s* (Boston: Twayne, 1987), pp. 29–47. Also see Martin Pernick, "Medical Professionalism," in Warren T. Reich, ed., *Encyclopedia of Bioethics*, vol. 3 (New York: The Free Press, 1978), pp. 1028–1034; Rossiter, *Women Scientists in America*, pp. 73–99; Charles Rosenberg, "Martin Arrowsmith: The Scientist as Hero," in Charles E. Rosenberg, ed., *No Other Gods: On Science and American Social Thought*, (Baltimore: Johns Hopkins University Press, 1976), pp. 123–131; JoAnne Brown, "Professional Language: Words That Succeed," *Radical History Review* 34 (1986): 33–51.

6. It is not clear that Hamilton ever actually served as third vice president. Four out of 266 presenters at the scientific sessions in 1915 were women (1.5 percent), only 2 were women in 1916 (both in the section on pathology and physiology). "Atlantic City Session," *JAMA* 53 (1914): 108–109; "Minutes of House of Delegates," *JAMA* 66 (1916): 1985–1987; "Preliminary Program of the Scientific Assembly," *JAMA* 64 (1915): 1708–1712; "Programs of the Sections," *JAMA* 66 (1916): 1520.

7. Regina Morantz-Sanchez, *Sympathy and Science: Women Physicians and American Medicine* (New York: Oxford University Press, 1985); Ellen More, "The Blackwell Medical Society and the Professionalization of Women Physicians," *Bulletin of the History of Medicine* 61 (1987): 603–628; Virginia Drachman, *Hospital with a Heart: Women Doctors and the Paradox of Separatism* (Ithaca, N.Y.: Cornell University Press, 1984); Mary Roth Walsh, *Doctors Wanted: No Women Need Apply* (New Haven: Yale University Press, 1977); Gloria Moldow, *Women Doctors in Gilded-Age Washington: Race, Gender, and Professionalization* (Urbana: University of Illinois Press, 1987).

8. More, "The Blackwell Medical Society," pp. 625–628; Virginia G. Drachman, "The Limits of Progress: The Professional Lives of Women Doctors, 1881–1926," *Bulletin of the History of Medicine* 60 (1986): 58–72; Drachman, *Hospital with a Heart*, pp. 151–195; Walsh, *Doctors Wanted*, pp. 263–267; Morantz-Sanchez, *Sympathy and Science*, chap. 10 and pp. 279–280, 336–339.

9. Penina Migdal Glazer and Miriam Slater, *Unequal Colleagues: The Entrance of Women into the Professions, 1890–1940* (New Brunswick, N.J.: Rutgers University Press, 1987), chaps. 1 and 6. Lynn Gordon, "Katharine Bement Davis," in Walter Trattner, ed., *The Biographical Dictionary of Social Welfare Reform* (Westport, Conn.: Greenwood Press, 1986); Jill Conway, "Women Reformers and American Culture, 1870–1930," *Journal of Social History* 5 (1972): 167–177. Jill Conway et al., "Introduction: The Concept of Gender," in *Daedalus* 116, no. 4 (1987): xxi–xxiv. G. J. Barker-Benfield, "'Mother-Emancipator': The Meaning of Jane Addams' Sickness and Cure," *Journal of Family History* 3 (1979): 395–420. Following Jill Conway, Barker-Benfield interprets the Addams typology as maternal heroism, the synthesis of selflessness and ambition. See pp. 396–399.

10. *Woman's Medical Journal* 25, no. 7 (1915): 159; Eve Fine, "Separate but Integrated: Bertha Van Hoosen and the Founding of AMWA," *JAMWA* 45, no. 5 (1990): 181–190, esp. p. 181.

11. See Bertha Van Hoosen, *Petticoat Surgeon* (Chicago: Pellegrini and Cudahy, 1947).

12. Van Hoosen, *Petticoat Surgeon*, p. 201; also in Walsh, *Doctors Wanted*, p. 213.

13. Mrs. George Bass, "The Relation of the Chicago Women's Club to the Mary Thompson Hospital," *Woman's Medical Journal* 25, no. 12 (1915): 279–280.

14. "Editorial: Amalgamation, Not Separation," ibid. 26, no. 5 (1916): 132–133.

15. *Woman's Medical Journal* 26, no. 1 (1916): 16–18; Minutes of Organization Meeting, Box 1, folder 2, AMWA/NYH. My thanks to Adele Lerner, archivist, for her help with the AMWA Collection.

16. Calculation is based on a $2 membership fee.

17. "Transactions of Annual Meeting of the Medical Women's National Association," *Woman's Medical Journal* 26 (1916): 164.

18. Bertha Van Hoosen, "Outline of Work for the Year," ibid. 26, no. 6 (1916): 159–160, 164. Isabelle T. Smart, "Report on Internships for Women," ibid. 27, no. 7 (1917): 162–163.

19. Ellen S. More, "'A Certain Restless Ambition,'" 636–660. Also see Kimberly Jensen, "Uncle Sam's Loyal Nieces: American Medical Women, Citizenship, and War Service in World War I," *Bulletin of the History of Medicine* 67 (1993): 670–690.

20. Arthur Dean Bevan, "The Problem of Hospital Organization, with Special Reference to the Coordination of General Surgery and the Surgical Specialties, *The Military Surgeon* 45, no. 2 (1919): 150–159; Rosemary Stevens, *American Medicine and the Public Interest*, 2nd ed. (New Haven, Conn.: Yale University Press, 1978), pp. 85–92, 124–128. Maj. Gen. M. W. Ireland, ed., *The Medical Department of the United States Army in the World War*, vol 1, by Col. Charles Lynch et al. (Washington, D.C.: U.S. Government Printing Office, 1923), p. 84.

21. More, "Rochester 'Over There.'"

22. Barbara J. Steinson, *American Women's Activism in World War I* (New York: Garland, 1982), p. 299; Filene, *Him/Her/Self*, p. 111; David Kennedy, *Over Here: The First World War and American Society* (New York: Oxford University Press, 1980), p. 30; Jean Bethke Elshtain, *Women and War* (New York: Basic Books, 1987), x–xiv. Cf. J. Stanley Lemons, *The Woman Citizen: Social Feminism in the 1920s* (Urbana: University of Illinois Press, 1973), pp. 14–15: "World War I was not just a boon to suffrage; the entire feminist movement received a boost."

23. Harvey Cushing, *From a Surgeon's Journal, 1915–1918* (Boston: Little, Brown, 1937), pp. 258, 267; George Washington Crile, *George Crile: An Autobiography*, ed. Grace Crile (Philadelphia: J. B. Lippincott, 1947), vol. 2, p. 359; GMB *Minutes*/NARC, vols. 1 and 2, Record Group 62, Box 426, pp. 2–4.

24. Seagrave was on her way to one of the National American Women's Suffrage Association's Women's Overseas Hospitals (see n. 30 below). *The Woman Citizen*, July 6, 1918, p. 114.

25. "Report of the American Women's Hospitals, June 6th to October 6th, 1917," p. 14, Box 1, folder 2, AWH/MCP. "The Brown-Gilmore Resolution, *Woman's Medical Journal* 27, no. 6 (1917): 149–150; "California Medical Women Urge Federal Recognition," ibid. 27, no. 10 (1917): 227–228; "A Most Interesting Report of Work of Colorado Women's War Service League," ibid. 28, no. 2 (1918): 39. Women, of course, were not the only group to suffer from exclusionary practices by the military; African American recruits, for example, were routinely segre-

gated. William Rothstein, *American Medical Schools and the Practice of Medicine* (New York: Oxford University Press, 1987), pp. 153, 363 n. 42.

26. Baker's ruling was not overturned until 1943. On Secretary Baker, see Daniel R. Beaver, *Newton D. Baker and the American War Effort, 1917–1919* (Lincoln: University of Nebraska Press, 1966), pp. 4–15; Kennedy, *Over Here*, pp. 96, 97; Frederich Palmer, *Newton D. Baker, America at War*, vol. 2 (New York: Dodd, Mead, 1981), pp. 34–35.

27. Letter from R. H. Frost to Eliza Mosher, September 25, 1917, reprinted in *Woman's Medical Journal* 27, no. 10 (1917): 226–227. Dr. Anita Newcomb McGee of Washington, D.C., had been employed as a contract surgeon during the Spanish-American War. She was hired to superintend the newly formed Army Nurse Corps. But McGee was not given a commission and she did not set a precedent for the inclusion of women in the Medical Reserves. Anita Newcomb McGee, "Can Women Physicians Serve in the Army?" ibid. 28, no. 2 (1918): 26, 27.

28. Mabel E. Gardner, "Bertha Van Hoosen," *JAMWA* 5, no. 10 (1950): 413, 414; McGee, "Can Women Physicians Serve in the Army?" pp. 26–28. "Opinion by Surgeon General Blue," *Woman's Medical Journal* 28, no. 2 (1918): 41.

29. MWNA Minutes, November 18, 1915, Box 1, folder 2, Bertha Van Hoosen, "The Medical Women's National Association," typescript, Box 5, folder 14, AMWA/NYH; Van Hoosen, *Petticoat Surgeon*, p. 210; *Woman's Medical Journal* 23, no. 7 (1913): 157; "National Association of Medical Women," ibid. 26, no. 1 (1916): 15; *Woman's Medical Journal* 28, no. 7 (1918): 157; Morantz-Sanchez, *Sympathy and Science*, pp. 248–249, 274–280; *American College of Surgeons, Third Year Book* (Chicago: American College of Surgeons, 1915). Cf. Jensen, "Uncle Sam's Loyal Nieces," pp. 670–690; idem., "The 'Open Way of Opportunity': Colorado Women Physicians and World War I," *Western Historical Quarterly* 27 (1996): 327–348.

30. The National American Woman's Suffrage Association (NAWSA) was the only other independent women's group to send a hospital unit, the Women's Overseas Hospitals, abroad. The NAWSA hospital and mobile dispensaries at Nancy in France managed 180 medical inpatients and 5,280 dispensary patients, delivered 17 babies, and performed 95 operations. Although the medical war work of the AWH and Women's Overseas Hospitals was similar, the NAWSA unit was more concerned with promoting woman suffrage than with professional advancement. Dr. Caroline Finley, director of obstetrics at the New York Infirmary for Women and Children, headed the original NAWSA unit. Finley told a reporter that she felt they were doing "a fine thing for suffrage." *The Woman Citizen*, May 4, 1918, p. 449, 450. In *The Woman Citizen*, p. 9, Lemons incorrectly conflated the AWH and NAWSA hospitals. [Lovejoy], "Notes Taken from 'The Woman Citizen,'" p. 5, Box 1, folder 4, AWH/MCP; "Letter from Dr. Laura Hunt to AWH Executive Committee, May 27, 1918, Box 22, folder 217, AWH/MCP. Also see idem., July 14, 1917, p. 121; February 2, 1918, p. 197; August 10, 1918, p. 213; September 21, 1918, p. 329; November 9, 1918, p. 491.

31. Bertha Van Hoosen, "Looking Backward," *JAMWA* 5, no. 10 (1950): 408; Van Hoosen, *Petticoat Surgeon*, p. 202.

32. Dan Joseph Singal, *The War Within: From Victorian to Modernist Thought in the South, 1919–1945* (Chapel Hill: University of North Carolina Press, 1982), pp. 1–36; Jackson Lears, *No Place of Grace: Antimodernism and the Formation of Mod-*

ern Culture, 1880–1920 (New York: Pantheon Books, 1981); Henry F. May, *The End of American Innocence: A Study of the First Years of Our Own Time, 1912–1917* (New York: Oxford University Press, 1979), pp. 6–30, 84, 108–110.

33. Letter from Rosalie Slaughter Morton to Dean Clara Marshall, April 30, 1898, Alumnae Archives and Special Collections on Women in Medicine, MCP Hahnemann University, Philadelphia.

34. Curriculum vitae for Rosalie Slaughter Morton, Envelope 2, MCP/MCP; Rosalie Slaughter Morton, *A Woman Surgeon* (New York: Stokes, 1937), pp. 42–45, 50, 107–117, 144.

35. Morton, *A Woman Surgeon*, pp. 23, 130–131, 178–191; Morton, curriculum vitae; Morantz-Sanchez, *Sympathy and Science*, pp. 146–147.

36. Morton, *A Woman Surgeon*, pp. 177–181, 197–213; Morton, curriculum vitae.

37. Morton, *A Woman Surgeon*, pp. 214–215.

38. Singal has described the ideology of professionalism among southern post-Victorians as a combination of noblesse oblige and heroic struggle. Part of Morton's struggle was to reshape this typology to fit a woman. See Singal, *The War Within*, pp. 27–32; Lears, *No Place of Grace*, p. 80; Christopher Lasch, *The New Radicalism in America: The Intellectual as a Social Type* (New York: Vintage Books, 1965), p. 62.

39. Morton, *A Woman Surgeon*, pp. 222–232.

40. Ibid., pp. 215–225, 228–229, 269–270.

41. The plan for a Women's Army General Hospital was never carried out. Ibid., p. 269. Also see Steinson, *American Women's Activism*, especially chap. 7.

42. "Report of the Medical Women's National Association, Second Annual Meeting, New York City, June 5th and 6th," *Woman's Medical Journal* 27, no. 6 (1917): 140–142, 147–148. Morton, *A Woman Surgeon*, pp. 271–272; Morton, "War Work," in *75th Anniversary Volume* (Philadelphia: Woman's Medical College of Pennsylvania, 1925), p. 365.

43. The brief account of Lovejoy's career contained in this and the following paragraphs is based on the following sources: passport application for Esther Pohl Lovejoy, 1946, and letter from Ernestine Strandborg to Esther Pohl Lovejoy, March 1, 1946, Box 2a, folder 19, AWH/MCP; "Lovejoy, Esther Pohl," in Barbara Sicherman and Carol Hurd Green, eds., *Notable American Women: The Modern Period* (Cambridge, Mass: The Belknap Press of Harvard University Press, 1980), pp. 424–426; telephone interview with Mrs. Estelle Fraade, April 1, 1987; Esther Pohl Lovejoy, *Certain Samaritans* (New York: Macmillan, 1927), pp. 8, 9.

44. Letter from Ernestine Strandborg to Esther Pohl Lovejoy, March 1, 1946, Box 2a, folder 19, AWH/MCP. Strandborg also teased Lovejoy for claiming to have washed her dirty linens in a German helmet on the voyage home from the war.

45. Compare Morton's reminiscences, replete with poetry and faded red roses, with Lovejoy's wisecracks about "a Pope with appendicitis" and the lack of workable plumbing. Morton, *A Woman Surgeon*, p. 268; Lovejoy, *Certain Samaritans*, pp. 16, 36; Kennedy, *Over Here*, pp. 178–225.

46. Morton, *A Woman Surgeon*, p. 1. Morton published portions of her mother's diary, describing her mother's Virginia courtship to the dashing Jack Slaughter, in the *Virginia Quarterly Review* 10 (1935): 61–81.

47. Letter from Rosalie Slaughter Morton to Marion Craig Potter, August 26, 1922, in Envelope 3, MCP/MCP.

48. Lovejoy, *Certain Samaritans*, pp. 16–36.

49. Van Hoosen, "Looking Backward," p. 408; Morton, *A Woman Surgeon,* pp. 188–89, 270.

50. Ironically, the American Red Cross traditionally was dominated by formidable women, including its founder, Clara Barton, and her successor, Mabel T. Boardman. On the eve of American participation in World War I, however, Boardman was supplanted by an all-male War Council of businessmen and financiers, to facilitate fund-raising. See Foster Rhea Dulles, *The American Red Cross: A History* (New York: Harper and Row, 1950), pp. 63–86, 138–141; John F. Hutchinson, *Champions of Charity: War and the Rise of the Red Cross* (Boulder, Colo.: Westview Press, 1996).

51. A modified version of the petition appeared in *Woman's Medical Journal.* In July, after Dr. Morton was appointed to the General Medical Board by Dr. Franklin Martin, he suggested that her Committee on Women Physicians circulate a modified version of the petition. About 500 signatures were collected, and the petition was presented to Martin in the late summer or early fall of 1917. Martin temporized, suggesting that the time had not yet arrived to present it to the surgeon general. AWH Executive Committee, Minutes, July 12, 1917, Box 1, folder 2, AWH/MCP; "The Brown-Gilmore Resolution," *Woman's Medical Journal* 27, no. 6 (1917): 149, 150; "Report of the American Women's Hospitals, June 6th to October 6th, 1917," p. 4, Box 1, folder 2, AWH/MCP.

52. War Service Committee, Executive Committee Minutes, June 9, 20, 21, and 28, 1917, Box 1, folder 2, AWH/MCP. Cf. AWH letterheads for September and November, 1917. The committee's stationery did not reflect the change for another four months.

53. Morton also knew Col. Jefferson Randolph Kean, director-general of the Red Cross Department of Military Relief. GMB *Minutes*/NARC, July 29, 1917, Record Group 62, Box 426, vol. 1, p. 151. Morton, *A Woman Surgeon,* pp. 134, 167–175, 269, 270; John M. Swan, *A History of United States Army Base Hospital No. 19* (Rochester, N.Y.: Wegman-Walsh Press, 1922) pp. 3, 4.

54. Morton proposed twelve women for the committee, from whom Dr. Martin selected the following: Drs. Marion Craig Potter, Caroline Purnell, Emma Culbertson, Caroline Towles, Adelaide Brown, Mary Lapham, Louise Strobel, Mary Parsons, Florence Ward, and Cornelia Brant. Morton had proposed Bertha Van Hoosen for the committee and was acutely embarrassed when Franklin Martin turned her down. Anxious to maintain harmony between herself and the MWNA president, Morton asked that, "in justice," Marion Craig Potter explain her good intentions to Van Hoosen. "Report of the American Women's Hospitals, June 6th to October 6th, 1917," pp. 13, 14, Box 1, folder 2, AWH/MCP. GMB *Minutes*/NARC, June 24, 1917, Record Group 62, vol. 1, Box 426, pp. 129–133. Morton, *A Woman Surgeon,* pp. 279–280; Rosalie Slaughter Morton to Marion Craig Potter, March 21, 1918, Box 2, folder 1, MCP/UR; Van Hoosen, *Petticoat Surgeon,* pp. 133–134.

55. M. Louise Strobel to Marion Craig Potter, November 24, 1917, Box 1, folder 21, MCP/UR, Minutes, Committee on Women Physicians, November 12, 1918, and May 15, 1918, Box 1, folder 21, MCP/UR; GMB *Minutes*/NARC, June 9, 1918, Record Group 62, Box 426, vol. 2, pp. 11, 12; Morton, *A Woman Surgeon,* pp. 284, 285.

56. "[AWH] Meeting of January 3, 1918," *Woman's Medical Journal* 28, no. 2

(1918): 30; Minutes, AWH Executive Committee Meeting, July 26, 1917, Box 1, folder 2, AWH/MCP.

57. Minutes, Open Meetings, Executive Committee, July 19 and 26, 1917, Box 23, AWH/MCP.

58. *Woman's Medical Journal* 27, no. 7 (1917): 164; Esther Pohl Lovejoy, *Women Physicians and Surgeons* (Livingston, N.Y.: The Livingston Press, 1939), p. 47; "[AWH] Treasurer's Report for October to December, 1917," *Woman's Medical Journal* 28, no. 2 (1918): 34, 35. See Morton, *A Woman Surgeon*, p. 286, for a somewhat rosier interpretation of those early months.

59. Jayne C. DeFiore, "Rosalie Slaughter Morton: Founder of American Women's Hospitals?" *Collections, Newsletter of the Archives and Special Collections on Women in Medicine* 16 (1987): 1, 2; Barker-Benfield, "Mother Emancipator," p. 396; Jill Conway, "Women Reformers," pp. 167–177.

60. Minutes, Executive Committee of the War Service Committee, MWNA, June 28 and July 12, 1917, Box 1, folder 2, AWH/MCP; Morton, "War Work," p. 262. *Woman's Medical Journal* 27, no. 6 (1917): 147; Rosalie Slaughter Morton, "The Battalion of Life," *The Forum*, April 1918 (reprint New York: American Women's Hospitals, 1918), pp. 3, 6, Envelope 2, MCP/MCP; Lovejoy, *Women Physicians and Surgeons*, p. 35.

61. Van Hoosen traveled to Washington to witness Morton's presentation of the document to the Red Cross on July 11. Whether or not she approved of the document, she seems to have done nothing to amend or correct it. According to the first AWH report, the committee merely sent it to Van Hoosen and generally kept "in touch" with her on an occasional basis. At the same time, the committee did take "weekly council" with Angenette Parry and Eliza Mosher, two MWNA officials based in New York. "Report of the American Women's Hospitals, June 6th to October 6th, 1917," pp. 10–11, Box 1, folder 2, AWH/MCP.

62. After the war, Morton founded the Serbian Education Committee and personally brought sixty Serbian students to the United States to receive a college education. Morton, *A Woman Surgeon*, pp. 170–174, 283; Emily Dunning Barringer to Marion Craig Potter, June 1, 1918, Box 2, folder 1, MCP/UR.

63. One particularly nasty misunderstanding arose during November and December 1917 over the credit and expenses incurred for the registration of women physicians. Executive Committee Minutes, Box 22, folder 218, AWH/MCP.

64. Lovejoy, *Women Physicians*, pp. 35, 38; AWH, "Annual Report, June, 1917 to June 1918," p. 8, Box 1, folder 3, AWH/MCP.

65. Unsigned memos dated 1917 and March 13, 1918, Records of the American Red Cross, Box 44, folder 1, National Archives Trust, Washington, D.C.; Gertrude A. Walker, *The American Women's Hospitals*, pamphlet, and Dr. Caroline Towles to Dr. Marion Craig Potter, August 7, 1917, Box 2, folder 1, MCP/UR.

66. It is not clear whether he was referring to physicians under the direction of his own Bureau for Women and Children or the total number of Red Cross physicians in France. *Woman's Medical Journal* 28, no. 6 (1918): 140, 141.

67. Rosalie Slaughter Morton to Marion Craig Potter, March 21, 1918, Box 2, folder 1, MCP/UR; Morton, *A Woman Surgeon*, p. 274; Dr. Belle Thomas to Dr. Bertha Van Hoosen, August 4, 1917, Box 2, folder 14, AWH/MCP. Also see Van Hoosen to Rosalie Slaughter Morton, September 10, 1917, ibid., where she writes that she hopes that Morton "can hold out."

68. In June, rather than a constitution, a revised set of "rules" for the War Service Committee was drawn up. AWH Executive Committee Minutes, June 3, 1918, Box 22, folder 217, AWH/MCP; *Woman's Medical Journal* 27, no. 2 (1917): 224; AWH Executive Committee Minutes, February 17, 1918, Box 22, folder 217, AWH/MCP; Van Hoosen to Morton, September 10, 1917, Box 2, folder 14, AWH/MCP; Morton to Van Hoosen, telegram, n.d., Box 1, folder 2, AWH/MCP.

69. Minutes, MWNA Council and Board of Directors meeting, June 10, 1924, pp. 29–30, Box 1, folder 6, AMWA/NYH.

70. Morantz-Sanchez, *Sympathy and Science*, p. 260; Lovejoy, *Women Physicians*, pp. 38–41; Morton, *A Woman Surgeon*, p. 288.

71. AWH campaign pamphlet, n.p., Box 1, folder 2, AWH/MCP. Also see, for example, "6000 Women Physicians Engage in Tremendous Task to Help Uncle Sam," *Washington Times*, January 20, 1918; "Drive to Create a Medical Corps of Women Begins," *New York Times*, March 24, 1918; and other miscellaneous clippings, Box 1, folder 3, MCP/UR; "The Campaign Is On!" *Woman's Medical Journal* 28, no. 2 (1918): 25; *Woman's Medical Journal* 28, no. 3 (1918): 61–62; Morton, *A Woman Surgeon*, pp. 288–291.

72. *Woman's Medical Journal* 28, no. 4 (1918): 70–79.

73. "Annual Report, June 1917 to June 1918," p. 10, Box 1, folder 3, AWH/MCP; also in March, NAWSA raised only about $75,000 for its hospitals, plus approximately $12,000 more later that year. See *The Woman Citizen*, March 16, 1918, p. 309, and August 24, 1918, pp. 250–251. In her autobiography Morton claimed, "Within ten days we had raised $300,000 in cash and in pledges promptly paid." *A Woman Surgeon*, p. 291.

74. Steinson, *American Women's Activism*, pp. 319–322.

75. By June 1918 the total number of AWH physicians working with the Red Cross in France had climbed to fifty-two. "Annual Report, June 1917 to June 1918," p. 4, Box 1, folder 3, AWH/MCP; *Woman's Medical Journal* 28, no. 4 (1918): 83.

76. A. J. P. Taylor, *The First World War* (London: Hamish Hamilton, 1963), pp. 164–170; Kennedy, *Over Here*, pp. 169, 170, 176, 177, 190.

77. In the same month the Red Cross also agreed to donate $35,000 worth of equipment to the NAWSA hospital. See above, n. 30.

78. AWH Executive Committee Minutes, February 18, 1918, Box 22, folder 218, AWH/MCP; *Woman's Medical Journal* 28, no. 4 (1918): 83.

79. Unsigned memorandum, March 13, 1918, Box 44, folder 1, Records of the American Red Cross, National Archives Trust, Washington, D.C.

80. AWH Executive Committee Minutes, February 17, April 11, and May 18, 1918, Box 22, folder 218, AWH/MCP; Elizabeth S. Hoyt to Mrs. C. M. Conger, April 8, 1919, Box 2, folder 1, MCP/MCP; *Woman's Medical Journal* 28, no. 7 (1918): 162; "Report of Chairman," ibid., 29, no. 8 (1919): 164.

81. Lovejoy, *Certain Samaritans*, p. 177. Unit No. 2 was established in the fall of 1918 in the town of La Ferté-Milon, a few miles north of the Marne and halfway between Luzancy and Blérancourt. Dr. Ethel Fraser of Denver directed the twenty-five-bed unit. It was established in a partially demolished building that, before the war, had housed the hospital known as the Hôtel de Dieu. American Women's Hospitals, *Bulletin* 1, no. 3 (1919): 15.

82. Letter from M. May Allen to Marion Craig Potter, December 3, 1918, Box

1, MCP/UR, letter from Barbara Hunt to the Executive Committee of the AWH, n.d., Box 8, folder 61, AWH/MCP; Hunt to Executive Committee, November 10, 1918; Box 8, folder 61, AWH/MCP. The rest of Hunt's partial inventory included 150 beds, plus sheets, drawsheets, and blankets, and in the laboratory, a small incubator, some tubes of agar, and some cultures of typhoid bacillus.

83. Dr. M. May Allen to Dr. Marion Craig Potter, December 3, 1918, Box 1, folder 6, MCP/UR. Drs. Hurrell, Allen, and Inez Bentley, Hurrells' assistant had all been members of the Blackwell Medical Society of Rochester, through which they knew Marion Craig Potter, Morton's close friend. These personal networks should be seen as the functional equivalent of the staffing patterns of the original base hospitals under Gorgas's plan. Prior personal acquaintance was seen as an advantage, facilitating cooperative work under difficult circumstances. More, "The Blackwell Medical Society"; Pract./BMS *Minutes*, book 3, September 13, 1906, MCP/UR.

84. Dr. Hazel Bonness, "Financial Report, July 1, 1918, to February 19, 1920," Box 8, folder 60, AWH/MCP.

85. Hurrell to AWH Executive Committee, February 10, May 3, and May 7, 1919, Box 8, folder 59, AWH/MCP.

86. AWH *Bulletin* 1, no. 2 (1919): n.p.; ibid., no. 3, n.p.; Hurrell to Executive Committee, August 20, 1919, Box 8, folder 59, AWH/MCP.

87. AWH *Bulletin* 1, no. 1 (1919): n.p.; ibid. no. 3, n.p.

88. Ibid., no. 2, n.p.

89. AWH press release, typscript, n.d., Box 2, folder 15, AWH/MCP.

90. Hurrell to AWH Executive Committee, February 22 and May 25, 1919, Box 8, folder 59, AWH/MCP.

91. M. Louise Hurrell to AWH Executive Committee, August 20, 1919, and Frances Cohen, AWH Commissioner, to AWH Executive Committee, August 7, 1919, Box 8, folder 60, AWH/MCP.

92. M. Louise Hurrell to Marion Craig Potter, March 14, 1919, Box 11, MCP/UR; Frances Cohen to AWH Executive Committee, August 7, 1919, Box 8, folder 60, AWH/MCP. As an example of Hurrell's humor, here is an excerpt from one of her letters to the AWH Executive Committee, a mise-en-scène from Luzancy: One evening a doctor returned from her dispensary route and, "as she seated herself for dinner, said she had just brought in a wretched old man for operation. The Director [Hurrell], aghast, said, 'But there are no more beds in the men's ward!' The doctor replied, as she began to eat, famishedly, 'Oh, that's all right. I had him put in the women's ward—he's blind.'" Of the chauffeurs, Hurrell wrote, they "are the doctors to these old autos, and better doctors than we are to our own old patients." Letter to the AWH Executive Committee, AWH/MCP, Box 8, folder 59, n.d.

93. Hurrell to AWH Executive Committee, May 25, 1919, Box 8, folder 59, AWH/MCP; Dr. Caroline Purnell, AWH Commissioner, to AWH Executive Committee, August 7 and 27, 1919, ibid., folder 60. Hazel Bonness to AWH Executive Committee, "Quarterly Report, July–September, 1919," ibid; Bonness to Esther Pohl Lovejoy, Chairman AWH Executive Committee, "Final Report," July 28, 1918, to January 31, 1920, ibid.

94. Bonness to Lovejoy, January 12, 1920, ibid.

95. Belle Thomas to Marion Craig Potter, July 13, 1918, Box 2, folder 1,

MCP/MCP; AWH Executive Committee Minutes, June 3, 1918, Box 22, AWH/MCP; Minutes, MWNA Annual Meeting, June 10 and 11, 1918, Box 1, folder 3, AMWA/NYH; Van Hoosen, *Looking Backward*, p. 408.

96. Telephone interview with Mrs. Estelle Fraade, assistant to Esther Pohl Lovejoy from 1937 to 1967 and executive director of the AWH from 1967 to 1979, April 1, 1987.

97. Emily Dunning Barringer to Marion Craig Potter, June 1, 1918, Box 2, folder 1, MCP/MCP.

98. MWNA Minutes, 10th annual meeting, June 9, 1924, Box 1, folder 6, AMWA/NYH.

6. New Directions

1. Patricia M. Hummer, *The Decade of Elusive Promise: Professional Women in the United States, 1920–1930* (Ann Arbor, Mich.: UMI Research Press, 1979). Two excellent studies are William H. Chafe, *The Paradox of Change: Women in the 20th Century* (New York: Oxford University Press, 1991), esp. pp. 22–51; and Nancy F. Cott, *The Grounding of Modern Feminism* (New Haven: Yale University Press, 1987).

2. These are the most conservative figures, taken from Cott, *The Grounding*, p. 350 n. 4. Nearly three-fifths were teachers, mostly in the elementary and secondary grades. Lois Scharf, *To Work and to Wed: Female Employment, Feminism, and the Great Depression* (Westport, Conn.: Greenwood Press, 1980), pp. 86–94; Chafe, *The Paradox*, pp. 70–71; Dorothy M. Brown, *Setting a Course: American Women in the 1920s* (Boston: G. K. Hall, 1987), p. 156; Mabel Newcomer, *A Century of Higher Education for American Women* (New York: Harper, 1959), Table 12, p. 179, cited in ibid., pp. 3–23, 151, 156; Susan Ware, *Beyond Suffrage: Women in the New Deal* (Cambridge, Mass.: Harvard University Press, 1981), p. 5. Cf. Sophonisba P. Breckenridge, *Women in the Twentieth Century* (1933; New York: McGraw-Hill, 1972), pp. 187–196. And see above, Chapter 4, Table 4.2.

3. Brown, *Setting a Course*, pp. 3–23; Blanche D. Coll, *Safety Net: Welfare and Social Security, 1929–1979* (New Brunswick, N.J.: Rutgers University Press, 1995), pp. 1–4; Ronald Numbers, *Almost Persuaded: American Physicians and Compulsory Health Insurance, 1912–1920* (Baltimore: Johns Hopkins University Press, 1978).

4. Richard A. Meckel, *Save the Babies: American Public Health Reform and the Prevention of Infant Mortality, 1850–1929* (Baltimore: Johns Hopkins University Press, 1990), pp. 108–111.

5. Jeffrey Paul Brosco, "Sin or Folly: Child and Community Health in Philadelphia, 1900–1930," Ph.D. diss., University of Pennsylvania, 1994, pp. 99–104, 115.

6. Sheila M. Rothman, "Women's Clinics or Doctors' Offices? The Sheppard-Towner Act and the Promotion of Preventive Health Care," in David J. Rothman and Stanton Wheeler, eds., *Social History and Social Policy* (New York: Academic Press, 1981), pp. 175–201, esp. pp. 179–181. Rothman superbly explicates the Children's Bureau's effort to maintain this distinction. Also see Molly Ladd-Taylor, *Raising a Baby the Government Way: Mothers' Letters to the Children's Bureau, 1915–1932* (New Brunswick, N.J.: Rutgers University Press, 1986).

7. Louis J. Covotsos, "Child Welfare and Social Progress: A History of the United States Children's Bureau, 1912–1935," Ph.D. diss., University of Chicago,

1976, pp. 91–96; Meckel, *Save the Babies*, pp. 202–203; Joseph Benedict Chepaitis, "The First Federal Social Welfare Measure: The Sheppard-Towner Maternity and Infancy Act, 1918–1932," Ph.D. diss., Georgetown University, 1968, p. 12.

8. Sydney Halpern, *American Pediatrics: The Social Dynamics of Professionalism, 1880–1980* (Berkeley: University of California Press, 1988), p. 86.

9. Frances Sage Bradley and Florence Brown Sherbon, *How to Conduct a Children's Health Conference*, U.S. Children's Bureau publication no. 23 (Washington, D.C.: U.S. Government Printing Office, 1917); Grace L. Meigs, "The Children's Year Campaign," *Transactions of the AMA Section on Diseases of Children* 69, no. 46 (1918): 52–53; both cited in Halpern, *American Pediatrics*, pp. 88, 90.

10. Brosco, "Sin or Folly," pp. 211–214.

11. Covotsos, "Child Welfare," pp. 118–122.

12. Ibid., pp. 127, 134–135.

13. John Duffy, *The Sanitarians: A History of American Public Health* (Urbana: University of Illinois Press, 1990), pp. 222–224, 233–235.

14. "Abstract of Action of Conference [of State and Provincial Health Authorities of North America] in Regard to the Reorganization of the Federal Health Activities (1919–1923)," in Box 19, folder 254, MME/SL. In 1937 a copy of this report was sent to Martha Eliot, assistant chief of the Children's Bureau, by one of the participating state health officers. See also Covotsos, "Child Welfare," p. 134; Chepaitis, "The First Federal Social Welfare Measure," pp. 113–119.

15. Quoted in Chepaitis, "The First Federal Social Welfare Measure," p. 118. Dorothy M. Brown, *Setting a Course*, pp. 49–54.

16. Haines was a former national secretary of the MWNA. *Bulletin of the Medical Women's National Association* 15 (1927): 11. See also Molly Ladd-Taylor, "'My Work Came Out of Agony and Grief': Mothers and the Making of the Sheppard-Towner Act," in Seth Koven and Sonya Michel, eds., *Mothers of a New World: Maternalist Politics and the Origins of Welfare States* (New York: Routledge, 1993), pp. 322–323, 329; Covotsos, "Child Welfare," p. 68; Rothman, "Women's Clinics," p. 185 n. 18.

17. Theda Skocpol, *Protecting Soldiers and Mothers: The Political Origins of Social Policy in the United States* (Cambridge, Mass.: The Belknap Press of Harvard University Press, 1992), pp. 506–511; *Bulletin of the Medical Women's National Association* 15 (1927): 11. "Report of Delegates to the General Federation of Women's Clubs National Meeting in Des Moines, 1920," *Medical Woman's Journal* 28, no. 7 (1921): 85; Meckel, *Save the Babies*, pp. 212–217; J. Stanley Lemons, *The Woman Citizen: Social Feminism in the 1920s* (Urbana: University of Illinois Press, 1973), pp. 123, 176; Molly Ladd-Taylor, *Mother-Work: Women, Child Welfare, and the State, 1890–1930* (Urbana: University of Illinois Press, 1994), p. 191 n.4; Sheila M. Rothman, *Woman's Proper Place: A History of Changing Ideals and Practices, 1870 to the Present* (New York: Basic Books, 1978) p. 303 n. 12.

18. Halpern's figures, based on a report by the White House Conference on Child Health and Protection, are somewhat higher, *American Pediatrics*, p. 86; Chepaitis, "The First Federal Social Welfare Measure," pp. 223–225; Lemons, *Woman Citizen*, p. 175.

19. Covotsos, "Child Welfare," p. 146; Chepaitis, "The First Federal Welfare Measure," pp. 378–379, Appendixes 5 and 6.

20. According to Richard Meckel, until 1916 no infant welfare station was

opened in a black neighborhood, even though infant mortality for African Americans between 1915 and 1920 was 65 percent higher than for whites. Meckel, *Save the Babies*, p. 142; Rothman, *Woman's Proper Place*, pp. 139–141, 303 nn. 12 and 14, 304 n. 16; Ladd-Taylor, "My Work Came Out of Agony and Grief," pp. 333–334.

21. Meckel, *Save the Babies*, p. 203; Lela B. Costin, *Two Sisters for Social Justice: A Biography of Grace and Edith Abbott* (Urbana: University of Illinois Press, 1983), pp. 146–149.

22. Harold C. Bailey, "Commentary," *American Journal of Obstetrics and Gynecology* 9 (1925): 725, cited in Charles R. King, "The New York Maternal Mortality Study," *Bulletin of the History of Medicine* 65, no. 4 (1991): 484–486; Costin, *Two Sisters*, pp. 146–149.

23. Costin, *Two Sisters*, p. 149; King, "New York Maternal Mortality Study"; Joyce Antler and Daniel M. Fox, "The Movement toward Safe Maternity: Physician Accountability in New York City, 1915–1940," *Bulletin of the History of Medicine* 50, no. 3 (1976): 569–595.

24. Martha M. Eliot, "The Control of Rickets: Preliminary Discussion of the Demonstration in New Haven," *JAMA* 85, no. 9 (1925): 656–663, quotation on pp. 656–657. Also see Brosco, "Sin or Folly," pp. 100–102.

25. *Medical Woman's Journal* 28, no. 7 (1921): 172, 177, 180; S. Josephine Baker, *Fighting for Life* (New York: Macmillan, 1939), pp. 47, 54, 55, 83, 84; J. B. Chepaitis, "Federal Social Welfare Progressivism in the 1920s," *Social Services Review* 46 (1972): 213–229; J. Stanley Lemons, "The Sheppard-Towner Act: Progressivism in the 1920s," *Journal of American History* 55 (1969): 776–786; Lemons, *Woman Citizen*, pp. 42, 123, 154–158, 165, 177 n. 4, 178 n. 22; Rothman, *Woman's Proper Place*, pp. 139–141; Ladd-Taylor, *Mother-Work*, pp. 167–206.

26. Lela B. Costin, "Women and Physicians: The 1930 White House Conference on Children," *Social Work* (March–April 1983): 108–114, quotation on p. 113.

27. Chepaitis, "The First Federal Social Welfare Measure," pp. 11–19; Covotsos, "Child Welfare"; Meckel, *Save the Babies*, pp. 109, 128, 200–219; Lemons, *Woman Citizen*, pp. 117, 123, 154–158, 177–78 n. 22.

28. *Medical Woman's Journal* 28, no. 7 (1921): 149–150, 172, 177, 180.

29. Lemons, *Woman Citizen*, p. 160; Ladd-Taylor, *Mother-Work*, p. 169.

30. Meckel, *Save the Babies*, pp. 197–198; Rothman, "Women's Clinics," pp. 190–193; Numbers, *Almost Persuaded*, pp. 107–108; H. H. Moore, *American Medicine and the People's Health . . . with Special Reference to the Adjustment of Medical Service to Social and Economic Change* (New York: D. Appleton, 1927), p. 621; James G. Burrow, *AMA: Voice of American Medicine* (Baltimore: Johns Hopkins University Press, 1963), p. 162.

31. Moore, *American Medicine*, pp. 507–508, 511–512; John M. Dodson, "The Growing Importance of Preventive Medicine to the General Practitioner," *JAMA* 81, (1923): 1427–1429. Dodson was the editor of the new journal *Hygeia*. See also Eugene Lyman Fisk and J. Ramser Crawford, *How to Make the Periodic Health Examination* (New York: Macmillan, 1927), pp. 11–13; Meckel, *Save the Babies*, pp. 189–198; and especially Rothman, "Women's Clinics," pp. 190–193.

32. D. O. Powell, "Haven Emerson," in Martin Kaufman, Stuart Galishoff, and Todd L. Savitt, eds., *Dictionary of American Medical Biography*, vol. 1, (Westport, Conn.: Greenwood Press, 1984), p. 231; Rothman, "Women's Clinics," pp. 191–193.

33. Cf. George Rosen, *The Specialization of Medicine with Particular Reference to Ophthalmology* (New York: Froben, 1944); Halpern, *American Pediatrics*, pp. 10–12, 44, 180 n. 35.

34. Halpern, *American Pediatrics*, pp. 80–83, 93–95.

35. Ibid., pp. 80–81, 91, 109. Thomas E. Cone, Jr., *History of American Pediatrics* (Boston: Little, Brown, 1979), pp. 202–203.

36. L. Emmett Holt, "American Pediatrics: A Retrospect and a Forecast," presidential address, American Pediatric Society, 1923, quoted in Cone, *History of American Pediatrics*, pp. 104, 158, 202–203; Harold Kniest Faber and Rustin McIntosh, *History of the American Pediatric Society, 1887–1965* (New York: McGraw-Hill, 1966), pp. 100, 113.

37. Rima D. Apple, *Mothers and Medicine: A Social History of Infant Feeding* (Madison: University of Wisconsin Press, 1987); Janet Golden, *A Social History of Wet Nursing in America: From Breast to Bottle* (New York: Cambridge University Press, 1996), p. 135; Halpern, *American Pediatrics*, pp. 62–69, 72–76.

38. Costin, *Two Sisters*, pp. 145–150.

39. Skocpol, *Protecting Soldiers and Mothers*, pp. 514–520; Covotsos, "Child Welfare," p. 153.

40. Skocpol, *Protecting Soldiers and Mothers*, pp. 518–522; Rothman, *Woman's Proper Place*, p. 125, cited in ibid., p. 515; Lemons, *Woman Citizen*, p. 123; Halpern, *American Pediatrics*, pp. 90–95. Also see Kriste Lindenmeyer, *"A Right to Childhood": The U.S. Children's Bureau and Child Welfare, 1912–46* (Urbana: University of Illinois Press, 1997), pp. 164–168.

41. Costin, *Two Sisters*, pp. 166–168. Of the 12 pediatricians in the American Child Health Association's Medical Service Division in 1927, 8 held leadership roles at the 1930 conference. Twenty-five (9 percent) of the 281 members of the conference committees were MWNA members, including S. Josephine Baker, Blanche M. Haines, Alice Hamilton, and Mary Riggs Noble. "American Child Health Association Medical Service Division," *Child Health Bulletin* 3 (1927): 134–138, quotation on p. 135; *White House Conference on Child Health and Protection: Directory of Committee Personnel, July 1, 1930* (Washington, D.C.: Department of the Interior, 1930).

42. Covotsos, "Child Welfare," p. 153, quoted in Theda Skocpol, *Protecting Soldiers and Mothers*, p. 517.

43. Costin, "Women and Physicians," pp. 108–114.

44. Letter from Dr. Haven Emerson to Dr. Edwards Park, November 21, 1930; letter from Dr. Martha Eliot to Dr. Edwards Park, November 29, 1930; letter from Dr. Edwards Park to Dr. Haven Emerson, December 1, 1930; letter from Dr. Haven Emerson to Dr. Edwards Park, December 4, 1930, all in Box 18, folder 250, MME/SL. Marion Hunt, "'Extraordinarily Interesting and Happy Years': Martha M. Eliot and Pediatrics at Yale," *Yale Journal of Biology and Medicine* 68 (1995): 159–170. My thanks to Toby Appel, director, History of Medicine Division, Yale Medical Library, for alerting me to the Hunt article.

45. Costin, *Two Sisters*, pp. 170–173, quoting from a supplement to the *U.S. Daily News*, sec. 11, vol. 5 (November 25–28, 1930).

46. Lemons, *Woman Citizen*, p. 165; Rothman, "Women's Clinics," pp. 179–181; Meckel, *Save the Babies*, pp. 217–219; Skocpol, *Protecting Soldiers and Mothers*, pp. 516, 535–536.

47. *Bulletin of the Medical Women's National Association* 38 (1932): 8–12; Costin, "Women and Physicians," pp. 108–114; Judith Lorber, *Women Physicians* (New York: Tavistock Publications, 1984), p. 23.

48. "Minutes of Annual Meeting," typescript, June 9, 1924, Box 1, folder 6, AMWA/NYH.

49. Lemons, *Woman Citizen*, 165.

50. Apple, *Mothers and Medicine*, pp. 105, 128.

51. "Report of the Committee on Public Health," *Bulletin of the Medical Women's National Association* 14 (1926): 21. Regina Morantz-Sanchez, *Sympathy and Science: Women Physicians and American Medicine* (New York: Oxford University Press, 1985), p. 303. See also Meryl Sali Justin: "Men, Women, and Women Physicians: The Medical Women's National Association and the Medical Profession, 1915–1945," senior thesis, typescript, Harvard University, March 20, 1980, p. 31. My thanks to Dr. Eugenia Marcus for allowing me to read her copy of this paper.

52. *Bulletin of the Medical Women's National Association* 13 (1926): 16–17; 14 (1926): 23; 17 (1927): 12.

53. Ellen S. More, "The American Medical Women's Association and the Role of the Woman Physician, 1915–1990," *JAMWA* 45, no. 5 (1990): 170; Skocpol, *Protecting Soldiers and Mothers*, pp. 519–521.

54. *A History of the National Federation of Business and Professional Women's Clubs, Inc., 1919–1944* (New York: National Federation of Business and Professional Women's Clubs, 1944), pp. 29, 30, 74. Also see Margaret Elliott and Grace E. Manson, "Earnings of Women in Business and the Professions," *Michigan Business Studies* 3, no. 1 (1930): 132, 168.

55. Lemons, *Woman Citizen*, pp. 201–204.

56. Carole R. McCann, *Birth Control Politics in the United States, 1916–1945* (Ithaca, N.Y.: Cornell University Press, 1994), pp. 25, 45–55. See also Cott, *The Grounding of Modern Feminism*; Skocpol, *Protecting Soldiers and Mothers*.

57. *Bulletin of the Medical Women's National Association* 21 (1928): 12, 22.

58. "Report of 16th Annual Meeting of the Medical Women's National Association, held in Detroit, Mich., June 22–24, 1930," *Medical Woman's Journal* 37, no. 7 (1930): 193–199.

59. Joyce M. Ray and F. G. Gosling, "American Physicians and Birth Control, 1936–1947," *Journal of Social History* 18, no. 3 (1985): 399–411, quotation on p. 400.

60. McCann, *Birth Control Politics*, pp. 76 n. 46, 77, 90 n. 103, 139–142, 150–152, 165. On Dr. Chinn, who graduated in 1892, see Darlene Clark Hine, "Co-Laborers in the Work of the Lord: Nineteenth-Century Black Women Physicians," in Ruth J. Abram, ed. *"Send Us a Lady Physician:" Women Doctors in America, 1835–1920* (New York: W. W. Norton, 1985), p. 112.

61. See Susan L. Smith, *Sick and Tired of Being Sick and Tired: Black Women's Health Activism in America, 1890–1950* (Philadelphia: University of Pennsylvania Press, 1995), esp. pp. 149–167, on which the following account is largely based. Also see Margaret Jerrido, "Dorothy Celeste Boulding Ferebee (1898–1980)," in Darlene Clark Hine, Elsa Barkley Brown, and Rosalyn Terborg-Penn, eds., *Black Women in America: An Historical Encyclopedia*, vol. 1, (Bloomington: Indiana University Press, 1994), pp. 425–426.

62. Paula Giddings, *When and Where I Enter: The Impact of Black Women on Race and Sex in America* (New York: Bantam, 1988), pp. 75–76, 95–118, 199–230.

63. Smith, *Sick and Tired*, pp. 56–66.

64. On Dorothy Ferebee, see Merze Tate, "Interview with Dr. Dorothy Boulding Ferebee (December 28, 31, 1979)," in Ruth Edmonds Hill, ed., *The Black Women Oral History Project*, vol. 3, (Westport, Conn.: Meckler, 1991), pp. 436–481. Vanessa Gamble, "Physicians, Twentieth Century," in Hine, Brown, and Terborg-Penn, eds., *Black Women in America*, vol. 2, pp. 926–928.

65. Smith, *Sick and Tired*, pp. 66–68, 149–167, quotation on p. 161.

66. Ibid., pp. 161–166; Tate, "Interview," pp. 443–450.

67. Patricia Evridge Hill, "'Carrying Health to the Country:' The Mountain Medical Service of the American Women's Hospitals," *Collections, Newsletter of the Archives and Special Collections on Women in Medicine* 31 (1997): 1–4. See also idem., "Invisible Labours: Mill Work and Motherhood in the American South," *Social History of Medicine* 9, no. 1 (1996): 235–251.

68. "Minutes of Annual Meeting," June 9, 1924, typescript, Box 1, folder 6, AMWA/NYH; *Women in Medicine* 1939 (March): 18–19; "Minutes of MWNA Annual Meeting," *Bulletin of the Medical Women's National Association* 41 (1933): 10; *Women in Medicine* 1944 (January): 13, 14; ibid., 1945 (January): 20; P. Preston Reynolds, "Hospitals and Civil Rights, 1945–1963: The Case of *Simkins v Moses H. Cone Memorial Hospital*," *Annals of Internal Medicine* 126, no. 11 (1997): 898–906. Van Hoosen's protest is cited in Justin, "Men, Women, and Women Physicians," pp. 28, 84 n. 13.

69. A survey by Dr. Margaret Craighill of women physicians who were board certified by 1940 showed that of a total of 388 women (2.7 percent of all those certified), 88 were board certified in pediatrics, 22.6 percent of women specialists. The first specialty board was established for obstetrics-gynecology in 1930, the pediatrics board in 1933. See also Howard A. Pearson, ed., *The Centennial History of the American Pediatric Society, 1888–1988* (New Haven, Conn.: Yale University Press, 1988), pp. 131–132; Faber and McIntosh, *History of the American Pediatric Society*, hand count of members, pp. 148 ff.; Margaret D. Craighill, "An Analysis of Women in Medicine Today," 1940, typescript, Box 1, folder 18, AMWA/NYH; Halpern, *American Pediatrics*, p. 171 n. 18; Carol Lopate, *Women in Medicine* (Baltimore: Johns Hopkins University Press, 1968), p. 121, Table 8-1, and p. 193, Appendix 1; American Medical Association, *Women in Medicine in America: In the Mainstream*, (Chicago: AMA, 1991), p. 14, Table 3.6; Rosa Lee Nemir, "Women Physicians Chairing Pediatric Departments in American Medical Schools, 1980," *JAMWA* 36, no. 6 (1981): 183–194. My thanks to Dr. Nemir for alerting me to this source.

70. Mary Roth Walsh, *Doctors Wanted: No Women Need Apply* (New Haven, Conn.: Yale University Press, 1977), pp. 81, 82; Halpern, *American Pediatrics*, p. 180 n. 35; Samuel X. Radbill, "Hospitals and Pediatrics, 1776–1976," *Bulletin of the History of Medicine* 53, no. 2 (1979): 286–291; Rickey Hendricks, "Feminism and Maternalism in Early Hospitals for Children: San Francisco and Denver, 1875–1915," *Journal of the West* 32 (1993): 61–69.

71. Summarized in Faber and McIntosh, *History of the American Pediatrics Society*, p. 130.

72. Like Dr. Martha Wolstein, Dr. Dorothy Anderson (1901–1963) was a pediatrician at Babies Hospital in New York. In a "classic paper" delivered to the APS in 1938, she used clinical and pathological analysis to describe cystic fibrosis of the pancreas. Wolstein, who was of an earlier generation, presented her first paper to the APS in 1922 and was elected a member in 1930. See also Cone, *History of American Pediatrics,* pp. 218, 224 n. 2; Faber and McIntosh, *A History of the American Pediatrics Society,* p. 190; D. H. Anderson, "Cystic Fibrosis of the Pancreas and Its Relation to Celiac Disease: A Clinical and Pathological Study," *American Journal of Diseases in Children* 56 (1938): 344.

73. Cf. Gerhard Sonnert, with Gerald Holton, *Who Succeeds in Science? The Gender Dimension* (New Brunswick, N.J.: Rutgers University Press, 1995), pp. 11–13; Penina Migdal Glazer and Miriam Slater, *Unequal Colleagues: The Entrance of Women into the Professions, 1890–1940* (New Brunswick, N.J.: Rutgers University Press, 1987), p. 14.

74. The following discussion of Drs. Martha Eliot (1891–1978) and Ethel Dunham (1883–1969) is drawn primarily from the following sources: archival materials in MME/SL, especially Oral History Project Interview, OH-1/Eliot, pp. 33–46, 61–74, 403, 435–437, and correspondence, Boxes 2, 3, 18 and 19; Marion Hunt, "Extraordinarily Interesting," pp. 159–170; William M. Schmidt, "Dunham, Ethel Collins," in Barbara Sicherman et al., eds. *Notable American Women: The Modern Period* (Cambridge, Mass.: Harvard University Press, 1980), pp. 212–213; Grover F. Powers, "Edwards A. Park, Yale Professor 1921–1927," *Journal of Pediatrics* 41 (1952): 651–659.

75. Pearson, *The Centennial History,* p. 137; Powers, "Edwards A. Park," p. 658. Also see letter from Martha Eliot to her mother, November 9, 1921, MME/SL. "Ned" Park's letters to Eliot over the years blended irreverence, professional politics, and great warmth. Frequently he signed off, "Give my love to Ethel."

76. Schmidt, "Dunham, Ethel Collins," p. 212; Hunt, "Extraordinarily Interesting," pp. 159–165, esp. p. 162; Powers, "Edwards A. Park," p. 656–657.

77. Faber and McIntosh, *History of the American Pediatrics Society,* pp. 173, 182; Schmidt, "Dunham, Ethel Collins," p. 212.

78. M. Eliot, "The Control of Rickets," *JAMA* 85 (1925): 656–663; Hunt, "Extraordinarily Interesting," p. 160.

79. When Secretary of Labor Frances Perkins was looking for Abbott's successor, Park and half a dozen other pediatrics leaders began concerted lobbying on behalf of Eliot. Oral History Project Interview, OH-1/Eliot, pp. 73, 74, 343–351; letter from committee of the APS to Secretary of Labor Frances Perkins, June 19, 1934, Box 3, folder 152; letter from Katherine Lenroot to Louis J. Covotsos, [November?] 1974, Box 3, folder 153, all in MME/SL. Cf. Schmidt, "Dunham, Ethel Collins," pp. 212–213; Hunt, "Extraordinarily Interesting," pp. 168–169.

80. For information about Dr. Jackson (1895–1977), I have relied on: Sara Lee Silberman, "Pioneering in Family-Centered Maternity and Infant Care: Edith B. Jackson and the Yale Rooming-In Research Project," *Bulletin of the History of Medicine* 64, no. 1 (1990): 262–287; Edith B. Jackson, "The Development of Rooming-In at Yale," *Yale Journal of Biology and Medicine* 25 (1953): 484–494; Morris A. Wessel and Frederic M. Blodgett, "Edith B. Jackson, M.D., and Yale Pediatrics," *Connecticut Medicine* 26, no. 7 (1962): 438 ff., reprint courtesy of the Yale Medical School

Archives; and the Edith Banfield Jackson Papers, MC 304, Finders' Aid, Schlesinger Library, Radcliffe College, Cambridge, Massachusetts, pp. 1–3. My thanks (again) to Toby Appel for making these materials available to me.

81. Jackson, "The Development of Rooming-In," pp. 487, 491; Silberman, "Pioneering," pp. 270, 274.

82. Jackson, "The Development of Rooming-In," p. 487.

83. Silberman, "Pioneering," p. 282.

84. Ibid., p. 285. For an excellent discussion of hospital-based birthing practices during the 1950s, see Judith W. Leavitt, *Brought to Bed: Childbearing in America* (New York: Oxford University Press, 1986).

85. On the career of Helen Taussig (1898–1986), see Helen Taussig, "Little Choice and a Stimulating Environment," *JAMWA* 36, no. 2 (1981): 43–44. I have relied primarily on the account given in Sherwin B. Nuland, *Doctors: The Biography of Medicine* (New York: Alfred A. Knopf, 1988), pp. 422–456. My thanks to Dr. Chester Burns for background on the term "heart station."

86. Nuland, *Doctors*, pp. 435–436; Pearson, *The Centennial History*, p. 137.

87. According to Dr. Mary Ellen Avery, formerly physician-in-chief of Boston Children's Hospital and a student and friend of Dr. Taussig, Taussig was dyslexic as well as partially deaf. She compensated by lipreading and by augmenting her stethoscope with an "electronic amplifier." Interview with Dr. Mary Ellen Avery, March 27, 1997, Children's Hospital, Boston.

88. Nuland, *Doctors*, pp. 436–444. Maude Abbott, who held a nontenured position on the medical faculty at McGill, was the author of *Atlas of Congenital Cardiac Disease* (1936) and the acknowledged international expert on the subject. For the pathbreaking career of Vivian Thomas, see Vivian Thomas, *Pioneering Research in Surgical Shock and Cardiovascular Surgery* (Philadelphia: University of Pennsylvania Press, 1985), cited in Nuland, *Doctors*, p. 508.

89. Nuland, *Doctors*, pp. 447–452.

90. On Virginia Apgar (1909–1974), see Robert J. Waldinger, "Apgar, Virginia," in Sicherman et al., eds., *Notable American Women*, pp. 27–28.

91. Cone, *History of American Pediatrics*, p. 238; L. Joseph Butterfield, "Virginia Apgar and the Apgar Score: Kudos and a Correction," *JAMA* 277, no. 22 (1997): 1762.

92. Dorothy Reed Mendenhall, *Unpublished Memoir* (ca. 1934), excerpted in Jill Ker Conway, ed., *Written by Herself: Autobiographies of American Women: An Anthology* (New York: Random House, 1992), pp. 188, 191–193.

7. Resisting the "Feminine Mystique," 1938–1968

1. Betty Friedan, *The Feminine Mystique*, (1963: reprint, New York: Dell Publishing Co., 1974), pp. 11–27; epigraph is quoted by Friedan on p. 23. On the postwar "adjustment" of professional women to the necessity and desirability of leaving the workforce for home and child rearing, see Margaret Rossiter, *Women Scientists in America: Before Affirmative Action, 1940–1972* (Baltimore: Johns Hopkins University Press, 1995), pp. 27–49.

2. In 1938 AMWA established the Committee on Aid to Medical Women in Distress, directed by Dr. Rita Finkler of New Jersey. It secured eleven affidavits to

enable women physicians and their families to emigrate to America and contributed a small amount of money to help them relocate. Nevertheless, by Atina Grossman's tentative estimate, only about one-third of the refugees were able to practice medicine in this country. Atina Grossmann, "New Women in Exile: German Women Doctors and the Emigration," in Sibylle Quack, ed., *Between Sorrow and Strength: Women Refugees of the Nazi Period*, (Washington, D.C.: German Historical Institute, and New York: Cambridge University Press, 1994), pp. 215–238; *Women in Medicine* 1939 (January): 20, (October): 21; "Refugee Report," June 1940, typescript, Box 11, folder 16, AMWA/NYH.

3. *Bulletin of the Medical Women's National Association* no. 37, (1932): 16. The issue was first revived by Bertha Van Hoosen's Committee on Medical Opportunities for Women, ibid., no. 36 (1932): 18; letter of August 2, 1935, Box 5, folder 12, and minutes of Board of Directors meeting, December 1, 1935, Box 5, folder 13, AMWA/NYH. On Maffett, see Rossiter, *Women Scientists . . . Before Affirmative Action*, pp. 20–24.

4. *Women in Medicine* 1939 (January): 20, (February): 18–20, (October): 23, 24. Lt. Col. Clara Raven, "Achievements of Women in Medicine, Past and Present: Women in the Medical Corps of the Army," *Military Medicine* 125 (1960): 105–111, esp. p. 108; "Resolution to AMA House of Delegates," June 10, 1940, typescript, Box 5, folder 15, AMWA/NYH; James G. Burrow, *AMA: Voice of American Medicine* (Baltimore: Johns Hopkins University Press, 1963), pp. 282–285.

5. Box 5, folders 15, 17, AMWA/NYH.

6. Burrow, *AMA*, pp. 282–285; Minutes, midwinter board meeting, December 7, 1941, typescript, Box 1, folder 17, AMWA/NYH; Karen Berger Morello, *The Invisible Bar: The Woman Lawyer in America* (Boston: Beacon Press, 1986), pp. 134–135; Susan M. Hartmann, "Kenyon, Dorothy," in Barbara Sicherman and Carol Hurd Green, eds., *Notable American Women: The Modern Period* (Cambridge, Mass.: Harvard University Press, 1980), pp. 395–397.

7. Dorothy Kenyon, "The Law of the Land," *Women in Medicine* 1943 (January): 7–10, 17–21; letter from Lt. Col. Francis M. Fitts to Emily Dunning Barringer, February 10, 1942, Box 5, folder 28, AMWA/NYH.

8. Kenyon, "Law of the Land"; letter to the editor from Dr. E. C. McCullough, Lt. Col., U.S. Army M.C., Retd., typescript, n.d., Box 5, folder 27, AMWA/NYH.

9. In 1943 the AMA also withdrew its opposition to admitting women into the Medical Reserves. Rossiter, *Women Scientists . . . Before Affirmative Action*, pp. 12–13; Burrow, *AMA*, pp. 282–285.

10. *Women in Medicine* 1943 (July): 12; *House Committee on Military Affairs, Appointment of Female Physicians and Surgeons in the Medical Corps of the Army and Navy: Hearings before Subcommittee No. 3 of the Committee on Military Affairs, House of Representatives, on H.R. 824 . . . and H.R. 1857*, 78th Cong., 1st sess., March 10, 11, and 18, 1943 (Washington, D.C.: U.S. Government Printing Office, 1943), pp. 1–101.

11. *House Committee on Military Affairs, Appointment of Female Physicians and Surgeons in the Medical Corps of the Army and Navy*, esp. 4–7, 15–16, 31, 44, 46. *Women in Medicine* 1943 (April): 9, (July): 15, (October): 15–16. Emily Dunning Barringer, *Bowery to Bellevue* (New York: W. W. Norton, 1950), pp. 240–241.

12. *Women in Medicine* 1943 (July): 15 ff.; ibid. 1946 (January): 21; Carol Lopate,

Women in Medicine (Baltimore: Johns Hopkins University Press, 1968), p. 193, Appendix 1; Mary Roth Walsh, *Doctors Wanted: No Women Need Apply* (New Haven, Conn.: Yale University Press, 1977), p. 186, Table 5.

13. *War Demands for Trained Personnel,* proceedings of a conference held at the Mayflower Hotel, Washington, D.C., March 20, 21, 1942 (New London, Conn.: Institute of Women's Professional Relations, Connecticut College, 1942), pp. 52–57.

14. Rossiter, *Women Scientists . . . Before Affirmative Action,* pp. 52–55.

15. Surgeon General's Consultant Group on Medical Education, *Physicians for a Growing America* (Washington, D.C.: U.S. Government Printing Office, 1959); Paul Starr, *The Social Transformation of American Medicine* (New York: Basic Books, 1982), p. 364.

16. Starr, *Social Transformation,* p. 364; *Physicians for a Growing America,* p. 13.

17. Ellen S. More, "The American Medical Women's Association and the Role of the Woman Physician, 1915–1990," *JAMWA* 45, no. 5 (1990): 175.

18. R. Wheeler, L. Candib, and M. Martin, "Part-Time Doctors: Reduced Working Hours for Primary Care Physicians," ibid., 45, no. 1 (1990): 47–54.

19. Anson Rabinbach, *The Human Motor: Energy, Fatigue, and the Origins of Modernity* (New York: Basic Books, 1990), pp. 3–4, 70–71.

20. Ellen More, "Doctors or Professors? Late Victorian Physicians and the Culture(s) of Professionalism," *Canadian Review of American Studies* 23 (1993): 125–148.

21. Samuel Haber, *Efficiency and Uplift: Scientific Management in the Progressive Era* (Chicago: University of Chicago Press, 1966); Susan Reverby, *Ordered to Care: The Dilemma of American Nursing, 1850–1945* (New York: Cambridge University Press, 1987), pp. 143–158.

22. "The Bed-Day," *American Medicine* 1902 (April): 626.

23. Edward T. Mormon, "Introduction," in *Efficiency, Scientific Management, and Hospital Standardization* (New York: Garland, 1989).

24. Quoted in J. Feinglass and J. W. Salmon, "Corporatization of Medicine: The Use of Medical Management Information Systems to Increase the Clinical Productivity of Physicians," *International Journal of Health Services* 20 (1990): 233–252.

25. Rosemary Stevens, *In Sickness and in Wealth: American Hospitals in the Twentieth Century* (New York: Basic Books, 1989), pp. 71–91.

26. William Pepper, *Higher Medical Education: The True Interest of the Public and of the Profession* (Philadelphia: Lippincott, 1884), Appendix 1, Table 3, pp. 46–47.

27. Kenneth M. Ludmerer, *Learning to Heal: The Development of American Medical Education* (New York: Basic Books, 1985); Gerald E. Markowitz and David Rosner, "Doctors in Crisis: Medical Education and Medical Reform during the Progressive Era, 1895–1915," in Susan Reverby and David Rosner, eds., *Sickness and Health in America,* (Philadelphia: Temple University Press, 1979), pp. 185–205.

28. "Abstract of the Final Report of the Commission on Medical Education," *JAMA* 99 (1982): 2206–2207.

29. John Z. Bowers and V. W. Lippard, "Introduction," in V. W. Lippard and E. F. Purcell, eds., *Case Histories of Ten New Medical Schools* (New York: The Josiah Macy, Jr. Foundation, 1972), pp. 7–9; Paul Starr, *Social Transformation,* pp. 359–361; Stevens, *In Sickness,* pp. 216–218.

30. Russell C. Maulitz and Daniel M. Fox, "The Bureau of Health Professions:

An Analytical History," unpublished manuscript, esp. pp. 13–20. My thanks to Dr. Maulitz for sending me this interesting study.

31. Roscoe A. Dykman and John M. Stalnaker, "Survey of Women Physicians Graduating from Medical School, 1925–1940," *Journal of Medical Education* 32 (1957): 3–38; L. M. Powers, H. Wisenfelder, and R. C. Parmelle, "Preliminary Report: Practice Patterns of Women and Men Physicians," October 14, 1966, pp. 5–8, Box B: "New" folder, "Research Committee figures," AMWA/NYH.

32. More, "The American Medical Women's Association," pp. 165–180.

33. Dykman and Stalnaker, "Survey of Women Physicians," pp. 18–24, 32, 40, quotation on p. 32. Mary Roth Walsh, *Doctors Wanted: No Women Need Apply* (New Haven, Conn.: Yale University Press, 1977), p. 253; Barbara H. Kehrer, "Factors Affecting the Income Levels of Men and Women Physicians: An Exploratory Analysis," *Journal of Human Resources* 11 (1976): 526–545, esp. 540–541.

34. "Minutes of the Midyear Meeting of the Board of Directors, American Medical Women's Association," *JAMWA* 15 (1960): 373–388; "Special Membership Meeting," ibid., 389–395.

35. AMWA membership also declined again after 1943. Box A: "New," loose folder, "Finance 1957," and loose papers, "Membership Report 1951–59," AMWA/NYH.

36. William H. Chafe, *The Paradox of Change: American Women in the 20th Century* (New York: Oxford University Press, 1991), pp. 154–161. See also the excellent discussion in Rossiter, *Women Scientists . . . Before Affirmative Action*, pp. 27, 34, 40–41.

37. "Program, November 1957," Box A, "New," AMWA/NYH. Allison Hepler argues, however, that women physicians took seriously the occupational health risks of housework. "The Pathology and Hygiene of Housework," paper given at the American Association for the History of Medicine, May 10, 1996, Buffalo, New York.

38. A. J. Spear, "Legislative Report," Box A: "New," loose folder, "Finance 1958," AMWA/NYH.

39. Wolfle is quoted approvingly in Jessie Bernard, *Academic Women* (New York: New American Library, 1964; 2nd printing, 1974), p. 49.

40. "Membership Report 1959," typescript in loose papers, revised constitution, 1958, Box A, "New," AMWA/NYH. Interview with Dr. Bertha Offenbach, Newton, Massachusetts, June 28, 1989; "Minutes of Midyear [board] Meeting," 15 (1960): 373–389; "Special Membership Meeting," ibid., 389–395.

41. *Physicians for a Growing America*, p. 24, emphasis added.

42. Bowers and Lippard, "Introduction," p. 7. Data on female enrollment in medical schools, 1965–1985, courtesy of the AAMC, was funded by the Office of the Vice President for Academic Affairs, UTMB, formerly Dr. George Bryan, to whom I am most grateful. Also see Lopate, *Women in Medicine*, Appendix 1, p. 193; *Minorities and Women in the Health Fields* (Washington, D.C.: U.S. Department of Health and Human Services, 1984), p. 104; John Z. Bowers, "Wife, Mother, Physician," *JAMWA* 22, no. 10 (1967): 760–764; John E. Chapman, "Toward a Common Denominator in the Equation toward a Medical Education," ibid. 24, no. 7 (1969): 561–565, esp. p. 565.

43. Alfred P. Ingegno, "The Case Against Female M.D.s," Part 4, *Medical Economics* 38 (December 4, 1961): 41–48, quotation on p. 48. Ingegno did concede that "at present, I think women doctors are a luxury we can afford."

44. J. Robert Buchanan, "The Selection of Medical Students," *JAMWA* 24, no. 7 (1969): 555–560; Lopate, *Women in Medicine*, p. 111.

45. Joan Cassell, *The Woman in the Surgeon's Body* (Cambridge, Mass.: Harvard University Press, 1998), p. 140.

46. Interview with Dr. Sally Abston, UTMB, Galveston, October 31, 1991. I am most grateful to Dr. Abston for agreeing to be interviewed.

47. Interview with Dr. Ruth Lawrence, University of Rochester School of Medicine and Dentistry, Rochester, New York, May 27 and 29, 1986. All subsequent quotations from Dr. Lawrence are taken from these interviews unless otherwise noted. My thanks to Dr. Lawrence for agreeing to be interviewed. Ruth A. Lawrence, *Breast-feeding: A Guide for the Medical Profession*, 4th ed. (St. Louis: Mosby, 1994).

48. Interview with Dr. Ruth Lawrence, May 29, 1986. Dr. Lawrence, interestingly, was interviewed twenty years earlier by Carol Lopate, perhaps because Lawrence had contributed a study of underemployment among women doctors in Rochester for the first Macy Foundation conference. See Lopate, *Women in Medicine*, pp. 180–181.

49. Biographical information was compiled from the author's interview with Dr. Mary Ellen Avery, Boston Children's Hospital, March 27, 1997, and from Avery's curriculum vitae (December 1996). I am gratefully indebted to Dr. Avery for the gracious donation of her time and copies of some of her personal files for use in this project.

50. M. E. Avery and J. Mead, "Surface Properties in Relation to Atelectasis and Hyaline Membrane Disease," *American Journal of Diseases of Children* 97 (1959): 517–523; Janice S. Kahn, "Hyaline Membrane Disease: A Sociological Study of Discovery and Change," Ph.D. diss., Boston University, 1990, esp. chap. 6, courtesy of Dr. Avery.

51. Judith Lorber, *Women Physicians: Careers, Status, and Power* (New York: Tavistock, 1984), pp. 49–63.

52. AMWA, "Medical Womanpower," transcript of panel discussion at AMWA summer convention, June 24, 1962, pp. 10, 11.

53. Ibid., pp. 11, 12.

54. Ibid., pp. 64, 68.

55. Curriculum vitae of Dr. Rosa Lee Nemir, personal communication, July 7, 1989.

56. The possibility that child care responsibilities could be shared by both working parents was rarely broached at the time. Lopate, *Women in Medicine*, pp. 162–163.

57. Memo from Dr. Rosa Lee Nemir to AMWA executive board, January 28, 1964, Box B: "New," folder: "AMA/AMWA Liaison," AMWA/NYH.

58. Minutes of meeting on survey of women physicians, July 17, 1965, Box A: "New"; agenda, meeting of advisory committee on "Survey of Women Physicians Graduating from Medical School 1930–1960," August 28, 1965, Box B: "New"; folder, "Research Committee—Figures; report on combined research study on "Survey of Women Physicians Graduating from Medical School 1930–1960," June 28, 1966, Box B: "New"; all in AMWA/NYH.

59. L. Powers, H. Wiesenfelder, and R. C. Parmalee, *Preliminary Report: Practice*

Patterns of Women and Men Physicians, October 14, 1966, pp. 5–8, Box B: "New," folder, "Research Committee figures," AMWA/NYH.

60. Telephone interview with Ms. Maxine Bleich, president of Ventures in Education and formerly vice president of the Josiah Macy, Jr. Foundation during Dr. Bowers's tenure, November 13, 1990. Bleich emphasized that Bowers was more committed to increasing the role of racial and ethnic minorities than women because, for the most part, the former were economically disadvantaged. I am grateful for Ms. Bleich's assistance.

61. Donna Haraway, *Primate Visions: Gender, Race and Nature in the World of Modern Science* (New York: Routledge, 1989), p. 103; "Josiah Macy, Jr. Foundation," in *The Foundation Center Sourcebook Profile* (1988), n.p.

62. The foundation sponsored two conferences in 1968, one titled Fuller Utilization of the Woman Physician (cosponsored by the Women's Bureau), and one called The Future of Women in Medicine. The last of this series, the Macy Conference on Women in Medicine, was held in 1976 and resulted in a published report by Dr. Elizabeth McAnarney. See *A Fifty Year Review and Report for the Year 1980* (New York: Macy Foundation, 1980), pp. 36–37; and annual reports for 1966, 1967, 1968, and 1976, Josiah Macy, Jr. Foundation. I am indebted to Thomas H. Meikle, president, the Josiah Macy, Jr. Foundation, for permission to consult these documents, and to Carlos M. Monteagudo, former administrative assistant to the president, for making them available.

63. John Z. Bowers, "Women in Medicine: An International Study," *New England Journal of Medicine* 275, no. 7 (1966): 362–365. I am indebted to the archivist of the Rockefeller Foundation for access to materials in the John Zimmerman Bowers Collection, Record Group 1.2, Series 103, Box 1, folder 1, and Series 200A, Box 120, folder 1059, and Box 160, folder 1452, Rockefeller Foundation Archive, North Tarrytown, New York.

64. Lopate, *Women in Medicine*, p. 24. J. Z. Bowers, "Forward," in ibid., pp. 5–8. M. Bunting, "Introduction," in ibid.

65. Quoted in ibid., pp. 20, 21.

66. AMWA second workshop, Washington, D.C., Thursday, November 3, 1966, pp. 2, 16, 17, Box A: "New"; letter from Peter S. Bing of President Johnson's National Advisory Commission on Health Manpower to President Margaret J. Schneider, October 26, 1966, Box A: "New," AMWA/NYH.

67. Lopate, *Women in Medicine*, pp. 142–143.

68. AMWA second workshop, pp. 19–20, 68–98, 174–177; "Membership Report, AMWA Meetings—1960s," Box 115C, folder "Interim-Sausalito—1969," AMWA/NYH.

69. In 1963 and 1964, the American Women's Hospitals donated $500 to PPFA to photocopy its *Manual of Contraceptive Practice* for use by women physicians overseas. Dr. Katharine Brooks Merrifield to Mrs. F. Workman, PPFA, March 21, 1959; Mary Steichen Calderone to Ruth Hartgraves, May 17, 1963, both in Box 10, folder 179, MSC/SL.

70. In 1994 SIECUS changed its name to the Sexuality Information and Education Council of the United States. I am grateful to Francesca Calderone-Steichen for informing me of the change.

71. Mary S. Calderone, "Sexual Health and Family Planning," 1968 Bronfman

lecture to the American Public Health Association cited in Mary Steichen Calderone, autobiographical essay, typescript, pp. 1–13, in personal communication to the author, May 25, 1984; Virginia Woolf, "Notes for a 1931 Speech before the London and National Society for Women's Service," quoted in Hermione Lee's excellent *Virginia Woolf* (New York: Alfred A. Knopf, 1997), pp. 590–591.

72. I have relied on the following sources for biographical material on Mary Steichen Calderone: interview with Mary S. Calderone, February 2, 1984, Rochester, New York; Francesca Calderone-Steichen, personal communication, December 5, 1997; David Mace, "Mary Steichen Calderone: Interpreter of Human Sexuality," in Leonard S. Kenworthy, ed., *Living in the Light: Some Quaker Pioneers of the 20th Century*, vol. 1 (Kennett Square, Pa.: Friends General Conference and Quaker Publications, 1984), pp. 75–87; Mary Steichen Calderone, "Mary Steichen Calderone," in Lynn Gilbert and Gaylen Moore, ed., *Particular Passions* (New York: Clarkson N. Potter, 1981), pp. 255–263; Mary Steichen Calderone, typescript abstract for *Encyclopedia of American Woman*, pp. 1–9, in Carton 1, folder 1, MSC/SL; SIECUS, "Biographical Data for Mary Steichen Calderone, M.D., M.P.H., 1975," ibid.; curriculum vitae for Mary S. Calderone, 1983–84, ibid.; "Calderone, Mary S(teichen)," *Current Biography* 1967 (November): 5–8, ibid.; letter from Mary S. Calderone to Dr. George W. Corner, Sr., August 17, 1981, ibid. Also see the very fine biography, Penelope Niven, *Steichen* (New York: Clarkson N. Potter, 1997). Many thanks to Francesca Calderone-Steichen, daughter of Mary Steichen Calderone, for her comments and assistance.

73. Obituary of Dr. Frank A. Calderone, *New York Times*, February 24, 1987, p. A25.

74. Interview with Mary S. Calderone, February 2, 1984; letter to Mary S. Calderone from the Valley Stream, New York, PTA, February 15, 1946, Box 12, folder 206, MSC/SL; typescript of talk on "sex education, especially of the young child," undated but annotated as "probably between 1948–50," Box 13, folder 222, MSC/SL.

75. McCann, *Birth Control Politics*, pp. 62, 76–77; Ray and Gosling, "American Physicians," pp. 401–402.

76. In 1946 83 percent of all clinic patients seeking contraceptive services were seen in "extramural" clinics of the kind sponsored by PPFA; only 17 percent were served by clinics located in hospitals or public health departments. Many of the physicians at these extramural clinics were women. See also Ray and Gosling, "American Physicians," pp. 403, 405; quotation on p. 403.

77. Quoted in Mace, "Interpreter of Human Sexuality," p. 79; Calderone, *Particular Passions*, p. 257.

78. Calderone, *Particular Passions*, p. 257.

79. Interview with Mary Steichen Calderone; Mace, "Interpreter of Human Sexuality," p. 79.

80. Participating were Alfred Kinsey, noted sex researcher; Dr. John Rock, a Catholic physician and developer of the first birth control pills; Dr. Hilla Sheriff, director of the Maternal and Child Division of the South Carolina Board of Health; and Dr. Alan Guttmacher, chief of obstetrics-gynecology at Mt. Sinai Hospital in New York City. Guttmacher became president of PPFA in 1962. Mary Steichen Calderone, ed., *Abortion in the United States* (New York: Hoeber-Harper, 1958), pp. 6–13; letter from Mary Steichen Calderone to Dr. Helen Wallace, February 3,

1959, Box 3, folder 43, MSC/SL, in which she labels abortion a "disease of society." For PPFA's controlled clinical trials, see especially Box 5, folders 69 and 70, MSC/SL; also see Marcia L. Meldrum, "'Simple Methods' and 'Determined Contraceptors': The Statistical Evaluation of Fertility Control," *Bulletin of the History of Medicine* 70, no. 2 (1996): 266–295.

81. "Confidential: Basis for a Proposed Resolution," March 1958; letter from Mary Steichen Calderone, Medical Director, PPFA, to William M. Schmidt, Harvard University School of Public Health, July 27, 1959; letter from Martha May Eliot to J. W. R. Norton, October 14, 1958; letter from Mary Steichen Calderone to Mrs. Nathan Elias (Dr. Leona Baumgartner), June 2, 1959; letter from Mary Steichen Calderone to Dr. Helen Wallace, February 3, 1959, all in Box 3, folder 43, MSC/SL. Also see "APHA Committees, 1958–1959," *American Journal of Public Health* 49, no. 3 (1959): 376–403; "Association News," *American Journal of Public Health* 49, no. 6 (1959): 814–815; "Policy Statements," *American Journal of Public Health* 49, no. 12 (1959): 1702–1704.

82. "Board of Trustees: Committees of the Board," *JAMA* 194, no. 4 (1965): 145; "The Control of Fertility," ibid., 230–238.

83. Friedan, *The Feminine Mystique*, pp. 17–21.

84. Letter from Mary Steichen Calderone to Alan Guttmacher, March 12, 1963; letter from Alan Guttmacher to Mary Steichen Calderone, March 18, 1963; letter from Mary Steichen Calderone to Mrs. George Gillespie, June 27, 1963; Mary Steichen Calderone to Alan Guttmacher and John Cotton, memorandum of resignation, February 18, 1964, all in Box 12, folder 205, MSC/SL.

85. "Calderone, Mary S(teichen)," *Current Biography*, pp. 5–8; John G. Rodgers, "Dr. Mary Calderone: Sex Educator," *Parade*, June 18, 1967, both in Box 1, folder 1, MSC/SL. Mary S. Calderone, "Above and Beyond Politics: The Sexual Socialization of Children," in Carole S. Vance, ed., *Pleasure and Danger: Exploring Female Sexuality* (Boston: Routledge and Kegan Paul, 1984), pp. 131–137.

86. "The Grandmother of Modern Sex Education," *Newsday*, February 22, 1966, Box 17, folder 286, MSC/SL; "Sex Education: Assault on American Youth," *American Education Lobby*, n.d., Box 14, folder 233, MSC/SL; "SIECUS," *Dan Smoot Report* 15, no. 11 (1969): 1; "Dr. Mary Calderone Latest Target of John Birch Society," *Congressional Record: Senate*, September 19, 1969, pp. S11013–S11017, in Box 14, folder 234, MSC/SL; "Cradle-to-Grave Intimacy," *Time*, September 7, 1981, p. 69; Complaint No. 48–81, National News Council, October 5, 1981, Acc. No. 82-M129, Box 1, folder 8, MSC/SL.

87. Membership report, Box 115C: "AMWA Meetings—1960s," folder "Interim—Sausalito—1969," AMWA/NYH. Lopate, *Women in Medicine*, p. 17.

88. Lopate, *Women in Medicine*, pp. 19–21.

89. "Annual Meeting," *JAMWA* 23 (1968): 56–87.

90. Letter from Louis T. Milic to Dr. Camille Mermod, July 17, 1970, Box B: "New," loose papers, AMWA/NYH.

91. Alice D. Chenoweth, "President's Message," *JAMWA* 23 (1968): 283, 284.

92. "Annual Meeting," *JAMWA* 24, no. 2 (1969).Cf. "Report on the 79th Annual Meeting of the AAMC," ibid., no. 4 (1968): 178–181; "1960–1972 Membership Figures," Box B: "New," AMWA/NYH; *Minorities and Women in the Health Fields* (Washington, D.C.: U.S. Department of Health and Human Services, 1984), Table 53, p. 120.

8. Medicine and the New Women's Movement

1. For the epigraphs, see E. Grey Dimond, "I: The Future of Women Physicians," *JAMA* 249, no. 2 (1983): 207–208; Marilyn Heins, "II: Medicine and Motherhood," ibid., 209–210. Mary Roth Walsh, *Doctors Wanted: No Women Need Apply* (New Haven, Conn.: Yale University Press, 1977); Walsh's subtitle, *Sexual Barriers in the Medical Profession, 1835–1973*, alerts her readers that this is a book straightforwardly about discrimination against women physicians. See also Carol Lopate, *Women in Medicine* (Baltimore: Johns Hopkins University Press, 1968); Lopate, writing at the behest of the Macy Foundation (see Chapter 7), was well aware of discrimination. But at the time, as she also knew, women physicians had not yet learned to "demand all that they might in the work situation [because] they often [felt] insecure about their bargaining power" (p. 186).

2. "Statement of Frances S. Norris, M.D., before the Special Subcommittee on Education Re: Sec. 805, H.R. 16098, concerning Discrimination against Women in Medical Schools," pp. 1–4; letter from WEAL to the Honorable Elliott Richardson, Secretary, Department of Health, Education and Welfare, October 5, 1970, both in WEAL Fund Collection, Carton 4, folder 157, Schlesinger Library, Radcliffe College, Cambridge, Massachusetts; *Women's Equity Action League, et al. v. Lauro F. Cavazos, Secretary of Education*, 279 U.S. App. D.C. 34; 879 f.2d 880; 1989 U.S. App. LEXIS 9729. My thanks to Kayhan Parsi for finding the previous citation. Cf. Margaret W. Rossiter, *Women Scientists in America: Before Affirmative Action, 1940–1972* (Baltimore: Johns Hopkins University Press, 1995), pp. 297, 370–374; Sheila Tobias, *Faces of Feminism: An Activist's Reflections on the Women's Movement* (Boulder, Colo.: Westview Press, 1997), pp. 71–92, 111, 122–125.

3. "Statement of Frances S. Norris," pp. 1–4.

4. Title IX stipulates that "no person in the United States shall, on the basis of sex, be excluded from participation in, be denied the benefits of, or be subjected to discrimination under any education program or activity receiving federal financial assistance." Margaret A. Campbell, M.D. [pseud.], *"Why Would a Girl Go into Medicine?"* (Old Westbury, N.Y.: Feminist Press, 1973), pp. 5, 19–20, 100. Sara M. Evans, *Born for Liberty: A History of Women in America* (New York: Free Press, 1989), pp. 290, 291; Rossiter, *Women Scientists . . . Before Affirmative Action*, pp. 374–376, 525 n. 31. Despite legislative progress, WEAL pursued its case and in 1989 was granted standing to bring it forward as a class action suit. But on June 26, 1990, the U.S. Court of Appeals decided against WEAL. By this time, according to a decision written by Ruth Bader Ginsburg, the original antidiscrimination suit had outgrown its mandate, having "cast the district court [site of the original lawsuit] as nationwide overseer or pacer of procedures government agencies use to enforce [educational] civil rights." The court called for the plaintiffs to begin addressing this issue through individual rather than class action suits.

5. In 1969 and 1970, four medical schools (SUNY Albany, Emory, Yale, and Loyola) still expressed a preference for male applicants. At the same time, 4.5 percent of the AAMC-listed schools—at the height of the civil rights movement—admitted to a preference for whites. Quotation from Campbell, *"Why Would a Girl,"* p. 9. Mary Roth Walsh, "Women in Medicine since Flexner," *New York State Journal of Medicine* (June 1990): 302–308, esp. p. 306; Carol Lopate, *Women in Medicine*, pp. 69–74; William G. Rothstein, *American Medical Schools and the Practice*

of Medicine (New York: Oxford University Press, 1987), p. 290, makes the inference of no "overt discrimination."

6. Walsh, "Women in Medicine," p. 306.

7. Robert F. Jones, "Women and the MCAT: An Overview of Research in Progress," in *AAMC Women in Medicine Research Symposium Abstracts*, October 28, 1984, Chicago. Reprint courtesy of Dr. Mary F. Schottstaedt, UTMB Women's liaison officer for 1984–85. A comparison of MCAT summary statistics for 1993 shows a narrowing of those differences. See Leah J. Dickstein, Daniel P. Dickstein, and Carol C. Nadelson, "The Status of Women Physicians in the Workforce," typescript prepared for the Council on Graduate Medical Education, April 8, 1994, p. 32.

8. The author was a member of the admissions committee at UTMB, Galveston. See also Robert C. Davidson and Ernest L. Lewis, "Affirmative Action and Other Special Consideration Admissions at the University of California, Davis, School of Medicine," *JAMA* 278, no. 14 (1997): 1153–1158; Darlene L. Shaw et al., "Influence of Medical School Applicants' Demographic and Cognitive Characteristics on Interviewers' Ratings of Noncognitive Traits," *Academic Medicine* 70, no. 6 (1995): 532–536; Carol L. Elam and Mitzi M. S. Johnson, "An Analysis of Admissions Committee Voting Patterns," ibid. 72, no. 10 (1997): S72–S75.

9. Wartime increases in the percentage of women accepted to medical school illustrate this phenomenon. For a discussion of the extent and variability of discrimination against women applicants to medical schools in the early twentieth century, see Regina Morantz-Sanchez, *Sympathy and Science: Women Physicians in American Medicine* (New York: Oxford University Press, 1985), pp. 249–254.

10. Janet Bickel and Phyllis R. Kopriva, "A Statistical Perspective on Gender in Medicine," *JAMWA* 48, no. 5 (1993): 141–144; Marilyn Heins and Jane Thomas, "Women Medical Students: A New Appraisal," ibid. 34, no. 11 (1979): 408–415; Walsh, "Women in Medicine," p. 306.

11. Rothstein, *American Medical Schools*, pp. 282–288; Janet Bickel et al., *Women in U.S. Academic Medicine Statistics, 1996* (Washington, D.C.: AAMC, 1996), Table 4. Between 1968 and 1974, according to Allen Singer of the AAMC, these increases were especially notable in that they surmounted unusual pressures on admissions from a temporary glut of Vietnam-era male applicants. Allen Singer, "The Effect of the Vietnam War on Numbers of Medical School Applicants," *Academic Medicine* 64 (1989): 567–573. "Women in Surgery—A Statistical Profile," *AWS Connections* 4 (Winter 1998): 4; Joan Cassell, *The Woman in the Surgeon's Body* (Cambridge, Mass.: Harvard University Press, 1998), pp. 67–76.

12. Beulah K. Cypress, *Characteristics of Visits to Female and Male Physicians: The National Ambulatory Medical Care Survey*, series 13, no. 49 (Washington, D.C.: U.S. Department of Health and Human Services, 1980), pp. 1–40, quotation on p. 14, Table S.

13. Carol Nadelson and Malkah T. Notman, "The Woman Physician," *Journal of Medical Education* 47 (March 1972): 176–183, quotation on p. 180; Heins and Thomas, "Women Medical Students," p. 408. For a student's perspective, see Perri Klass, *A Not Entirely Benign Procedure* (New York: G. P. Putnam's Sons, 1987).

14. See the discussion of Dr. Mary Howell, associate dean for student affairs, Harvard Medical School, later in this chapter.

15. Bickel began her involvement with medical education as the "all-purpose"

admissions/financial aid/student affairs officer at the newly opened Brown University Medical School. She was at Brown until 1976, when she moved to the AAMC Group on Student Affairs. Interview with Janet Bickel, July 16, 1990; Joseph A. Keyes, Jr., "Forward," in Janet Bickel and Renee Quinnie, *Building a Stronger Women's Program: Enhancing the Educational and Professional Environment,* 2nd ed. (Washington, D.C.: AAMC, 1993), pp. 2, 6–10; memorandum, Janet Bickel to Sarah Slavin, July 6, 1990; "Chapter 3: Deciding Whether and Where to Apply to Medical School," in *Medical School Admission Requirements,* 1991–92 (Washington, D.C.: AAMC, 1991), pp. 29–34; all sources courtesy of, and with thanks to, Janet Bickel. Also see, "AAMC Project Committee on Increasing Women's Leadership in Academic Medicine," *Academic Medicine* 71, no. 7 (1996): 801–810.

16. Telephone interview with Phyllis Kopriva, November 15, 1990; Allen Turner, "Healthy, Irrepressible Perspective," *Houston Chronicle,* June 29, 1997, p. 1D.

17. Kopriva interview, November 15, 1990; *Ad Hoc Committee Report on Women Physicians to the Board of Trustees* (Chicago: AMA, n.d.), pp. 1–42, courtesy of Phyllis Kopriva. My appreciation to Ms. Kopriva for the donation of her time and assistance. House of Delegates data are from a study commissioned by the AMA in 1992 in AMA Policy 530.974 of the *Policy Compendium,* sent to me courtesy of Samuel Wolinsky of the Chicago *Sun Times.* My thanks to Mr. Wolinsky. The AMA resolution was quoted in Kathryn E. McGoldrick, "Editor's Page: Shattering the Glass Ceiling," *JAMWA* 49, no. 1 (1994): 3.

18. American Medical Association, *Women in Medicine in America: In the Mainstream* (Chicago: AMA, 1991), Table 3.1, p. 11; Dickstein et al., "The Status of Women Physicians," Table 10, p. 39; *Minorities and Women in the Health Fields* (Washington, D.C.: U.S. Department of Health and Human Services, 1984), Table 48, p. 112.

19. Janet Bickel, Aarolyn Galbraith, and Renee Quinnie, *Women in U.S. Academic Medicine Statistics 1994* (Washington, D.C.: AAMC, 1994), Table 5; "AAMC Project Committee on Increasing Women's Leadership," pp. 799–811, Table 1; "Women Applicants Level Off, Faculty Continue Growth," *Academic Physician and Scientist* 1995 (January): 5; Janet Bickel et al., *Women in U.S. Academic Medicine Statistics 1998* (Washington, D.C.: AAMC, 1998), p. 2.

20. Marlys H. Witte, Arnold J. Arem, and Miguel Holguin, "Women Physicians in U.S. Medical Schools: A Preliminary Report," *JAMWA* 31, no. 5 (1976): 211–213, quotation on p. 212; K. Farrell et al., "Women Physicians in Medical Academia," *JAMA* 241 (1979): 2808–2812, esp. 2809.

21. The conference was organized by Drs. Lila Wallis and Arlene Scadron, presidents of the Women's Medical Associations of New York City and New York State, respectively; "Conference on Women in Medicine, Goals for Today and Tomorrow," *JAMWA* 34 (1979): 441; interview with Lila Wallis, New York City, March 30, 1989. My thanks and appreciation to Dr. Wallis.

22. A. L. Scully, "President's Report," *JAMWA* 34 (1979): 440. See Box: "Correspondence, Early '70s," folder: "Junior Memberships 1972–74"; Box: "Meetings, '60s and '70s," folder: "Proceedings of AMWA 58th Annual Meeting," esp. pp. 115–119, AMWA/NYH.

23. Interview with Dr. Bertha Offenbach, Newton, Massachusetts, June 28, 1989. Telephone interview with Dr. Rosa Lee Nemir, New York City, July 5, 1989;

interview with Dr. Carol Nadelson, Boston, June 28, 1989; interview with Dr. Suzanne Hall, New York City, July 5, 1989. My thanks to all of the above for generously sharing their time and interests with me.

24. Interview with Dr. Carol Nadelson. Interview with Dr. Eugenia Marcus, Newton, Massachusetts, June 27, 1989. Interview with Dr. Lila Wallis. Telephone interview with Dr. Leah Dickstein, September 8, 1989. My great appreciation to Drs. Marcus and Dickstein for their time and encouragement.

25. Constitution, Box: "Minutes, Correspondence c. 1981," folder "Minutes of the AMWA House of Delegates," November 5–6, 1981, p. 4, AMWA/NYH; Mary I. Bunting, "Afterthoughts by Way of Introduction," in Carolyn Spieler, ed., *Women in Medicine—1976: Report of a Macy Conference* (New York: Josiah Macy, Jr. Foundation, 1977), pp. 1–7; Janet Bickel and Renee Quinnie, *Women in Medicine Statistics 1990* (Washington, D.C.: AAMC, 1990), Tables 1, 10.

26. Interview with Dr. Susan Stewart, AMWA president, New York City, July 30, 1990; interview with Eileen McGrath, AMWA executive director, McLean, Virginia, July 16, 1990. I am most grateful to Dr. Stewart and Ms. McGrath for their contributions of time and assistance.

27. Bickel et al., *Women in U.S. Academic Medicine Statistics 1997*, Table 4; J. Sybil Biermann, "Women in Orthopedic Surgery Residencies in the United States," *Academic Medicine* 73, no. 6 (1998): 708–709.

28. The AWS was founded in 1981, the AWP in 1983. Joyce A. Majure, *1994 Pocket Mentor* (Westmont, Ill.: Association of Women Surgeons, 1994); "DSM-IIIR: Controversial Diagnoses Still Not Resolved," *News for Women in Psychiatry* 5, no. 1 (1986): 1, 2; interview with Dr. Leah Dickstein.

29. Ian Fisher, "The No-Complaints Generation," *New York Times Magazine*, October 5, 1997, p. 71.

30. Interviews with Eileen McGrath, Dr. Susan Stewart, and Dr. Suzanne Hall.

31. Judith Lorber, "The Limits of Sponsorship for Women Physicians," *JAMWA* 36, no. 11 (1981): 329–338. Also see Lorber, *Women Physicians: Careers, Status and Power* (New York: Tavistock, 1984).

32. Steven R. Daugherty, DeWitt C. Baldwin, Jr., and Beverley D. Rowley, "Learning, Satisfaction, and Mistreatment during Medical Internship," *JAMA* 279, no. 15 (1998); 1194–1199; Sharyn A. Lenhart and Clyde H. Evans, "Sexual Harassment and Gender Discrimination: A Primer for Women Physicians," *JAMWA* 46, no. 3 (1991): 77–82, quotation on p. 78; Melissa Schiffman and Erica Frank, "Harassment of Women Physicians," *JAMWA* 50, no. 6 (1995): 207–211; Lois Margaret Nora, "Sexual Harassment in Medical Education: A Review of the Literature," *Academic Medicine* 71, no. 1 (1996) supplement, S113–S118; Adriane Fugh-Berman, "Tales Out of Medical School," *The Nation*, January 20, 1992, pp. 1, 54–56.

33. Between 1940, when the first woman was board certified in neurosurgery, and 1995, only sixty-nine women were certified in the specialty. Rosa Lynn Pinkus, "Politics, Paternalism, and the Rise of the Neurosurgeon: The Evolution of Moral Reasoning," *Medical Humanities Review* 10, no. 2 (1996): 43 n. 54.

34. According to Lenhart and Evans, "Sexual Harassment," p. 78, the term *micro-inequities* was coined by Mary Rowe of MIT. See Mary Rowe, "Barriers to Equality: The Power of Subtle Discrimination to Maintain Unequal Opportunity," *Employee Responsibilities and Rights Journal* 3 (1990): 153–163.

35. Frances K. Conley, "Breaking the Glass Ceiling," *Harvard Medical Alumni*

Bulletin (Winter 1992–93); 38–44. I am indebted to Dr. Mary Ellen Avery of Harvard University for providing me with this source. Also see Jane Gross, "Female Surgeon's Quitting Touches Nerve at Medical Schools," *New York Times*, July 14, 1991, p. 8; "Headliners: On Her Terms," ibid., September 8, 1991; Courtney Leatherman, "Stanford Said to Ask Physician to Quit Neurosurgery Post," *Chronicle of Higher Education*, March 4, 1992, p. A17; Mary Roth Walsh, "Before and after Frances Conley," *JAMWA* 47, no. 4 (1992): 119–122; Sharon Lenhart, "Gender Discrimination: A Health and Career Development Problem for Women Physicians," *JAMWA* 48, no. 5 (1993): 155–159; Lenhart and Evans, "Sexual Harassment," pp. 78–79; P. Caplan, *Lifting a Ton of Feathers: A Woman's Guide to Surviving in the Academic World* (Buffalo, N.Y.: University of Toronto Press, 1993), cited in "AAMC Project Committee," p. 804 n. 18.

36. Sherrie H. Kaplan, et al., "Sex Differences in Academic Advancement," *New England Journal of Medicine* 335, no. 17 (1996): 1282–1289; Bonnie J. Tesch et al., "Promotion of Women Physicians in Academic Medicine: Glass Ceiling or Sticky Floor?" *JAMA* 273, no. 13 (1995): 1022–1025, esp. p. 1024; Lawrence C. Baker, "Differences in Earnings between Male and Female Physicians," *New England Journal of Medicine* 334, no. 15 (1996): 960–964; Mike Mitka, "Female Physician Income Lags Despite Experience," *American Medical News*, December 16, 1991. The Kaplan et al., study focused on 126 U.S. academic departments of pediatrics; Tesch et al. surveyed a cross-section of medical school departments over an eleven-year period ending in 1990; Baker found that in 1990, among male and female physicians in practice nine years or less, income gaps were accounted for by specialty choice, practice setting, and hours worked. But for physicians in practice more than ten years, men earned more than women. Also see Natalie Angier, "Among Doctors, Pay for Women Still Lags," *New York Times*, January 12, 1999, p. D7.

37. Bickel, et al., *Women in U.S. Academic Medicine Statistics 1994* pp. 4–5, Tables 7, 8; Bickel et al., *Women in U.S. Academic Medicine Statistics 1997*, Figure 1; "AAMC Project Committee on Increasing Women's Leadership," p. 803.

38. Janet Bickel of the AAMC refers to "cumulative career disadvantages," based on work by sociologists Jonathon R. Cole, Harriet Zuckerman, and Burton Singer. These authors coined the term *theory of limited differences* to refine Robert K. Merton's theory of "cumulative advantage," which, in examining the productivity of scientists, emphasizes a supposedly "rational" initial distribution of institutional resources, implying a level playing field. Merton's theory thus can not account for the effects of gender bias. Also see Janet Bickel, "Maximizing Professional Development of Women in Academic Medicine," in *Association of Women Surgeons Newsletter* 9, no. 3 (1997): 4, 7. I am grateful to Janet Bickel for alerting me to the work of Cole et al., some of which was presented as "The Productivity Puzzle" at a Josiah Macy, Jr. Conference on Women in Science in 1984. For a fuller exposition, see Jonathon R. Cole and Burton Singer, "A Theory of Limited Differences: Exploring the Productivity Puzzle in Science," in Harriet Zuckerman, Jonathon R. Cole, and John T. Bruer, eds., *The Outer Circle: Women in the Scientific Community* (New York: W. W. Norton, 1991), pp. 277–310; Harriet Zuckerman, "Accumulation of Advantage and Disadvantage: The Theory and Its Intellectual Biography," in C. Mongardini and S. Tabboni, eds., *L'opera di R. K. Merton e la sociologia contemporeana* (Genoa: Edzioni Culturali Internationali Genova, 1989), cited in Gerhard Sonnert

and Gerald Holton, "Career Patterns of Women and Men in the Sciences," *American Scientist* 84 (1996): 63–71 (see Introduction, n. 19, above).

39. Kaplan et al., "Sex Differences in Academic Advancement," p. 1288. For an earlier, more optimistic assessment, see Katherine G. Nickerson, et al., "The Status of Women at One Academic Medical Center," *JAMA* 264, no. 14 (1990): 1813–1817. In a subsequent publication, the two lead authors of this article noted that a disproportionate number of women faculty in their study were on the clinical rather than the tenure track. Nancy M. Bennett and Katherine G. Nickerson, "Women in Academic Medicine: Perceived Obstacles to Advancement," *JAMWA* 47, no. 4 (1992): 115–118, esp. p. 117.

40. "The Matriculation of Women at Harvard Medical School: A History of Conflict and Debate," *JCSW Newsletter* 10, no. 1 (1995): 1–8: quotation on p. 1. Figures are derived from a comparison of the following sources: "Engendering Gender Awareness," *Harvard Medical Alumni Bulletin* (Winter 1992/93): 39–40; press release and letter to Dean Daniel Tosteson from Drs. Mary Ellen Avery, Elizabeth Hay, Alice S. Huang, Lynne Reid, and Priscilla Schaffer, July 19, 1990, from the personal files of Dr. Mary Ellen Avery, to whom I am most grateful; American Medical Association, *Women in Medicine in America*, Table 3.16, p. 18.

41. Interview with Mary Howell, M.D., Ph.D., (1932–1998) by Regina Morantz, Medical College of Pennsylvania Oral History Project on Women in Medicine, April 15, 1978, pp. 45–47, typescript at Schlesinger Library, Radcliffe College, Cambridge, Massachusetts; Margaret A. Campbell, M.D. [pseud.], *"Why Would a Girl Go into Medicine?" Medical Education in the United States: A Guide for Women* (Old Westbury, N.Y.: The Feminist Press, 1973), pp. 3, 71; Wolfgang Saxon, "Mary Howell, A Leader in Medicine, Dies at 65," *New York Times*, February 6, 1998, p. A25. My appreciation to Dr. Gert Brieger for providing me with a copy of Howell's obituary. Cf. Elissa Ely, "Remembrance," *Harvard Medical Alumni Bulletin* 72 (Summer 1998): 54–55. Thanks, too, to Dr. Lucy Candib for providing Dr. Ely's written memorial to Howell.

42. Howell's observations add another layer of complexity to Rosabeth Moss Kanter's description, noted in the introduction, of the effects of varying degrees of underrepresentation on the power of a subgroup, such as women. Campbell, *"Why Would a Girl*, pp. 24, 29; interview with Mary Howell, pp. 22, 47; Saxon, "Mary Howell"; Mary C. Howell "What Medical Schools Teach About Women," *New England Journal of Medicine* 291, no. 6 (1974): 304–307.

43. As quoted in Jane G. Schaller, "Women and the Future of Academic Pediatrics," *Journal of Pediatrics* 118, no. 2 (1991): 318; see also Judith Lorber, "The Limits of Sponsorship." Figures are taken from Bickel et al., *Women in U.S. Academic Medicine Statistics 1994*, Table 9.

44. Figures are courtesy of Jeanne Benetti, U.S. Census Bureau, personal communication. See also Marian Gray Secundy, "To Have a Heritage Unique in the Ages: Voices of African American Female Healers," in Ellen Singer More and Maureen A. Milligan, *The Empathic Practitioner: Empathy, Gender, and Medicine* (New Brunswick, N.J.: Rutgers University Press, 1994), pp. 222–236, esp. 235 n. 11.

45. To some extent, the situation in medicine is merely a reflection of larger societal trends. See Dorothy J. Gaiter, "The Gender Divide: Black Women's Gains

in Corporate America Outstrip Black Men's," *Wall Street Journal*, March 8, 1994, pp. A1–A4. For 1992 the exact statistical breakdown for minority applicants was: African Americans, 509 men / 782 women; Hispanics, 240 men / 178 women; Native Americans, 55 men / 48 women. "Minorities in Medicine," *Science* 258, no. 13 (1992): 1087. Cf. Timothy Ready and Herbert Nickens, "Black Men in the Medical Education Pipeline: Past, Present, and Future," *Academic Medicine* 66, no. 4 (1991): 181–187, especially Figure 1, p. 182.

46. In 1991, citing "stagnation" in the aggregate numbers of minority applicants to medical schools, AAMC president Robert G. Petersdorf announced a new initiative spearheaded by the AAMC, the Project 3000 by 2000 Health Professions Partnership Initiative, designed to increase the number of enrolled minority students in health professions degree programs to 3,000 by the year 2000. Robert G. Petersdorf, "Not a Choice, an Obligation," address presented at the annual meeting of the AAMC, November 10, 1991; "Minorities in Medicine," *Science* 258 (November 13, 1992): 1; Jordan J. Cohen, "Some Good News about Diversity: Public Funding for Project 3000 by 2000," *Academic Medicine* 72, no. 9 (1997): 775.

47. African Americans represented only 4.7 percent of graduates in 1997. According to the AMA's Web page, "Data on Minority Physicians," the AMA has collected "race/ethnicity" data for approximately two-thirds of American physicians. It estimates that, to date, 2.6 percent of this number are African Americans and 4.6 percent are Hispanic Americans. In 1997, two African American women were deans of American medical schools, Anna C. Epps, at Meharry Medical College, and Barbara Ross-Lee, at Ohio University College of Osteopathic Medicine. All figures exclude Asians for the sake of an accurate comparison, since in 1975 no tabulations were made for the Asian/Pacific Islander category. In 1995, Asians comprised 15.6 percent of medical graduates. After 1980, the figure for Hispanic medical graduates included graduates of medical colleges in the Commonwealth of Puerto Rico. The increasing representation of Hispanic and Latino medical graduates is quite recent—too recent to allow for any extended historical treatment at this time. Between 1975 and 1980 it increased from 1.1 percent to 3.4 percent; in 1994, Hispanic and Latina female medical graduates accounted for 2.2 percent. Ironically, Dr. Antonia Coella Novello's appointment in 1990 as the first Hispanic woman surgeon general may have helped mask this dearth. Any future history of women in American medicine will want to give specific attention to their contributions. See also Ruth E. Zambrana, "The Underrepresentation of Hispanic Women in the Health Professions," *JAMWA* 51, no. 4 (1996): 147–152, esp. Table 2; and the *JAMA* "education numbers" 236, no. 26 (1976): 2962, Table 6; 276, no. 9 (1996): 717; and 278, no. 9 (1997): 747; "Minorities in Medicine," p. 1087; and Vanessa Northington Gamble, *Making a Place for Ourselves: The Black Hospital Movement* (New York: Oxford University Press, 1995), pp. 183, 191. My thanks to Dr. Sara Clausen for assisting with the search for these statistics.

48. Priscilla Ferguson Clement, "Managing on Their Own: Ailing Black Women in Philadelphia and Charleston, 1870–1918," in Barbara Bair and Susan E. Cayleff, eds., *Wings of Gauze: Women of Color and the Experience of Health and Illness* (Detroit: Wayne State University Press, 1993), pp. 180–190. Also see Diane Price Herndl, "The Invisible (Invalid) Woman: African-American Women, Illness, and Nineteenth-Century Narrative," *Women's Studies* 24 (1995): 553–572.

49. Gamble, *Making a Place for Ourselves*, pp. 190–193; P. Preston Reynolds,

"Hospitals and Civil Rights, 1945–1963: The Case of *Simkins v Moses H. Cone Memorial Hospital*," *Annals of Internal Medicine* 126, no. 11 (1997): 898–906. The key legal decisions in the campaign to integrate hospitals came in 1964, when, in *Simkins v. Moses H. Cone Memorial Hospital*, black physicians and dentists success-fully sued to integrate all hospitals. As Gamble points out, the ironic effect of these decisions was also to hasten the demise of most black hospitals.

50. This point is well made in the following works: Stephanie J. Shaw, *What a Woman Ought to Be and to Do: Black Professional Women Workers during the Era of Jim Crow* (Chicago: University of Chicago Press, 1996), pp. 13–14, 65–67; Nancy C. Talley-Ross, *Jagged Edges: Black Professional Women in White Male Worlds* (New York: Peter Lang, 1995), pp. 72, 75, 98, 108; and Secundy, "To Have a Heritage," pp. 222–236.

51. James Jones was a professor of educational psychology at nearby University of Arkansas at Pine Bluff, and later at Texas Southern University in Houston. See also Lydia E. Brew, *Edith: The Story of Edith Irby Jones, M.D.* (Houston: Lucious New, Jr., 1986), pp. 11–16; Norma Martin, "Medical Marvel: Dr. Edith Jones Has Blazed a Few Trails," *Houston Chronicle*, August 29, 1993, pp. 1, 2C; "Edith Irby Revisited," *Ebony* 10 (July 1963): 52–59; Charles Whitaker, "Breakthroughs Are Her Business," *Ebony* (June 1986): 89–96. My thanks to Vanessa Gamble for provid-ing me with copies of many of the foregoing sources for Dr. Edith Irby Jones.

52. Martin, "Medical Marvel"; Brew, *Edith*, pp. 50–53.

53. Leonard Laster, *Life after Medical School* (New York: W. W. Norton, 1996), pp. 277–304, quotations on pp. 282, 283.

54. Secundy, "To Have a Heritage," pp. 223, 225, 227, 231.

55. Marilyn J. Greer, Vernon W. Fields, and Ellen More, "Career Goals and Life Plans of Women Physicians," typescript draft, August 1996, pp. 1–49. Also see Marilyn Jane Greer, "Women Physicians' Specialty Choices," Ph.D. diss., Univer-sity of Texas Health Science Center, Houston, 1994. My thanks to Dr. Marilyn Greer for agreeing to collaborate on the study and for her statistical and survey-de-sign expertise. I am also grateful to the personnel of the Office of Research and Academic Affairs, University of Texas Health Science Center, Houston, and the Institute for the Medical Humanities, UTMB, Galveston, for splendid staff sup-port. I would also like to thank Dr. Judith Cadore for helping to organize a pretest subgroup, and the five physicians who, along with Dr. Cadore, completed the pre-test. Finally, we were ably assisted by students Gerald Waddy and Victoria Neidell, to whom I am most grateful. This project was approved by the Institutional Review Boards of both institutions: #OSP-93-330 (UTMB); #HSC-SPH-93-070 (UTHSCH).

56. Frequency distributions did not carry "significance," any difference in re-sponses between the two cohorts that equaled or exceeded 20 percent.

57. Rothstein, *American Medical Schools*, pp. 304–305, 316–320.

58. Vanessa Northington Gamble, "On Becoming a Physician: A Dream Not Deferred," in Evelyn C. White, ed., *The Black Women's Health Book: Speaking for Ourselves*, (Seattle: Seal Publications, 1990), pp. 52–64.

59. These specialties are listed in order of the number of women residents in each. Notably, women account for 61 percent of all pediatrics residents and 57 percent of residents in obstetrics-gynecology. *JAMA* 276, no. 9 (1996): Appendix 2.

60. Gang Xu et al., "A National Study of the Factors Influencing Men and

Women Physicians' Choices of Primary Care Specialties," *Academic Medicine* 70, no. 5 (1995): 398–404.

61. This points, of course, to one of the tragic implications of the 1996 case *Hopwood v. State of Texas*, in which the U.S. Fifth Circuit Court of Appeals ruled to bar the use of racial diversity as a factor in college admissions within the Fifth Circuit Court's District (a decision the U.S. Supreme Court declined to overrule). According to a study by the AAMC released in November 1997, in states where affirmative action programs have been ruled illegal (namely, California, Texas, Mississippi, and Louisiana), applications from minority students declined by 17 percent and acceptances by 27 percent. "Fewer Minorities Entering Nation's Medical Schools," *Galveston Daily News*, November 2, 1997, p. A3; James J. Young, "Affirmative Action in Medical School Admissions: The Texas Experience," *Academic Medicine* 72, no. 7 (1997): 616–617; Vera B. Thurmond and Louis L. Cregler, "Specialty Choices and Practice Locales of Black Graduates from a Predominantly White Medical School," *Academic Medicine* 68 (1993): 929–930; Miriam Komaromy et al., "The Role of Black and Hispanic Physicians in Providing Health Care for Underserved Populations," *New England Journal of Medicine* 334, no. 20 (1996): 1305–1310; Kevin A. Schulman et al., "The Effect of Race and Sex on Physicians' Recommendations for Cardiac Catheterization," *New England Journal of Medicine* 340, no. 8 (1999): 618–626; Ernest Moy and Barbara Bartman, "Physician Race and Care of Minority and Medically Indigent Patients," *JAMA* 273, no. 19 (1995): 1515–1520; Gang Xu et al., "Factors Influencing Primary Care Physicians' Choice to Practice in Medically Underserved Areas," *Academic Medicine* 72, no. 10 (October Supplement): S109–S111.

62. Stephen M. Keith et al., "Effects of Affirmative Action in Medical Schools," *New England Journal of Medicine* 313, no. 24 (1985): 1519–1525; Sarah Brotherton, "The Relationship of Indebtedness, Race, and Gender to the Choice of General or Subspecialty Pediatrics," *Academic Physician and Scientist* 70 (1995): 149–151. At the time of the study, about 54.3 percent of black and 68.4 percent of other participants were married; around 50 percent of both cohorts were responsible for the care of one to three children. The complete survey results are available, on request, from myself and from Marilyn Greer.

63. All quotations are from an interview of Dr. Judith Martin Cadore, May 20, 1998, Friendswood, Texas. I am greatly indebted to Dr. Cadore for agreeing to be interviewed for this project.

Conclusion

1. The following discussion of family leave is drawn from a much longer panel presentation, "Family Leave, Physician Productivity, and the Dilemma of Difference," presented by myself, Janet Bickel, and Dr. Leah Dickstein at the annual meetings of the AAMC (1991) and the Society for Health and Human Values (1991). I offer my gratitude and appreciation for the insights and assistance of my copresenters. Martha C. Minow, *Making All the Difference* (Ithaca, N.Y.: Cornell University Press, 1990), pp. 20–21, 57.

2. Emilie H. S. Osborn, Virginia L. Ernster, and Joseph B. Martin, "Women's Attitudes toward Careers in Academic Medicine at the University of California, San Francisco," *Academic Medicine* 67, no. 1 (1992): 59–62, esp. p. 60; Phyllis L.

Carr et al., "Relation of Family Responsibilities and Gender to the Productivity and Career Satisfaction of Medical Faculty," *Annals of Internal Medicine,* 129 (1998): 532–538; Christine Laine, "On Being Dr. Mom," ibid., pp. 579–580, quotation on p. 580; Gerhard Sonnert and Gerald Holton, "Career Patterns of Women and Men in the Sciences," *American Scientist* 84 (1996): 63–71; Nan Stone, "Mother's Work," *Harvard Business Review* (September–October 1989): 50–56.

3. Sarena D. Seifer, Barbara Troupin, and Gordon D. Rubenfeld, "Changes in Marketplace Demand for Physicians," *JAMA* 276, no. 9 (1996): 695–699; Judith Lorber, "A Welcome to a Crowded Field: Where Will the New Women Fit In?" *JAMWA* 42, no. 5 (1987): 149–152; J. M. Davis, "Too Many of Us?" *Postgraduate Medicine* 77 (1985): 22–24; E. J. Volpintesta, "Physician Glut: Myth or Reality?" ibid., p. 24; Philip R. Kletke, William D. Marder, and Anne B. Silberger, "The Growing Proportion of Female Physicians: Implications for U.S. Physician Supply," *American Journal of Public Health* 80, no. 3 (1990): 300–304.

4. The following are representative samples from the literature: Kevin Grumbach and Philip R. Lee, "How Many Physicians Can We Afford?" *JAMA* 265, no. 18 (1991): 2369–2372; Mary Jo Lanska, Douglas J. Lanska, and Alfred A. Rimm, "Effect of Rising Percentage of Female Physicians on Projections of Physician Supply," *Journal of Medical Education,* 59 (1984): 849–855; Sylvia Hurdle and Gregory Pope, "Physician Productivity: Trends and Determinants," *Inquiry* 26 (1989): 100–115.

5. *Ad Hoc Committee Report on Women Physicians to the Board of Trustees* (Chicago: AMA, 1981), p. 38. Quotation from "Report of the AMA Board of Trustees on Parental Leave (No. I-91)," presented by Joseph T. Painter, typescript, n.d., p. 2. Cf. American Medical Association, *Women in American Medicine: In the Mainstream* (Chicago: AMA, 1991), p. 17; Norman B. Kahn, Jr., and Richard B. Addison, "Comparison of Support Services Offered by Residencies in Six Specialties, 1979–80 and 1988–89," *Academic Medicine* 67, nos. 3 (1992): 197–202; Emina H. Huang and Olga Jonasson, "A Pregnant Surgical Resident? Oh My!" *JAMA* 265, no. 21 (1991): 2859–2860.

6. Etan Milgrom, M.D., "Parent or Resident or Both: A Father's Dilemma," *Hospital Physician* 25 (1989): 52. My thanks to Janet Bickel for this reference.

7. For a feminist elaboration of this perspective, see Lucy M. Candib, *Medicine and the Family* (New York: Basic Books, 1995).

8. Paula J. Clayton et al., "Mood Disorder in Women Professionals," *Journal of Affective Disorders* 2 (1980): 37–46; Amos Welner et al., "Psychiatric Disorders among Professional Women," *Archives of General Psychiatry* 36 (1979): 169–173. For a metanalysis of physician suicide studies, see Sari Lindeman et al., "A Systematic Review on Gender-Specific Suicide Mortality in Medical Doctors," *British Journal of Psychiatry* 168 (1996): 274–279. My thanks to Dr. Ruth Levine for alerting me to these sources.

9. Carr et al., "Relation of Family Responsibilities," p. 535.

10. In 1995 61 percent of male and 53 percent of female physicians worked in office-based practices. American Medical Association, *Physician Characteristics and Distribution in the U.S., 1996–97* (Chicago: AMA, 1997), pp. 61, 62, Tables B-11, B-11A.

11. Philip R. Kletke, David W. Emmons, and Kurt D. Gillis, "Current Trends in Physicians' Practice Arrangements: From Owners to Employees," *JAMA* 276, no. 7

(1996): 555–560; Martin L. Gonzalez, ed., *Socioeconomic Characteristics of Medical Practice 1997* (Chicago: AMA, 1997), pp. 21–30 and p. 57, Table 4.

12. Gang Xu, et al., "A National Study of the Factors Influencing Men and Women Physicians' Choices of Primary Care Specialties," *Academic Medicine* 70 (1995): 398–404; Wendy Levinson et al. for the American College of Physicians, "Position Paper: Parental Leave for Residents," *Annals of Internal Medicine*, 111 (1989): 1035–1038; Leah J. Dickstein, Daniel P. Dickstein, and Carol C. Nadelson, "The Status of Women Physicians in the Workforce," typescript prepared for Council on Graduate Medical Education, April 8, 1994. Table 11, p. 40.

13. Francis Kwakwa and Olga Jonasson, "The Longitudinal Study of Surgical Residents, 1993–1994," *Journal of the American College of Surgeons* 183, no. 5 (1996): 429; Joan Cassell, *The Woman in the Surgeon's Body* (Cambridge, Mass.: Harvard University Press, 1998), pp. 114–117, Karen Sandrick, "The Residency Experience: The Woman's Perspective," *Bulletin of the Amerian College of Surgeons* 77, no. 8 (1992): 10–17.

14. Cassell, *The Woman in the Surgeon's Body*, p. 67; Kathleen E. Ellsbury et al., "The Shift to Primary Care: Emerging Influences on Specialty Choice," *Academic Medicine* 71, no. 10 (1996): S16–S18, quotation on p. S18. According to this study, women were more likely to choose primary care, but "gender per se" had no independent predictive value.

15. Keith D. Lillemoe et al., "Surgery: Still an 'Old Boys' Club?" *Surgery* 116, no. 2 (1994): 255–261.

16. Molly Carnes, "Balancing Family and Career: Advice from the Trenches," *Annals of Internal Medicine* 125, no. 7 (1996): 618–620.

17. I am grateful for the support of the office of Dr. George Bryan, then vice president for academic affairs at UTMB, for sponsoring my attendance at the workshop in 1991.

18. For reference to the cartoon, see Sharyn A. Lenhart and Clyde H. Evans, "Sexual Harassment and Gender Discrimination: A Primer for Women Physicians," *JAMWA* 50, no. 6 (1995): 78, 82 n. 10.

19. It is suggestive—though no more than that—that when the editor of the journal *Academic Physician and Scientist* posted a query to readers for an article titled "Academics Find Solutions to Career Stress and Burnout" ([September/October 1998]: 1, 4–5), her only respondents were women. She wondered whether burnout affects women disproportionately or if they are just more willing to talk about it.

20. Judith Lorber, *Women Physicians: Careers, Status, and Power* (New York: Tavistock, 1984); Judith Lorber, "Can Women Physicians Ever Be True Equals in the American Medical Profession?" *Current Research on Occupations and Professions* 6, no. 1 (1991): 25–37.

21. Debatable perceptions of a physician "glut" and actual declines in physicians' real income during the 1970s and 1980s may also have fueled some of the resistance noted by Lorber. See Lorber, "Can Women Physicians Ever Be True Equals?" quotation on p. 34. Also see Frederic W. Hafferty, "Physician Oversupply as a Socially Constructed Reality," *Journal of Health and Social Behavior* 27 (1986): 358–369, esp. p. 360; Richard Restak, "We Need More Cheap, Docile Women Doctors," *Washington Post*, April 27, 1986, p. C1.

22. A. E. Eyler, Daniel W. Gorenflo, and Kathleen Musser, "Physician Gender and Productivity in a Managed Care Setting," *Journal of Women's Health* 5, no. 3

(1996): 221–224. This study concluded that when measured according to panel patients seen per hour, and controlling for specialty, there was no significant difference in productivity according to gender of physician.

23. Klea D. Bertakis et al., "The Influence of Gender on Physician Practice Style," *Medical Care* 33, no. 4 (1995): 407–416; N. Lurie et al., "Does the Sex of the Physician Matter?" *New England Journal of Medicine* 329, no. 7 (1993): 478–482; Carol S. Weisman and Martha Ann Teitelbaum, "Physician Gender and the Physician-Patient Relationship: Recent Evidence and Relevant Questions," *Social Science and Medicine* 20, no. 11 (1985): 1119–1127; Robert M. Arnold, Steven C. Martin, and Ruth M. Parker, "Taking Care of Patients: Does it Matter Whether the Physician Is a Woman?" *Western Journal of Medicine* 149, no. 6 (1988): 729–733. Women generally fare better than men in comparisons of malpractice suits. Cf. James Morrison and Peter Wickersham, "Physicians Disciplined by a State Medical Board," *JAMA* 279, no. 23 (1998): 1889–1893. My thanks to Prof. Joal Hill for this reference.

24. Lorber, "Can Women Physicians Ever Be True Equals?" p. 35.

25. Fay Jarosh Ellis, "Academic Careers Need Not Compete with Family Life," *Academic Physician and Scientist* (January/February 1998): 1, 4–5; "Deborah E. Powell, M.D.," ibid., pp. 6–7; letter to the editor from Wendi Neckameyer, in ibid. (March/April 1998): 3.

26. Pamela Charney, "Update in Women's Health," *Annals of Internal Medicine* 129 (1998): 551–558; Eliot Marshall, "A 5-Year Initiative Slowly Takes Shape," *Science* 278 (1997): quotation on p. 1558. Marshall was quoting cardiovascular specialist Marianne Legato of Columbia University College of Physicians and Surgeons.

27. Ellen Schur and JoDean Nicolette, "Medical Students and the Future of Women's Health," *JAMA* 277, no. 17 (1997): 1406–1407; Sue V. Rosser, *Women's Health: Missing from U.S. Medicine* (Bloomington: Indiana University Press, 1994); Lila Wallis, "Women's Health: Developing a New Interdisciplinary Specialty," *Journal of Women's Health* 1 (1992): 107–108; Janet B. Henrich, *Academic and Clinical Programs in Women's Health* (Washington, D.C.: Council on Graduate Medical Education, 1994); Glenda D. Donoghue, ed., *Women's Health in the Curriculum: A Resource Guide for Faculty* (Philadelphia: National Academy on Women's Health Medical Education, 1996); *Women in Biomedical Careers: Dynamics of Change* (Washington, D.C.: NIH Office of Research on Women's Health, 1992); Marc Nelson, JoDean Nicolette, and Karen Johnson, "Integration or Evolution: Women's Health as a Model for Interdisciplinary Change in Medical Education," *Academic Medicine* 72, no. 9 (1997): 737–740.

28. Philip C. Carling et al., "Part-time Residency Training in Internal Medicine: Analysis of a Ten-Year Experience," *Academic Medicine* 74, no. 3 (1999): 282–284. The part-time residents were "evaluated by faculty to be superior to full-time trainees in both clinical competence and humanistic skills." Quotation on p. 2.

29. For example, a study of the Department of Internal Medicine at Johns Hopkins found that one-third of male faculty, as well as two-thirds of female faculty, were dissatisfied with meetings scheduled on Saturdays or after 5 P.M. on weekdays, because they conflicted with family responsibilities. Linda P. Fried et al., "Career Development for Women in Academic Medicine," *JAMA* 276, no. 11 (1996): 898–905, esp. p. 899. See also Milgrom, "Parent or Resident or Both,"

p. 52; Leah J. Dickstein, "Female Physicians in the 1980s: Personal and Family Attitudes and Values," *JAMWA* 45, no. 4 (1990): 122–126; Janet Bickel, *Medicine and Parenting* (Washington, D.C.: AAMC, 1991); Sarah E. Brotherton and Susan A. LeBailly, "The Effect of Family on the Work Lives of Married Physicians," *Journal of the American Medical Women's Association* 48, no. 6 (1993): 175–181. However, cf. "Women vs. Men Pediatricians: A 'Homework' Disparity," *Pediatric Management* 3 (July 1993): 51 (source courtesy of Dr. Alma C. Golden); R. Chris Wray, Jr., "Editorial: Pregnancy and Plastic Surgery Residency," *Plastic and Reconstructive Surgery* 91, no. 2 (1993): 344–345.

30. John D. Stobo, Linda P. Fried, and Emma J. Stokes, "Understanding and Eradicating Bias against Women in Medicine," *Academic Medicine* 68, no. 5 (1993): 349; Linda P. Fried et al., "Career Development for Women," pp. 898–905. For faculty women in the sciences, see "A Study on the Status of Women Faculty in Science at MIT," *The MIT Faculty Newsletter* 11, no. 4, Special Edition (1999): 1–3; online at http://web.mit.edu.fnl.women/women.html.

31. Edward J. Volpintesta, letter to the editor, *JAMA* 279 (1998): 1609; Mark Linzer, "Leaders or Lemmings?" ibid., p. 341; Dennis H. Novack, et al., "Calibrating the Physician: Personal Awareness and Effective Patient Care," ibid. 278 (1997): 502–509, quotation on p. 505.

Index

Abbott, Edith, 159–160
Abbott, Grace, 76, 152, 154, 159–160, 174
Abbott, Maude, 178, 179
Abortion, 35–36, 103, 209–210
Abortion in the United States, 210
Abston, Sally, 194–195
Academic careers, 170–181, 232, 249–250, 257
Academic Physician and Scientist, 255
Adamson, Sarah Reed. *See* Dolley, Sarah Adamson
Addams, Jane, 73, 90
African American mothers, 77–78, 85–86, 153
African American women physicians, 4–6, 12, 100, 110, 165–168; medical associations, 236; community service, 236–237, 242; racism and, 239–240; gender issues and, 240; specialization, 241–242; individual profiles, 237–239, 242–247
Aldridge, George, 38, 83
Alexander, Virginia, 6, 110
Allen, M. May, 49, 66
Almshouses, 105
Alpha Kappa Alpha (AKA), 166
American Academy of Medicine (AAM), 72–73
American Association for the Study and Prevention of Infant Mortality (AASPIM), 72, 73, 75, 82

American College of Surgeons (ACS), 111, 118–119, 120
American Committee for Devastated France (ACDF), 140, 144–145
American Medical Association (AMA), 39–40, 50, 79, 122; Code of Ethics, 40; Council on Medical Education, 98, 107, 111; public health and, 123, 129, 156; Sheppard-Towner Act and, 155–156; family planning and, 209–212; Department of Women in Medicine, 223–224
American Medical Women's Association (AMWA), 124, 191; World War II and, 182–186; "Medical Womanpower," 200–201; policy changes, 204–205; Regional Conference and Workshop on Women in Medicine, 225; medical students in, 225–227. *See also* Medical Women's National Association
American Pediatric Society (APS), 112, 157, 170, 173
American Public Health Association (APHA), 174, 209–211
American Red Cross, 26, 127, 129, 134–137
American Women's Hospitals (AWH), 11, 67, 127–128, 169; Red Cross and, 134–137; Morton and, 134–140; objectives, 138–139; Unit No. 1, 140–146
America's Resources, 187, 192
Anthony, Susan B., 16, 25, 37, 38, 56, 66, 83

Separatism, 9, 43, 65, 96–97, 123
Settlement houses, 76, 84, 90–94
Sex education, 164, 207, 212
Sex Information and Education Council
    of the United States (SIECUS), 205,
    212
Sheppard-Towner Act, 11, 72, 148,
    151–155, 161, 181; AMA and, 155–156
Sherbon, Florence Brown, 150
Sherman-Ricker, Marcena, 84, 86
Simpkins, Modjeska, 166
Singer, Burton, 8
Slaight, Mary, 62
Smith, Clement, 199
Smoot, Reed, 152
Snoke, Albert W., 176
Soble, Nathan, 63
Social feminism, 44, 120
Social housekeeping, 44, 67
Social reform, 44, 55–56, 62
Social Security Act (Title V), 161, 174
South, Lillian H., 123
Sparkman, John, 185
Specialization, 6, 31, 49, 54–55, 111, 170,
    227; earnings, 230–231; African
    American women physicians and,
    241–242; childbearing issues and,
    252–255. *See also* Obstetrics/gynecology;
    Pediatrics
Specificity, 16
Spencer, R. H., 186
Spock, Benjamin, 191
Stalnaker, John M., 190, 193, 202
*Standards of Prenatal Care: An Outline for the
    Use of Physicians*, 154
Stanford University, 229, 230
Stark, Mary, 52, 56, 59
Starr, Ellen, 90
Starr, Paul, 39–40, 122
Stastny, Olga, 164
State boards of health, 151
State medicine, 75
Staupers, Mabel K., 165, 166
Stevens, Rosemary, 118
Stevenson, Sarah Hackett, 39, 123
Stimson, Henry, 185
Stone, Hannah, 165, 209
Strauss, Nathan, 74
Suffrage, 37–38, 56, 66, 71, 132
Surgeons, 118–119, 194–195, 227–228

Talks to My Patients (Gleason), 19
Taussig, Helen, 171, 177–179, 198

Tayler-Jones, Louise, 162, 163
Terrell, Mary Church, 86
Thomas, Vivian, 179
Thompson, Lou, 165
Thoms, Herbert, 176
Title V, 161, 174
Title VI, 216–217
Title VII, 216–217
Title IX, 218–219
Tracy, Martha, 71–72, 79, 81, 183
Turner, Harriet, 38, 52, 61, 83

University of Michigan, 51, 100–101, 108
University of Pennsylvania, 31–32, 52
University of Texas Medical Branch
    (UTMB), 194, 244

Vandervall, Isabella, 110
Van Hoosen, Bertha, 24, 49, 104, 168, 169,
    181; internships and, 105, 108, 111; AMA
    and, 122, 124, 125; war service and, 128,
    130–131, 132, 138

Wald, Lillian, 74–75, 76, 90, 150, 159, 160
Walker, Catherine, 86
Walker, Gertrude, 137
Walker, Mary, 146
Walsh, Mary Roth, 190, 216, 218–219
Waters, Ralph M., 180
Watts, Virginia Jane, 100
Weigel, Louis, 50, 60
Welch, William, 112
Whipple, Electra, 65
Whipple, George, 207
White House Conference on Child Health
    and Protection (1930), 158–159
White House Conference on the Care of
    Dependent Children, 73, 76
White-Thomas, Cornelia, 49, 63, 88, 118,
    120
Whittemore, Ruth, 179
Whittmore, Emma, 85
*"Why Would a Girl go into Medicine?"
    Medical Education in the United States: A
    Guide for Women* (Howell), 233
Wilbur, Ray Lyman, 158, 160
Wile, Julius, 115, 116
Will, Elsa (Leveque), 92, 93
Willard, Frances, 25, 37
Williams, Anna Wessel, 79, 121
Williams, J. Whitridge, 154
Williams, Mary E., 167
Winslow, C.-E. A., 72

CPSIA information can be obtained at www.ICGtesting.com
Printed in the USA
LVOW091453310712

292332LV00005B/1/P